THE DEFENSE INDUSTRY
IN THE POST-COLD WAR ERA

CORPORATE STRATEGIES
AND PUBLIC POLICY PERSPECTIVES

TECHNOLOGY, INNOVATION, ENTREPRENEURSHIP AND
COMPETITIVE STRATEGY SERIES
Series Editors: John McGee and Howard Thomas

Published

STEFFENS
Newgames: Strategic Competition in the PC Revolution

BULL, THOMAS & WILLARD
Entrepreneurship: Perspectives on Theory Building

SANCHEZ, HEENE & THOMAS
Dynamics of Competence-Based Competition

DAI
Corporate Strategy, Public Policy and New Technologies

BOGNER
Drugs to Market

Forthcoming titles

PORAC
The Social Construction of Markets and Industries

SCHULZ & HOFER
Creating Value with Entrepreneurial Leadership and Skill-Based Strategies

Other titles of interest

DAEMS & THOMAS
Strategic Groups, Strategic Moves, and Performance

DENNING
Making Strategic Planning Work in Practice

DOZ
Strategic Management in Multinational Companies

McNAMEE
Developing Strategies for Competitive Advantage

TSE
Marks and Spencer

Related journals—sample copies available on request
European Management Journal
International Business Review
Journal of Retailing and Consumer Services
Long Range Planning
Scandinavian Journal of Management

THE DEFENSE INDUSTRY IN THE POST-COLD WAR ERA

CORPORATE STRATEGIES AND PUBLIC POLICY PERSPECTIVES

edited by

Gerald I. Susman and Sean O'Keefe

1998

Pergamon

An Imprint of Elsevier Science

Amsterdam – Lausanne – New York – Oxford – Shannon – Singapore – Tokyo

388,4762330973
D313

ELSEVIER SCIENCE Ltd
The Boulevard, Langford Lane
Kidlington, Oxford OX5 1GB, UK

First edition 1998

Library of Congress Cataloging-in-Publication Data
The defense industry in the post-cold war era : corporate strategies
  and public policy perspectives / edited by Gerald I. Susman and Sean
  O'Keefe.
      p.   cm. – – (Technology, innovation, entrepreneurship, and
  competitive strategy series)
    Includes bibliographical references (p.   ) and index.
    ISBN 0-08-043356-1 (hardcover)
    1. Defense industries––United States.   2. Economic conversion–
  –United States.   I. Susman, Gerald I.   II. O'Keefe, Sean.
  III. Series
  HD9743.U6D377   1998
  338.4'76233'0973– –dc21
                                                                98–36744
                                                                     CIP

British Library Cataloguing in Publication Data
A catalogue record from the British Library has been applied for.

ISBN: 0-08-043356-1

⊗ The paper used in this publication meets the requirements of ANSI/NISO Z39.48-1992 (Permanence of Paper).
Printed in The Netherlands.

CONTENTS

EDITORS' PREFACE

The idea for this book began in Fall 1993 when we agreed to co-teach a course on the Management of Innovation at Penn State. One of us (Susman) was a professor of management with a special interest in the management of technology and the other (O'Keefe) had just joined the Penn State faculty as a professor of finance after having served in the Bush Administration as Comptroller of the Department of Defense and Secretary of the Navy. Our interest in co-teaching this course was recognition that our backgrounds blended nicely and that we could learn much from each other. This also was the time when interest was surging on how defense-related firms would cope with the dramatic decline in the defense budget. Many firms would merge, be acquired, liquidate or survive in a weakened state. Other firms, however, would stumble temporarily, but ultimately flourish because they recognized that their strong technology base was a core competence and a potential source of competitive advantage. We saw a wonderful opportunity to observe how companies integrated their corporate and technology strategies. The time-period during which these companies could develop and implement successful strategies was brief, so it was almost as if we could observe strategy formulation and implementation under "time-lapse photography" conditions. Companies could blossom or wilt before our eyes, as it were, and we had a unique opportunity to understand why.

  Obviously, defense-related companies do not exist in a vacuum. The Department of Defense also was formulating public policies to cope with the end of the Cold-War. Senior Pentagon officials had to decide to what extent the federal government should intervene to influence the structure of the defense industry in the post-Cold War era. They also had to decide what national security threats remained or might emerge in this new era and how they should respond to them. They had to make these decisions while acknowledging that the procurement budget was unlikely to increase very much, if at all, in the next decade. The significance and complexity of these decisions are challenging and intriguing in their own right. We saw a magnificent opportunity to ask the people who are responsible for making these decisions or are knowledgeable observers to comment about the issues involved in making these decisions. Again, timing was critical. The complexity, subtleties, and uncertainty involved in such decision-making would be lost in retrospective reconstructions if they were made only a few years hence.

Our respective backgrounds and the events of the day were perfectly matched for the venture that we had conceived. We planned to hold a symposium that explored corporate strategies and public policies that had shaped or were shaping the structure of the defense industry in the post Cold-War era. Before we proceeded very far, however, one of us (O'Keefe) left Penn State to accept an endowed professorship at the Maxwell School of Citizenship and Public Affairs at Syracuse University. What we thought originally was going to be a single university initiative was transformed into a collaboration between two great American universities. The Second Klein Symposium on the Management of Technology was held at Penn State on September 14–17, 1997. It focused primarily on corporate strategy. The Bantle Symposium on Industrial Diversification: The Public Policy Perspective was held at Syracuse University on October 26–28, 1997. Professors, researchers, policy analysts, and current and former government officials were invited to write papers for one or both of these events. We sought to balance the list of invitees to assure a diverse range of views and political perspectives on the subjects to be discussed. Also, senior executives at several defense-related companies were invited to make presentations at Penn State or Syracuse on their strategies for applying their existing physical and intellectual assets to commercial markets. This book is the product of all of these respective contributions.

One might think that organizing and convening two inter-related symposia would complicate greatly an already challenging task. In realty, what happened was that we were able to draw upon the best of both of our respective institutions and the talent and dedication of so many wonderful people. We have many people to thank at Penn State and Syracuse University for making our two symposia so successful.

We profoundly thank Robert and Judith Klein whose generosity made the Second Klein Symposium on the Management of Technology possible at Penn State. Their friendship and warmth are deeply cherished by Gerald Susman who holds the endowed professorship that bears their name. The Robert and Judith Klein Professorship of Management continues to be crucial to initiatives at Penn State to conduct research and extend knowledge of management of technology issues. The people at Penn State whose contributions should be recognized include Judy Sartore and Deb Ieraci of the Research Publications Center of the Smeal College of Business Administration. The Klein Symposium would not have run so smoothly were it not for their experience and wise judgment at many critical junctures in the planning of this event. Judy Sartore was indispensable as usual in preparing the manuscript for submission to the publisher. Her forthcoming and richly deserved retirement will leave a void that seems impossible to fill. Also, deserving of recognition for their help in preparing the manuscript are Shirley Kovach, Tammy Laukitis and Bobbi Jo Lesko of the Department of Management and Organization in the Smeal College of Business Administration. Finally, Mustafa Bayulgen is thanked for his assistance in preparing summaries of the presentations made by company executives.

The Bantle Symposium at Syracuse University was made possible by the generous contribution of Lou and Ginny Bantle. The symposium was a clear expression of their desire that business and government relations be advanced by exploring issues of common concern between the sectors. Sean O'Keefe is particularly grateful for their support and friendship as he is honored to be the Louis A. Bantle Professor of Business and Government Policy, named in memory of Lou's father. At the risk of understatement, the symposium wouldn't have happened were it not for the dedicated efforts of Anthony Tsougranis who endured the time commitments for this venture while working on his doctorate at the Maxwell School. Kathleen Millson, the Senior Administrator for the Syracuse-John Hopkins National Security Studies seminars, patiently shepherded the participants of the National Security Management Course through every moment of the symposium. Thereafter, her paper management skills and those of Retha Senke, in the Maxwell School Public Administration Department, were invaluable in helping to see this edited volume come to fruition.

Most importantly, we thank our wives, Elizabeth Susman, and Laura O'Keefe. Gerry thanks Liz simply for being the wonderful and loving person that she is. The task of completing this book was infinitely easier to endure because she was there to talk with at the end of some very long days. Liz, also an academic, has the patience and understanding to excuse an occasionally absent and frequently late spouse. Her graciousness sets a high standard that is tough for anyone to match. Comparably, Laura's understanding of Sean's work habits developed by his fifteen years of public service experience was put to the test during the conduct of this venture. Fortunately for Sean, she employed the full extent of that understanding and tireless support in return for his undying gratitude.

Gerald I. Susman
Sean O'Keefe

LIST OF CONTRIBUTORS

John A. Alic has served as an adjunct faculty member at the Johns Hopkins School of Advanced International Studies. He also has consulted for organizations including the Inter-American Development Bank and the U.S. Department of Commerce.

David J. Berteau is Corporate Vice President at Science Application International Corporation (SAIC). He served as Principal Deputy Assistant Secretary of Defense for Production and Logistics from 1990–1993. He was Chairman of the Defense Conversion Commission in 1992.

Mark A. Blodgett completed his Master's of Business Administration degree (with honors) at the Smeal College of Business Administration at The Pennsylvania State University in 1998. He previously served for twelve years in the U.S. Army. He is currently employed by PricewaterhouseCoopers as a management consultant.

David S.C. Chu is Director of RAND's Washington office and Associate Chairman of RAND's Research Staff. He served as Assistant Secretary of Defense (Program Analysis and Evaluation) from 1988–1993.

Robert Callum graduated in 1998 with master's degrees in International Relations and Public Administration from the Maxwell School of Citizenship and Public Affairs at Syracuse University. He is now a research analyst with the Center for Naval Analyses in Alexandria, VA.

Jonathan M. Feldman is a Research Fellow at the Centre for Innovation and Entrepreneurship at Linköping University in Linköping, Sweden.

Kenneth Flamm holds the Dean Rusk Chair in International Affairs at the Lyndon B. Johnson School of Public Affairs, University of Texas, Austin and is a Senior Fellow, Foreign Policy Studies, at the Brookings Institution.

Terrence Guay is Assistant Professor of Political Science and Director of Graduate Studies for the International Relations Program at the Maxwell School, Syracuse University.

John J. Hamre is Deputy Secretary of Defense. Prior to assuming his current post, he served as Under Secretary of Defense (Comptroller) from 1993–1997.

Maryellen R. Kelly is Associate Professor of Public Policy at the H. John Heinz School of Public Policy and Management at Carnegie-Mellon University. She is currently on leave to the U.S. National Institutes of Standards and Technology in Gaithersburg, Maryland.

Theodore M. Hagelin is Professor of Law and Director, Technology Transfer Research Center, Syracuse University, College of Law.

Paul G. Kaminski is Chairman and CEO of Technovation, Inc. He served as Under Secretary of Defense for Acquisition and Technology from 1994–1997. He has served as Chairman of the Defense Science Board and as a member of the Defense Policy Board.

Ann Markusen is Professor of Economics and Director of the Project on Regional and Industrial Economics at Rutgers, The State University of New Jersey, New Brunswick, New Jersey.

Michael Oden is Assistant Professor of Planning in the Department of Community and Regional Planning at the University of Texas, Austin.

Sean O'Keefe is the Louis A. Bantle Professor of Business and Government Policy in the Department of Public Administration at the Maxwell School of Citizenship and Public Affairs at Syracuse University. He served as Secretary of the Navy from 1992–93 and Comptroller and Chief Financial Officer of the Department of Defense from 1989–1992.

John W. Peterson is Manager of Technology Strategy at Lucent Technologies.

Erik R. Pages is Vice President at Business Executives for National Security (BENS). He served as Director of the Office of Economic Conversion Information (OECI) at the U.S. Department of Commerce.

Gene Porter is Senior Fellow at the Center for Naval Analyses. He has served in several posts within the Office of the Secretary of Defense.

Michael Radnor is Professor of Management at The Kellogg Graduate School of Management, Northwestern University.

William L. Shanklin is Professor of Marketing at the Graduate School of Management at Kent State University.

Gerald I. Susman is the Robert and Judith Klein Professor of Management, Chair of the Department of Management and Organization, and Director of the Center for the Management of Technological and Organizational Change at the Smeal College of Business Administration, The Pennsylvania State University.

Todd A. Watkins is Class of 1961 Associate Professor of Economics in the College of Business and Economics, Lehigh University.

Matthew C. Waxman is a consultant to the Rand Corporation.

John P. White served as Deputy Secretary of Defense from 1995–1997. He previously served as Director of Business and Government at the John F. Kennedy School of Government at Harvard University and as Deputy Director of the Office of Management and Budget in the Carter Administration.

CASE PRESENTATIONS

John Stuelpnagel, Director, R&D Operations, Northrop-Grumman Corporation

William T. Hanley, Chief Executive Officer, Galileo Corporation

James A. Koshak, President, U.S. Materials, Hexcel Corporation

Albert Smith, President, Electronic Systems Sector, Harris Corporation

Richard Flam, Chief Scientist, ORBIT/FR

John Manuel, Vice President, Domestic Business Development, Lockheed-Martin

Gerald Brasuell, Vice President and General Manager, Systron Donner Inertial Division, BEI Sensors and Systems Company

Richard J. Farrelly, Vice President, Corporate Development, Base Ten Systems, Inc.

James D. Scanlon, President, Lockheed-Martin Control Systems

John V. Sponyoe, President, Lockheed-Martin Federal Systems

PART ONE:

SETTING THE SCENE

I

INTRODUCTION: POST-COLD WAR CHALLENGES FOR GOVERNMENT AND INDUSTRY

GERALD I. SUSMAN AND SEAN O'KEEFE

PART ONE: SETTING THE SCENE

The Department of Defense (DoD) and firms in defense-related industries face the challenge of responding effectively to profound changes since the end of the Cold War. The world has emerged from an era in which the spectre of global confrontation by two superpowers has subsided. Concurrently, the national security policy of the United States has evolved to recognize the diminished prospect of global conflict, but emergence of a range of other threats to U.S. interests. A limited, albeit spirited debate has been engaged to redefine these national security policy objectives. Without clear resolution in this debate, the defense strategy to carry out policy lacks the desired clarity to efficiently achieve the objectives.

As Deputy Defense Secretary John Hamre suggests, strategy uncertainty complicates the development of clear priorities for weapon system performance requirements. As the Congress, the Administration, and the military services seek to reach agreement on these important strategy issues, the unsettled environment for management of systems tends to yield higher acquisition cost and lost opportunities for reducing the cost of operations.

Modernization of weapon systems is one of the central objectives of the current strategy advanced in the Quadrennial Defense Review. However, current plans for financing modernization require that the resources be

3

derived from extant defense activities, given that there is virtually no prospect for overall defense budget increases forecast for the foreseeable future. Former Deputy Defense Secretary John White indicates that the options are for DoD to identify efficiencies in designing and producing weapon systems, operating and maintaining existing assets, reducing infrastructure by closing more bases, and reducing force structure by cutting civilian and military personnel. While likely that the latter two options must be explored to achieve adequate resources for modernization, these issues are not treated extensively in this book. This edited volume covers a range of perspectives which pertain to the first two aspects of White's observations, with particular emphasis on the role of defense industry and government strategies to achieve such efficiencies.

The defense industry has weathered significant reductions in the procurement portion of the annual defense budget. Defense-related firms have responded to this dramatic decline in market demand much like commercial firms have anytime extensive market adjustments occur. They have diversified, retrenched, sold strategically weak divisions, merged with or acquired competitors. While the strategies have been similar, the methods have been a bit different.

Uniquely, defense-related firms have found the challenge of diversifying into commercial markets to be formidable because the commercial marketing and distribution channels are fundamentally different from those of the defense sector. Similarly, requirements that the defense industry be particularly sensitive to quality control, government specifications, and regulatory compliance has the effect of minimizing cost competitiveness in the commercial sector. Even smaller defense-related firms which are less encumbered with these unique overhead requirements often lack the capital to acquire the resources or skills they need to compete in the commercial sector, so must form alliances with firms that have complementary assets.

Compounding the challenge for the defense industry, the government maintains a strong interest in the strategic choices industry makes in grappling with defense budget cuts. Indeed, the public policies advanced can guide the overall industry trends. In recent years, the Defense Department actively promoted mergers and acquisitions by subsidizing some of the restructuring costs and relaxing anti-trust scrutiny. As defense industry consolidation neared completion in late winter 1998, the government raised new concerns as to whether the diminished competitive environment would be adequate to support innovation and continued cost reduction.

This book focuses on the challenges that defense-related firms face in enhancing their financial well-being and DoD faces in maintaining affordable national security in the post-Cold War era. The corporate strategies and public policies that each develops respectively sometimes have little or no impact on the other. Sometimes, however, they reinforce or undermine each other. This book explores the conditions that defense-related firms and DoD currently face and the future they respectively envision as well as the corporate strategies and public policies that each develops in response to these conditions and visions. The authors who contributed chapters to this book describe these corporate strategies and public policies, assess their respective strengths and weaknesses, and, where appropriate, endorse them or recommend alternatives. Finally, senior executives from ten small and large defense-related firms recount their experiences in diversifying successfully into commercial markets and the challenges they met or still face in planning and implementing their strategies effectively.

PART TWO: CORPORATE AND GOVERNMENT INTERESTS

Government interests are affected in many ways by defense resource availability. The sheer size of the defense budget has a profound impact on the economy and the ability to fund other priorities. The Department of Defense (DoD) and other government agencies provide economic adjustment assistance to communities that are heavily dependent on defense expenditures or facilitate and promote the sale of U.S. weapons to foreign countries. DoD is concerned about deterioration of the defense industrial base in the wake of defense budget reductions and our capability to mobilize quickly in a national emergency. DoD also is concerned about dependence on foreign sources of state-of-the-art technology and may take steps to assure domestic sources of such technology. Dual-use technologies can lead to creation of an integrated military-commercial industrial base and lower costs through increasing economies of scale. It also can lower costs by promoting insertion of commercial components into weapons systems. Maintaining U.S. military superiority with dual-use technologies also poses new challenges. Maximizing use of commercially viable technology and products eases access into a range of markets, but it also encourages advocacy for technology transfer and export controls. Such regulatory regimes tend to be difficult to construct and even more difficult to enforce. Of equal concern, dual-use technologies must be supported in the commercial market place. To the extent that public financing may be required, this could defeat the objective of using dual-use technology applications.

Diversification of defense-related companies into commercial markets also encourages an integrated industrial base, lowers costs, and facilitates transfer of technologies across military and commercial sectors. Consolidation also lowers costs, but may lead to a segregated defense and industrial base, and little transfer of technology across military and industrial sectors. The proper balance between government and industry in promoting these respective interests is controversial and different contributors to this book take different positions on it. The issue of balance is often in the background of discussions concerning how to reduce defense costs and enhance capabilities. We believe that this book makes a constructive contribution to the debate.

Chu and Waxman are cautious about developing policies that encourage defense contractors to pursue any specific competitive strategy. It is a risky endeavor and may produce unintended consequences. They urge anyone who proposes policies to be as clear as possible about policy targets and objectives. For example, targeting the defense industry for change is problematic since the definition of an industry player depends on whether a defense firm is defined by volume or percentage of defense to total sales. General Electric, for example, is the ninth largest defense contractor in volume of sales, but only 3.1 percent of its sales are defense-related. Also, policies have been proposed to maintain defense industrial base readiness, insure domestic supplies of critical components, prevent undue defense industry influence on U.S. foreign policy, and promote dual-use applications of technology. Chu and Waxman suggest that the objectives of such policies are not always clear and sometimes may conflict. They believe that the Department of Defense is more likely to achieve efficiency gains by targeting policy towards changes within its own domain via acquisition reform than by policies aimed at restructuring the defense industry.

Flamm analyzes several data bases to determine how much defense spending actually declined between 1985 and 1997. The most frequently cited statistic is a two-thirds decline in the procurement budget. However, this statistic is based on congressional appropriations. Flamm shows that actual spending outlays indicate a decline that is closer to 40 percent, which is about where expenditures were before the start of the Reagan build-up. If so, then should the federal government be concerned about eliminating excess capacity? Despite policies that have encouraged consolidation in the defense industry, measures of physical production and capacity utilization suggest that excess capacity remains in several industry sectors. Fewer companies doesn't necessarily mean less capacity. Flamm's analysis shows that the number of companies that performed R&D on aircraft, aircraft parts, and missiles already were declining during the build-up. This suggests that consolidation would have occurred in some industry sectors without government intervention. Flamm and other authors in this book suggest that large weapon systems may now require a minimum critical level of R&D just to perform the necessary front-end work. Nevertheless, government incentives which subsidized merger expenses may have accelerated the consolidation trend. Also, companies do not have sufficient incentive to lower overhead if they can recover it on cost-plus contracts. How many companies are the right number of companies is a tough question to answer. Fewer companies may lead to greater efficiencies, but the trade-off may be less innovative companies due to less competition.

Hagelin indicates that basic and applied research declined more than any other categories of the Research, Development, Testing and Evaluation (RDT&E) budget between 1994 and 1997. With the potential for dual-use technologies to be adopted quickly for military and commercial use by U.S. allies and adversaries, Hagelin sees a vicious circle developing in which DoD increasingly must spend more for basic and applied research to assure that the U.S. maintains military superiority through advanced technology. Also, U.S. laws regarding intellectual property rights and technology transfer tend to favor economic over military priorities, which exacerbates the problem. Hagelin also brings the interests of American universities into the picture. Universities are the recipients of a significant amount of basic and applied research dollars. While the National Institutes of Health and the National Science Foundation have increased their funding of university research, DoD has been a significant source of funding for electrical and mechanical engineering and materials science. These disciplines contribute significantly to the development of U.S. military superiority and to training of graduate students for careers in industry.

Guay and Callum indicate that the European perspective on acquisition reform is very different from that in the United States. For Europe, acquisition reform is primarily an industry restructuring issue. There clearly is much more excess capacity in Europe than in the United States. There are ten helicopter and warplane manufacturers among the EU countries compared to four in the United States. The question is not whether European governments have a role in restructuring the defense industry, but what the nature of their role is. Governments are parties to negotiating and/or sanctioning cross-national deals between companies on investment contributions and pro-rata returns. Trying to privatize or consolidate defense firms in Europe is as difficult politically as closing bases is in the United States. The European challenge is exacerbated by unemployment levels in some countries that are more than 11 percent.

Guay and Callum raise a point made in chapters by Flamm and Alic, which is that large weapon systems may now require a minimum critical level of R&D just to perform the necessary front-end work. An increasingly important issue in both Europe and the United States is how to spread fixed R&D costs over a sufficient number of units to break even. Minimum economies of scale means that no country can have a stand-alone defense industry and survive. The European defense companies must rely on exports to achieve such economies, but the United States has aggressively pursued weapons exports and has increased its world market share considerably. The fierce competition for foreign military sales has emerged at the same time, however, that demand has markedly declined and suppliers from former Soviet markets have entered the fray as well. Most European companies are unwilling to partner with U.S. companies for fear that they will become merely subcontractors. Achieving economies of scale through joint weapons projects among European companies requires negotiation of agreements on joint specifications and inter-operability across national borders, which is a difficult task even for companies within the same nation.

PART THREE: CORPORATE PERSPECTIVE

Corporate Strategies

The chapters in this section of the book concern the strategies that defense-related firms pursued when they were faced with significant reductions in the defense budget. Defense firms responded to the reduction the same way that any commercial firm would respond to a decline in sales in its primary market. They responded by shrinking their asset base and/or looking for new sources of revenue in related or unrelated markets. Defense-related firms responded in four possible ways. They (1) remained in defense, but diversified into related commercial businesses (diversification), (2) remained in defense, but sold non-essential assets (retrenchment), (3) exited defense by spinning-off defense-related businesses (spin-off), or (4) remained in defense, but acquired companies that served different defense-related markets (consolidation).

These four strategies are firm-level responses which concern business divisions or similar firm-level units that develop, make and sell products. These strategies are concerned with the current and future mix of such units and the paths that firms take to reach their future mix. When companies modify their mix, they may sell, acquire, build or convert existing facilities. The term "defense conversion" is often associated with issues that relate to the last option. Which option the firm chooses is not the primary focus of this section, but it is obviously very important to the managers and workers in communities in which the existing facilities are located. The diversification and consolidation strategies assume that the firm derives benefit from sharing resources between existing and newly acquired or developed units. At the firm level, these benefits may be economies of scale or scope or stimulation of new ideas. The term "dual-use" could be applied to benefits that the firm derives from sharing between units that are located almost anywhere along the value-chain, e.g., shared research and development or shared manufacturing. The outcome of such sharing is products that can serve most military and commercial needs and are cheaper to produce.

While firms pursue only one strategy at a time, they may pursue strategies sequentially. For example, firms that pursue spin-off may have started initially with diversification or consolidation. They may have pursued either of the latter strategies to bolster sagging revenues or to make their defense units more attractive for ultimate sale. They also may have chosen spin-off because other consolidating firms enticed them with premium prices for their defense units. Firms that pursue retrenchment are able to shrink their assets faster than their revenues decline. This strategy makes sense for firms that make tanks, munitions, and ships because commercial markets for their products are non-existent or stagnant. These firms may return the cash that was generated from asset sales to stockholders or, after pausing briefly, begin to pursue consolidation. Their initial choice to sell assets rather than leverage them makes them less willing or capable of choosing diversification as a successor strategy.

Both diversification and consolidation are growth strategies, but they pursue growth very differently. Diversification achieves growth by internal ventures and selected acquisition of businesses with high potential for synergy with existing businesses. Consolidation achieves growth by acquisition of businesses that serve different defense markets. Economies are achieved primarily by consolidating production of similar products into common facilities and by eliminating redundant managerial and support staff among the businesses acquired. Diversification is more likely than consolidation to transform completely the identity of a firm because of the new and different businesses that it enters. For example, Westinghouse, McCaw, and Monsanto were transformed so completely through diversification that they eventually sold off the core business on which their firms were founded.

Advocates of either diversification or consolidation use the term "core competence" to support their strategy choice. However, they seldom make explicit the different organizational processes to which this term applies. Companies can develop a core competence anywhere along the business value chain. Diversification into commercial markets requires developing a core competence in linking technology capabilities to customer needs, then acquiring and coordinating the resources needed to develop and commercialize a product. They must estimate accurately what customers want and deliver a reliable product to them in a timely manner and at an acceptable price. Consolidation requires exploiting an existing core competence in translating defense department specifications into a product that performs as required and developing and manufacturing the product in accordance with specified procedures and cost accounting practices. Each type of core competence is supported by organizational structure, procedures, and culture that are incompatible with those that support the other. This explains why the transition between these two types of core competence is so difficult. Defense-related companies usually prefer to remain exclusively within the defense industry or develop or acquire segregated commercial facilities.

Markusen, Oden and Pages indicate that consolidation was the strategy of choice for most large defense contractors. These firms were cash-rich and their top management saw significant opportunities to make substantial profits from consolidation. The businesses they acquired had mature production programs which generated significant cash-flow and required relatively small research and development expenditures. The Pentagon was not expected to initiate any new programs that required significant R&D expenditures for the foreseeable future so there was little incentive to undertake new internally generated defense-related research. The payoff for internally generated commercial research generally takes years to realize. Markusen suggests that Wall Street investment bankers, consulting firms and some academics encouraged defense contractors to choose easier paths by promoting the self-serving intellectual justification that defense firms are ill-equipped to diversify successfully into commercial markets.

Oden compared six diversifying firms with six consolidating firms and showed that sales growth and profitability were higher among the diversifying firms between 1989 and 1994. These data dispel the myth that defense-related firms cannot diversify successfully into commercial markets. However, these data more accurately suggest that firms that are already moderately diversified will continue to diversify successfully. Oden's six diversifying firms had a range of defense to total sales of 23–52 percent in 1989, while his consolidating firms had a range of 60–90 percent. Also, some of Oden's six consolidating firms had chosen retrenchment during 1989–1994, e.g., General Dynamics, so it is not surprising to see a decline of sales growth among these firms during this period. Nevertheless, these data show that experience with diversification counts. Those firms that develop the capability to diversify are more willing and able to pursue this strategy successfully when the opportunity arises.

Pages concludes that the most attractive opportunities to make money via consolidation are essentially over. The large firms that pursued consolidation have acquired most of the defense-related businesses that were for sale. The mature programs of the acquired businesses are at steady-state or winding down. A resurgence of growth of new programs within the defense sector is unlikely. Thus, companies that had pursued consolidation will likely turn now to diversification into commercial markets for growth opportunities. Many of these companies turned early to civilian government agencies for prospective customers, given similarities in the contracting relationship. These companies, however, can learn much from commercial firms that have pursued diversification.

As indicated earlier, firms can diversify by acquisitions and divestitures or by internal growth. Either path to diversification can transform the character of the firm. However, transformation by adding and deleting businesses differs significantly from the metamorphosis of an existing business via internal growth. Large defense-related firms, such as Westinghouse and Rockwell, transformed themselves essentially into commercial firms along the former path. Smaller defense-related firms seldom have the resources to pursue this path, so are more likely to diversify by internal growth. In Chapters 13–15, Radnor and Peterson, Shanklin and Susman and Blodgett suggest how challenging the pursuit of diversification via internal growth is. Radnor and Peterson indicate that firms need to reassess and realign their technology value-chain and business value-chain when shifting from defense to commercial markets. Susman and Blodgett describe the internal growth strategies which small and medium-sized infra-red sensor firms pursue in entering promising commercial markets. These firms pursue growth by expanding from performing a single function, e.g., manufacturing or research and development, into performing multiple functions, e.g., producing and marketing a product. They also may pursue growth by expanding the number of applications for which their products may be used.

The opportunities to make money probably would have led most large defense contractors to pursue consolidation anyway, but Markusen, Oden, and Pages make clear that Pentagon policies facilitated and accelerated the direction. These policies included relaxation of anti-trust provisions and introduction of incentives for mergers. The Clinton administration initiatives to encourage diversification by subsiding research and development which had potential dual-use applications were minuscule by comparison. The consequence of these policies has been a dramatic reduction in the number of large defense contractors. The remaining ones are giants with significant market share and political clout.

Markusen, Oden and Pages are generally critical of policies that have shaped the current structure of the defense industry. They believe that these policies have resulted in lost opportunities for defense firms to diversify into commercial markets, and are concerned that the few remaining large defense contractors will unduly influence perceptions of future national security threats and the weapons that will be needed to deal with them. Markusen calls for abolishing these policies, while Pages looks beyond them and recommends policies to capitalize on renewed defense firm interest in diversification.

Organization Structure and Dynamics

Kelley and Watkins extend the diversification question beyond prime contractors upon which most of Part Three has focused and look at second-tier subcontractors. Although 70 percent of all defense dollars went to the one-hundred largest prime contractors, approximately 60 percent of all DoD dollars were distributed by prime contractors to subcontractors for the purchase of components. Notwithstanding the issue of the undue influence of large contractors raised earlier, perhaps much of the effort to encourage diversification should be aimed at the subcontractor tier. Diversification at this level might allay a concern expressed by several contributors to this book, which is that consolidating prime contractors would seek to vertically integrate via acquisition of second-tier subcontractors or, a more likely scenario, that some of the larger second-tier subcontractors would start to acquire other second-tier subcontractors, thus starting a second wave of defense industry consolidation at this level.

Kelley and Watkins look at the plants of both prime contractors and subcontractors in what they call the "machine dominated goods" or MDG sector and conclude that prime contractors are already very diversified, at least at the plant level. Their data show that in most plants of prime contractors or subcontractors, 50 percent or more of their goods are shipped to commercial customers. These plants also have a greater percentage of programmable automation (PA) and more extensive networks of collaboration and information sharing with customers, suppliers and competitors than do commercial plants. If the latter is true, then future policy should focus as much on technology transfer and learning from the defense to the commercial sector as the other way around. DoD programs such as the Industrial Modernization and Incentive Program (IMIP) may have achieved specifically targeted goals in the defense sector more effectively than market forces achieved in the commercial sector, i.e., adoption of PA.

Kelley and Watkins' data suggest that policy makers need not be very concerned about maintenance of the defense industrial base. The number of firms in the MDG sector has remained stable over the past ten years. Even if the mix of products within their plants were to shift more to commercial sales, the MDG plants have the capability to return to defense production should surge demand require it. However, the presence of PA doesn't mean necessarily that the plants are exploiting flexibility very effectively. Jaikumar (1986) presented data that U.S. plants were using PA equipment on a more limited product mix and for longer product runs than were Japanese plants. Even if U.S. plants now are exploiting the inherent flexibility of PA technology, are they exploiting its static or dynamic capabilities? Static flexibility focuses on quick changeover to make existing parts, while dynamic flexibility focuses on quick transition to making new parts. One suspects that DoD policy makers were more concerned originally with static flexibility because of the significant cost reduction potential from producing parts in very small batches. Dynamic flexibility is more important to firms that seek to increase revenue from new products and new customers, which is the essence of diversification. In this case, dynamic flexibility must be tied to new product development capability, which Feldman discusses below.

The extent to which diversification should be measured at the plant level can cut two ways. Kelley and Watkins argue that their plant level data should ease concerns about defense-commercial segregation because segregation at the division or higher organizational level may be misleading. Just because a company has separate defense and commercial divisions doesn't mean that defense and commercial production isn't occurring within plants that both divisions share. However, if policy is directed at the capability to diversify across defense and commercial markets, then perhaps change should focus on the firm's capability to develop and commercialize new products. If a firm suffers cutbacks in defense contracts, it needs the capability to generate new business through an effective new product development process as much or more than it needs its plant level flexibility.

Kelley and Watkins readily admit that conclusions from the MDG sector may not be generalizable to other sectors. The MDG sector consists of firms from twenty-one SIC codes that fabricate components for use in defense or commercial sub-systems of aircraft, satellites and missiles. Production processes do not start to diverge very much until the assembly of subsystems and systems, which is more likely to take place at prime contractor than subcontractor plants. Even for assembly, many of the major systems and subsystems of aircraft, e.g., engines or wings, are designed to serve similar functions in their final products, thus do not differ dramatically. Thus, diversification within the MDG sector requires less of a transformation than for other products in other sectors. Also, the customers for fabricated components are not final consumers, so the manner in which customers are solicited and products are sold doesn't require as radical a transformation as might be required for defense firms that diversify into final consumer markets. Nevertheless, these plants represent a substantial percentage of DoD sales and value-added to production. They suggest that any diversification policy needs to focus carefully on where it can make the most difference and where such processes are less likely to occur naturally due to market forces.

Feldman illustrates the challenge of transforming a defense company's existing product development process to meet the requirements of commercial markets. The challenge exists because the firm is trying to transform an existing key organizational process, which makes it a case of defense conversion as well as diversification. Feldman illustrates the challenge with detailed documentation of Boeing-Vertol's experience with the design and development of rapid transit cars. The company's short-lived venture is often cited as an object lesson in the futility of diversification by defense firms into commercial markets. Yet Feldman demonstrates that Boeing-Vertol had greater engineering success than is generally realized. To be sure, Boeing-Vertol's first rapid transit cars, which had been designed for the cities of Boston and San Francisco, had so many maintenance problems that Boston refused delivery of the remaining cars on its original contract. San Francisco's maintenance problems were less serious, in part because their cars were delivered later, giving Boeing-Vertol more time to improve the design of the cars. Boeing-Vertol subsequently designed a car for the Chicago Transit Authority (CTA) which met all design specifications and performed very well. The differences between these projects are instructive.

The CTA case study demonstrates that defense firms can diversify successfully into non-defense markets when their customer is very knowledgeable about the product and its operating requirements and can provide detailed specifications to the designer. They also can be successful when their customer has considerable power to dictate these specifications and assure delivery of stringently tested products at a fixed price. The CTA staff was very experienced with rapid transit cars and knew in great detail what specifications were required to operate successfully in

Chicago. Their design specifications involved low risk options of materials and components, and tolerances that were appropriate to the operating needs of the cars. They had the power to veto any unnecessary design options promoted by Boeing-Vertol designers. Virtually none of these conditions applied to the Boston-San Francisco case. The customers could not give detailed specifications to Boeing-Vertol engineers and relied instead on consultants who hindered more than facilitated communication between the customers and Boeing-Vertol designers. Also, the schedule was too ambitious for the number of cars ordered, which encouraged the designers to cut corners on testing the cars before they were delivered to their customer.

Feldman identifies other conditions that led the Chicago cars to succeed and the Boston-San Francisco cars to fail. However, one might ask whether defense contractors must have customers who are knowledgeable and powerful enough to keep defense engineers out of trouble. Defense contractors are much more likely to find customers with these desirable attributes in civilian government projects than in commercial markets. If commercial customers do not have these attributes, then engineers must learn practices that are suitable for commercial markets and have these practices supported by appropriate institutional changes.

Aside from conditions that lead defense contractors to succeed in commercial markets, Feldman also raises a policy issue regarding the specific markets which defense firms should enter.

He indicates that one of the reasons why Boeing-Vertol abandoned the rapid transit cars was the uncertainty that the federal government would provide financial support for mass transportation. This raises the question of whether the government should actively encourage defense firms to enter markets which respond to public service needs. Over the years, traditional defense industry firms have explored other markets which involve government support or active participation on the assumption that defense contractors have the knowledge and capabilities to better respond to government influenced markets. The track record to demonstrate this assumption is mixed at best. A generation ago, the defense industry responded to the emerging market for alternative energy prompted by activist public policies. This experience exposed the problems associated with transitioning to consumer-based markets, and the limitations of applying marketing and distribution systems associated with the more limited defense sector. Others refer to the challenges posed by pursuing such a strategy.

Berteau's analysis of the IT service industry suggests that most defense-dependent IT service companies will survive even if they don't diversify into commercial markets. The demand for IT services is growing within the defense sector and, more generally, within federal, state and local governments. Demand will grow because IT services reinforce an ongoing trend to cut costs by rationalizing decision-making and information flows in all organizations, public or private, and by outsourcing in-house services. These trends can be expected to continue for quite some time. Furthermore, although few new weapons will be developed over the next decade, the focus of modernization of existing weapons will be information intensive.

If defense-dependent IT service companies were to diversify into commercial markets because the prospects for growth in commercial markets are greater than in government markets, then an interesting question is whether these companies' prior exclusive experience in contracting with DoD will put them at a competitive disadvantage relative to commercial companies that never had DoD contracts. How much do these defense-related companies have to unlearn in order to compete in commercial markets or must they acquire or merge with existing IT service companies that have commercial experience? As Berteau suggests, the prospects for growth in both commercial and government sectors are high enough to encourage many IT service companies to seek new business in both sectors. Both DoD and IT service companies will benefit to the extent that economies and learning between sectors is facilitated. This can happen through acquisition reform and removing any impediments toward standardization of software as the IT industry matures.

Acquisition reform would eliminate many of the reasons for IT service companies to segregate their defense and commercial businesses, but what else can government do to facilitate dual-use application of IT services? Berteau suggests that if defense officials could define performance requirements more clearly, then IT service companies could do a much better job of delivering their services. This raises the question as to whether it is inherently more difficult to define performance requirements for DoD services because of their complexity or the conditions under which such services are performed. Also, is the lack of clarity due to the poor selection and training of the DoD personnel who are responsible for articulating those requirements, or due to laws and regulations which prevent hiring and retaining the most knowledgeable and experienced personnel from private industry?

PART FOUR: GOVERNMENT PERSPECTIVE

As discussed earlier, the Department of Defense encouraged defense-related firms to pursue consolidation by relaxing Federal Trade Commission restrictions on mergers and acquisitions and by partially reimbursing overhead costs associated with restructuring. The latter initiative as well as many other government initiatives that affect firm behavior require Congressional approval. Nevertheless, Alic suggests that differences between the Congress, DoD and the military services on funding priorities and weapon systems requirements are so deeply embedded in institutional structure and practice that changes in rules and regulations alone will not produce the desired results. DoD also can lower costs and enhance capabilities by directly changing the way it operates internally and, indirectly, by changing the rules and regulations under which defense-related firms currently operate. Acquisition reform and outsourcing of non-critical services has achieved some of these objectives.

Alic raises interesting questions about where the focus of acquisition reform should be. As in commercial new product development, clarity and consensus regarding concept identification and development have perhaps the greatest leverage for cost reduction in downstream phases. However, clarity and consensus in these phases are obscured by existing rules and regulations that are imposed by the Congress and by competition between the military services for the weapons that each service wants. In the post-Cold War period, consensus on threats to national security is less likely than ever, impeded in part by vested interests from those who have benefited from past perceptions of threats.

In spite of the significant downturn in the procurement part of the defense budget, the research, development, test and evaluation (RDT&E) has remained virtually constant. For fiscal 1998, RDT&E is about 80 percent of expenditures for procurement. RDT&E expenditures may have remained nearly constant because of a continuing commitment to high technology warfare. It would be interesting to see whether there has been any shift in the high technology areas in which RDT&E dollars have been spent. Undoubtedly, there has been a shift towards greater expenditures in electronics. As Alic suggests, there would be significant savings if mechanisms existed to increase consensus on what weapons were needed and what their performance attributes should be. Some differences might be easier to reconcile if program officers had more discretion to negotiate with defense contractors about the operational implications of these performance attributes.

The implications of the high ratio of RDT&E to procurement expenditures for defense contractors are interesting also. If two-thirds of RDT&E expenditures go to private industry, as Alic indicates, then which companies are the recipients of these twenty billion-plus dollars? Are the big consolidators the recipients, e.g., Lockheed Martin or Raytheon, or are other prime contractors which chose different strategies? Who are the second-tier suppliers associated with these prime contractors? Can the prime contractors who receive these RDT&E dollars prosper with limited prospects for follow-on contracts? If they cannot, then is it possible for them to leverage the capabilities that they developed from their RDT&E contracts into commercial applications? Dual-use initiatives tend to focus on the design and manufacture of components and subsystems for defense and commercial applications. Can the knowledge and skill required to manage large-scale research and development projects be formalized sufficiently to transform them into candidates for dual-use application in the commercial sector?

David Berteau puts acquisition reform into perspective by offering an overview of the conditions that led to the creation of the Defense Conversion Commission in 1992. He identifies the political and economic forces that made acquisition reform possible at that time. Finally, he summarizes the Commission's recommendations, describes the actions that were taken, assesses the results that have been achieved, and outlines what he believes remains to be done.

Paul Kaminski, one of the architects of acquisition reform, discusses a variety of initiatives to reduce costs by encouraging DoD and defense-related firms to use commercial components and products in weapons systems and in operations and maintenance activities. Porter also suggests a similar set of initiatives. For example, program managers and acquisition officers should be required to use commercial components and products unless there is clear justification for a mil-spec. In cases such as software, decisions on open system architecture are a prerequisite to being able to insert commercial products into existing platforms. The Single Process Initiative is aimed at reducing redundant processes that proliferated as a result of lack of awareness by program managers that similar processes already existed elsewhere to accomplish the same purpose. Significant savings can be achieved in operations and maintenance by shifting attention from unit costs to ownership and maintenance costs. Finally, commercial accounting systems need to be adopted wherever possible so that the focus shifts from cost certification to cost savings.

Erik Pages discusses the outsourcing of non-critical services such as payroll processing, housing, travel, and utilities to private vendors. These services have been performed in-house traditionally because no private sector sources for them existed previously. DoD has the discretion to outsource, but the most troublesome issue is that DoD currently writes most service contracts in a manner which discourages private sector firms to bid on them. These contracts require vendors to assume all of the risk in return for relatively little reward. The decision to outsource should be accompanied by a job transition program to retrain or re-employ those whose jobs are displaced by outsourcing. Private vendors that contract to perform the formerly in-house services often hire these displaced workers.

Kaminski also discusses partnerships between government, industry and universities. Encouraging private industry to invest in specific basic research areas may be a way to leverage the use of scarce dollars in research areas with long-term interest to DoD. Kaminski also discusses teaming between large and small companies, which, in his view, combines large company "mass" with small company "velocity" or innovation capability. The Technology Reinvestment Program encouraged such teaming by favorably weighting proposals in which small and large companies were teamed in joint research projects. The TRP also funded government–industry partnerships which sought commercial applications of existing defense-related technologies. While some private firms were subsidized for developing a technology from which they profited—always a concern of free enterprise advocates—the government also benefited from the dual-use application of technology, which lowered costs through increased economies of scale.

PART FIVE: CASE PRESENTATIONS

Senior executives from eight defense-related companies were invited to the Klein Symposium and two senior executives from different units of a large defense-related company were invited to the Bantle Symposium to make presentations on their experiences in responding to the sales plunge that followed the dramatic defense budget cuts of the early 1990s. Five of the eight executives who spoke at the Klein Symposium were from small to medium sized companies which had sales that ranged from $10 million to $1 billion per year. The other five executives were from companies or units of large companies which had sales that ranged from $4 billion to $27 billion per year.

The five small to medium sized defense-related firms survived and prospered after the defense downturn by concentrating on their core competencies. They developed these competencies further by acquiring companies that had competencies that enhanced or complemented their own. Sometimes they entered markets in which their existing competencies were well suited to rapidly growing markets, e.g., wireless telecommunications, medical imaging, and automotive electronics. Galileo Corporation saw modest prospects for commercial market growth for infra-red sensors, but it drew upon its more generalized competencies in opto-electronics and successfully entered markets for remote medical sensor products. Foreign sales accounted for a significant portion of the sales growth of many of these firms. Success also was based on selecting customers that were similar to their defense customers, e.g., Base Ten Systems worked with the Food and Drug Administration and pharmaceutical manufacturers, or to selecting markets in which contracts with a few large customers led to large and relatively secure sales volume, e.g., BEI Sensors and Systems selected the automobile industry.

Both large and small firms made acquisitions for a wide variety of reasons. Northrop acquired Grumman and Westinghouse Defense Electronics in order to be less dependent on air frames. Hexcel acquired Ciba in order to enhance its presence in foreign markets. It acquired Hercules in order to backward integrate into carbon fibers and acquired Fiberite in order to forward integrate into satellite materials and prepregs for use in final consumer products. ORBIT/FR acquired Advanced Electro-magnetics to broaden its competence base and to obtain control over an asset which it viewed as important to future market growth. Lockheed Martin acquired launch vehicle and satellite manufacturing capabilities from General

Dynamics and General Electric, respectively, in order to enhance its position in the burgeoning military and commercial telecommunications market.

Many of the products that these firms developed originally for military applications were sold with minor modifications to commercial customers. For example, ORBIT/FR did this for measurement and test equipment for microwave antennas, as did Harris Corporation with deployable antennas for hand-held devices that communicate with the GEO satellite. Sometimes, the experience that led firms to develop and produce more refined and less expensive commercial products found their way back into military applications. Lockheed Martin developed a video game with Sega that was based on tank trainer technology. This resulted in a smaller system with improved screen resolution that now is being used to improve tank trainers. BEI Systron developed solid-state quartz gyros for the automobile industry that are now being sold to military customers for whom the gyros were developed originally. Acquisition reform, e.g., the Single Process Initiative, is making such dual-use applications and cross-sector learning possible. DoD officials have shown refreshing flexibility in accepting commercial quality standards, e.g., ISO 9001, in lieu of Mil-Q 9858A. BEI Systron has benefited from such flexibility. Base Ten Systems found that its experience with Mil-Q 9858 was very helpful in being able to meet FDA's stringent certification standards.

Most of these firms recognized early that selling to commercial customers was very different than selling to prime contractors or to DoD. Commercial customers focused on performance requirements, not on specifications. Also, they paid for products upon delivery, not when major design and production milestones were passed. Consequently, these firms set up new divisions and staffed them with people who had the requisite experience and skills to locate and sell to commercial customers or they partnered with companies that already had developed distribution and logistics channels to such customers. Harris Corporation successfully developed and launched several new products through forming "channel partnerships". Such partnerships are especially critical for defense-related firms that enter fragmented markets with a large number of relatively small customers. A large firm like Northrop-Grumman learned this lesson with its unsuccessful venture into law enforcement surveillance equipment. Northrop Grumman and the Control Systems Group and Federal Systems Group of Lockheed Martin did very well when they managed large-scale system integration projects for single large customers. A small company like Galileo, however, learned that reliance on a few large customers can leave it highly vulnerable to unexpected contract terminations.

Finally, these defense-related firms recognized early that successful competition in commercial markets requires constant effort at productivity enhancement through capital investment and employee training. BEI Systron invested heavily in automation to manufacture solid-state quartz gyros. Hexcel invested in a variety of state-of-the-art methods to eliminate steps from the manufacture of composite materials. Galileo significantly raised its productivity by involving teams of employees in the continuous assessment and re-engineering of its manufacturing processes.

REFERENCES

Jaikumar, R., "Post-Industrial Manufacturing." *Harvard Business Review*, pp. 69–76, November-December, 1986.

2

THE EVOLVING NATIONAL SECURITY AGENDA: THE SEARCH FOR PUBLIC CONSENSUS[1]

THE HONORABLE JOHN J. HAMRE
Deputy Secretary of Defense

This is one of the hardest audiences in terms of thinking about the content of my speech. Frankly, you have been thinking about these issues for a long time, so it is pretty hard to blow smoke at you. I am not going to try. I thought that I would take a few minutes to give you my evolving perspective on the Department of Defense.

I have made a few observations over the last few weeks. Number one, I was startled the other day by a press synopsis of a recent poll conducted by the University of Maryland. The poll, in examining support for the defense budget, found that 80 percent of the American public, when pressed in various ways, thought that we should make more cuts—80 percent. In a poll conducted about a year and a half earlier, surveyors named the six nations that the United States was most likely to face in a conflict: Iraq, Iran, China, North Korea, etc. Then they asked whether the United States should spend: (a) as much as the largest of those opponents spends, (b) as much as all of them put together, or (c) twice as much as all of them put together. Only 7 percent of the American public picked "c" Our budget is actually "c." First, let me say that a comparison of published budgets is not fair because one cannot compare, say, the Chinese defense budget with ours. But the basic perception is startling. The implication of that basic perception is that we have not adequately explained to the American public that the costs of maintaining the world's finest military are substantial. That is observation number one.

Observation number two. I am surprised at the lack of debate in the country over defense issues. For example, there has been little public debate on NATO enlargement. This is noteworthy in that we, as a government, are going to change a fundamental premise of our security policy as it has been conducted over the last half century, and we are not engaging in a thorough public debate on the issue. Another example of the lack of rigorous national debate is the Quadrennial Defense Review. I would like to think the QDR is such a solid product that we do not need to talk about it. The truth is that it is a solid product, but it needs more debate, because debate on issues of this magnitude helps us as a nation to galvanize our thinking about our current and future goals. Observation number three. In the absence of this debate, we have a default consensus on the defense budget. There has been little discussion nationally to propose alternatives that differ dramatically from the status quo. Competing political interests, which counter-balance each other, have determined the defense budget rather than a vigorous, informed debate about priorities and requirements at the national level.

With those observations, let me try to draw a few conclusions. Number one. It is extremely difficult in this environment to postulate a strategic direction, even though the Department has done so in the QDR. It would be difficult for anybody. When few people are listening—since the American public is not engaged—it is very difficult to propose a dramatic change of direction. So this tends to lead toward strategies that embrace virtually everything because there is no clear path. In a couple of weeks, the National Defense Panel will be issuing its report, and it will say we really ought to be focusing on the very different worlds that may emerge 20 years from now. Well, that is a great idea, and I'd like to pick which room I get when I arrive in Heaven, too. We have got to live through the very dramatic and interesting challenges we face over the next two to three months, let alone the next two to three years. So in this kind of environment it is understandable—maybe not right, but understandable—that one's strategic direction will be informed largely by a desire to conserve that which is definitely valuable right now. It is very difficult to branch off into dramatic new directions.

Number two. In this environment, we are going to be tugged in many different ways by Congress—too often over second- and third-order issues. With that frame of reference, it is very difficult to hold a strategic focus because the institution of Congress is so important to the Department of Defense. It is the crucible that turns an individual idea into a national commitment.

Number three. In this environment, it is more important than ever to use the tools, the resources, and the people that we have to formulate our strategic long-term thinking. We have to do that. If there is no clear outside threat to guide our thinking right now, and there is no clear consensus inside our country, then we have to get it right ourselves. We must have a disciplined internal process to think about our future. I noticed when I was speaking earlier about the QDR that some eyes were rolling. Everyone, of course, is entitled to his or her own view. What the QDR did was bring all of us together into a wrenching and difficult process where we thought in fairly detailed ways about our future. I would posit that no other organization in the world spends as much time studying its future as ours. And I would argue that we have thought about every possible strategic surprise. I do not believe, for example, that some super-duper beam weapon—a weapon whose implications we have not been studying in a fair amount of detail or through modeling or simulating—is going to suddenly emerge from some laboratory. I do not believe a threat is going to emerge that we have not thought about in detailed and systematic ways. That does not mean we have all the answers about what to do about a terrorist organization that uses chemical or biological weapons in a civil setting like the Tokyo subway. We do not have a comprehensive solution to that vexing problem, but we sure as hell are thinking about it. In this environment it is enormously important for this institution to think in a very systematic and disciplined way about our future.

You may not like elements of the QDR, and you may think that we missed huge things. I would be delighted to engage in that discussion. But I honestly believe that we have thought about this in a very disciplined way. It is possible that we got it wrong. Human beings can fail. Even in such a large and diverse organization as ours one still must come to a single recommendation to inform the decision that the Secretary of Defense makes. But I would venture to say that we have thought about these issues in a more disciplined way than anyone else. That is not to be smug or say that other views are unimportant, but no one should suggest that we came to any easy quick answers about our future.

Let me wrap up by posing some questions and offering some preliminary answers. One of the first questions it is fair to ask is how real is this revolution in military affairs? My personal view is that no other military in the world is looking so systematically at the interaction of technology and organizations as ours. We are now in a qualitatively different posture vis-a-vis any conceivable opponent. We will be in a position to largely shape the way any engagement evolves tactically once we get there. When it comes to dealing with an environment that emerges, particularly if it involves some of these terrifying new potential weapons like chemical and biological weapons, we still have significant challenges. But I would argue that our opponents do too. It is not clear to me how they would gain a strategic advantage by using those weapons. Certainly they would gain some tactical advantage, but not necessarily a strategic advantage. So it is not clear to me why a thinking, rational government—setting aside terrorist organizations—would choose such weapons. Where you will see the most promise in the revolution of military affairs is going to be in the application of simulation technology and organizational decision-making. Tactically, especially for battlefield advisors and senior individuals, that is going to be the breakthrough that will qualitatively take us to the next step, the next evolution of our remarkable military capability.

Second, what is this revolution in business affairs? I think there is a lot of encouraging but isolated evidence that as a business entity the department in many ways has done some pioneering work. But it is very uneven. It tends to be internal with respect to organizations rather than on a broader enterprise-wide basis. It clearly must be the area we emphasize the most over the next couple of years because it is the only area where we will find any real growth in our budget. Frankly, finding greater purchasing power in our existing budget must become the centerpiece of our efforts over the next couple of years. As I've rambled on, I hope I have given enough provocative thought to stimulate some questions on your part.

NOTES

1. Keynote address at Bantle Symposium on Industrial Diversification: The Public Policy Perspective, Maxwell School of Citizenship and Public Affairs, Syracuse University, October 26, 1997.

3

DoD IN THE 21ST CENTURY: THE ROLE OF THE PRIVATE SECTOR[1]

JOHN P. WHITE

Reforming the elements of the Department of Defense which provide support to operations may be one of the most critical functions for the future. If you think about areas in which American business ought to participate more fully, this is a major opportunity. We are foregoing somewhere between $10 and $20 billion a year, which is getting effectively chewed up inside the Department because we have too much infrastructure, are too inefficient, and waste a great deal of money. The private sector ought to, in my judgment, take a much stronger role in terms of leadership in the debate about what we ought to be doing to reform the Department in the area of support functions.

To effectively engage in this debate it is important to set the perspective. First of all in terms of the world that we now live in, the demographics are such that we know as we enter the next century we will be living in a significantly different world. Just look at where birth rates and migration rates are in different parts of the world. We know, and we didn't have to be reminded of it recently by the capital markets, that we live in a very economically interdependent world and we live in a world in which there is a great deal of change. I would suggest that most of that change on balance in the political realm, is in our favor, not against us. That is to say, over the last decade the movement has been more away from people who in fact were armed with confrontational policies against the United States and what we stand for, to those who are in a more conciliatory mood, or even want to move aggressively closer to us.

One of the things that you see at the top of the Defense Department is the number of people, particularly heads of state and ministers of defense, who insist on coming to see you every time they are in town. Now that is because people, and this is not just our traditional friends—but across the spectrum from all over the world—perceive correctly, I would submit, that the Department of Defense, plays a very pivotal and major role in terms of the outcome of a whole array of events around the world. My most cherished proposal to Secretary Perry was that, instead of having all of these interminable parades (we would easily have between three and four visitors a week) that if their countries were below a certain G.D.P. you would get a video instead of a parade. It would have saved us a lot of time.

If I turn to the security forum issues, my argument today is that, relatively, we are well off. We do not have major threats around the world—in the sense of a global threat. We may have those in the future, but if so, I'm of the school that says we will see some of that coming—whether or not we are wise or foolish in terms of how we respond to rising threats—our history has shown that we have done that both ways—we will see. We have regional threats, as we are all aware, but those regional threats are less than they were a decade ago, with the exception of maybe Korea, which continues to worry me every day. I think it is the thing that the Secretary and Deputy Secretary ought to worry about most. And then, of course, we have crises and contingencies, which keep us busy around the world, but in fact are manageable. Finally we have a whole set of symmetric challenges: loose nukes, biochemical threats, terrorist threats or information war threats, infrastructure threats. I would submit that we don't know how these are going to manifest themselves or how much trouble they are going to cause. But we do know that terrorism and information warfare are very fundamental threats and clear vulnerabilities.

All of this requires, in my judgment, a significant amount of innovation from where we are today. And that innovation has to come, in large measure, from the Department of Defense. Turning from the world at large to current policy, I think our policies are well-known and well-articulated, certainly in this administration, both in terms of economic policy and political policy. I will submit that in large measure, when there is a change of administration that change will be one of emphasis not one of fundamentals. Now that brings us to the national security and DoD policy.

First turn to the Quadrennial Defense Review (QDR) because that sets the parameters for effective private sector involvement. First of all, it seems to me, in the QDR we faced three fundamental challenges. The first was that the world had changed in a great many operational ways that we hadn't anticipated before and we had to make adjustments for that change. The QDR was enormously complicated by the fact that we had a second overarching problem—a very major imbalance in terms of our resource allocations. We had gotten too much up-front in terms of the current expenditures relative to what we ought to be putting into the investment accounts and we had to deal with that resource misallocation, unless we wanted to forego what I considered our responsibility to make sure that we were funding the future. Which means, roughly a procurement budget of about $60 billion at the end of the program years. The third issue is the uncertainty of all of this as it relates to planning. That is, we have to do something as we go forward with respect to how we deal with uncertainty, and we can translate that into the revolution in military affairs and a lot of other labels. But the issue is fundamental. We do not know where we are going, nor should you expect that we would—not having been there before. The world has changed a great deal and we have this inherent tension as we did in the QDR, between trying to understand that we ought to live with a great deal of uncertainty and that military planners need to be given some direction and specificity in terms of what it is they are supposed to plan for.

Those are the three fundamental issues. The first issue has to do with what we have done with shaping and responding and other changes in the strategy which require a good deal of innovation. I would argue that in fact there are very significant changes of that sort in the QDR. They were necessary changes. They are not fully implemented, but they clearly will be important and will make a difference.

Let me turn to the second resource misallocation issue. The magnitude of the problem is somewhere between $15 and $20 billion a year. We have a shortfall or a soaking up of resources to the tune of $15–20 billion a year in our current accounts, in our operational account, that we would prefer to have spent on investment. That's a big number. It's a number that has persisted now for several years despite our best efforts. We have not been successful to date in fundamentally changing that number. We continually and embarrassingly miss our estimates for the program years. For example, in the 1995 budget we estimated a 1998 investment budget of $54 billion and we came in at $42.6 billion. I will tell you we started that process at about $39 billion and it took heroic efforts to get over $42 billion which is a big difference from where we were. In my view, the most important problem is the unprogrammed operations. This usually means people coming in at the end of the program year and telling you "I've got 'n' billion dollars worth of maintenance that I didn't know I had," "I've got 'n' billion dollars worth of steaming hours that I didn't know I had," and so on and so forth. There are some contingencies in there but most of these are management issues and not surprises in the sense of contingencies. The second element, which is also equally large, has to do with the problem of unrealized savings. "I told you I was going to save 'x', I didn't make it—I only saved one-third of 'x'". Again, these are management issues. They have to do with our inability to manage the process, particularly the process of change, in a way that allows us to predict going in where we will come out.

Now you can, of course, affect that by making more conservative predictions. Managers will tell you that you will get a smaller surprise, but you will also get a significantly lower response in terms of what you are trying to do. So you have a very basic incentive problem here in terms of what it is you do, and then you have these surprises or shortfalls, which are very significant. Then there are new program needs. New program needs are going to occur. The inherent nature of the national security business is that you are going to have surprises. And that means you are going to have new programs arise whether they are national missile defense, or theater defense, or whatever they might be. Some of them will be significantly more expensive, in the billions, than you would have anticipated. All of these, once again, are management problems. These are problems of how you run the Department. There is only one place to get the money and that is out of the investment accounts. As a result of getting it out of your investment accounts, you wreak havoc with a whole set of other programs—drive up their costs through changes in quantities, through changes in planning factors and increase technical risk in those programs. And those compound the threat that you have of the $15–20 billion a year shortfall.

Much of this is caused by the fact that we haven't taken down the infrastructure. The budget from 1985 until the forecast for 2003 will go down about 40 percent, the end strength about 36 percent, base structure only about 26 percent, yet some 61 percent of all our people work in infrastructure. We proposed solutions to the problem, in large measure, in the QDR track. We are going to cut some manpower—military and civilian. We are going to adjust the modernization account some, but not a great deal. Most importantly, we are going to try to change the support operations by cutting the size of the infrastructure program, by reengineering and by increasing the outsourcing and hopefully saving somewhere on the order of $7 billion. If we can also persuade the Congress to do a couple more rounds of base closures, which are sorely needed, maybe we can get another $3 billion. So now with some estimates putting us at $10 billion, it's still necessary to look for another $5–10 billion. So we are a very long way from solving our problem and, on top of that, we have big investment risks, contingency risks and other factors that we have forgotten about along the way as we do these calculations. Hence the determination is that we need a significantly larger set of infrastructure changes, and I will return to that because that is the heart of the argument.

If we do not provide a way to find these savings internally, because the budget is not going to get any larger, we are going to underfund investment relative to the consensus of the senior management in the Department and it is going to be very significant. You don't have to compound $10 billion a year for very long to realize that you are talking about very major changes to what you are able to provide your fighting forces in 15 years. By the time you realize that you didn't get it right it is obviously too late.

So this is a very, very fundamental, critical problem. It is as if you were operating a company, spent no money effectively on research or new product development, and only sold your current products. You will do well in the short-run and in the long-run you will go out of business.

The third issue that I mentioned had to do with planning uncertainty and the inherent nature of the problem we face and the tension between needing to know what is necessary in order to plan and the recognition that we don't know where we are going. If you go back through the Commission on Roles and Missions report, you will see that we spent a good deal of time worrying about this issue. I think it still deserves a lot of attention and it again has to do with innovation and changing the way you do things. It has to do with making the Department more flexible, more responsive and so on. I think those kinds of changes are critical if you are going to solve the first two problems. You cannot solve the problems with respect to changing the operations and being more responsive to the problems of management performance if you don't change the character of the way you are doing business in the Department. So they are all of a piece and, therefore, fundamentally important.

We have some pretty good context for this. It is not as if these are problems that we just discovered; these are problems that we've had for quite some time. They are clearly more obvious to us in this era than in the prior era and rightly so. But Goldwater–Nichols, I would argue, generally provides a format in which to deal with them. The kinds of planning that has been going on in the Department—Vision 2010, the Roles and Missions report—there are a whole set of things that speak to these issues. This is not all new or mysterious but I would submit that we need something substantially more. While there have been a number of actions taken, we do need a much more fundamental change. We really do need a revolution in business affairs. We do need to change the way the department does business. And the private sector provides us with the perfect model. The private sector has lived through a similar problem and has solved that problem in quite an incredible way and provides us with a great deal of information in terms of the kinds of things that have to be done if we are to be successful. Again, it seems to me that we will not be successful if we only do it through public management. We have to have the support of the private sector. Not just its business support, but its political support.

Let me talk just a bit about what is going on in the private sector. First of all, in the early 1980s we had big problems. We had lost our way. We were not investing enough. Our innovation rates were down. Profit margins were down. We were losing a huge amount of market to our global competitors. People had discovered alternative models of doing business, such as in Japan, and explained to us patiently that these models were not only better than ours, but in fact were going to destroy the American free-enterprise system because they were so much more attendant to what was needed in the world. And many American commentators accepted those views. Some, importantly, did not. The American business community decided that it couldn't live with the result of what was being forecasted. Samuel Johnson said, "There is nothing that clears your mind like learning you will be hanged in the morning." In this case we had a magnificent transformation of American business. How did we do it? We did it by focusing on our core competencies, by ruthlessly cutting our costs, by stressing innovation, by outsourcing damn near everything that is not central to our business, by flattening our organizations and reaching out to our customers and finding out what it is that they really want, and by embracing change as a cultural way of behaving as opposed to the traditional way of behaving—as in looking continually for some form of stability, when in fact you are not going to have stability. American business did this in the 1980s and, as a result, was poised, when the world economy turned around, to exploit that in ways that no other major economies have been so poised in the 1990s. U.S. industry got it right. The Japanese did not get it right. The Europeans did not get it right any more than they have gotten it right in terms of their defense industry problems. I want to caution us against any hubris, but we should understand that fundamentally our model worked better than any other and as a result we are the low cost producers of steel and autos and chips and software and so on. It was not easy. It took a lot of wrenching and took a lot of headlines about downsizing and caused a lot of pain to a lot of people. But not nearly as much pain as if we had not gone through this fundamental transformation—and we're still going through it. It has become a way of life, as it must.

I would submit to you that this is the model that DoD ought to adopt. It is not doing business like a business. Now DoD is not a business, but we must accept modern management practice and implement that practice. We have a lot of precedence for it. We do a number of these things as we go through our reforms in the Department, but we do not do nearly enough of them in order to reap the kinds of benefit that we need if we are going to be successful. I think DoD ought to follow the example of American business. We ought to focus on our core competencies . . . joint combat operations. We ought to embrace change. We ought to push forward on innovation. We ought to cut costs. We ought to outsource everything that is not core to what it is that we are doing. And we ought to do all of this very, very aggressively. If we do that, we will be well on the way to providing the wherewithal to create the investment level that we will need in the future. If we do not do that, I think we are doomed, first of all, to underfund the national security of the future. Secondly, and more fundamentally, I think this is all going to catch up with us politically. That is to say, the American people are going to wake up and say "$250 billion—wow, that is a lot of dough and tell me again what you guys are doing with that?"

And our answer is going to be "Well, we got a lot of these bases and depots and we are a little slow, but we'll catch up after awhile." That is an unacceptable answer.

I think there are a whole set of quite specific programs that we ought to both expand and implement.

Acquisition reform—we are not there yet. We have made some progress, but not there yet. And it is going to take persistence. Secretary Bill Perry once said to me, "This is the third time I've tried acquisition reform. This time it is beginning to work". Personnel reductions—again, another 39–40 percent from 1989 on into the future of the program. It forces us to do these other things, such as base closures, and relief from being required to do 60 percent of our maintenance in public depots. There is a whole laundry list of changes that we need from the Congress, but first we need to get through the major ones. We need to change OMB Circular A-76. OMB Director Frank Raines has demonstrated some sympathy to these issues and has demonstrated that he is willing to make significant changes. We need to retrain DoD managers so that they are more skilled at managing this outsourcing process and managing these contracts, not simply managing an in-house workforce. We need to create specific programs to reach out to non-DoD industry. We are no longer the dominant buyer that we once were in this marketplace.

Much of the technology that is very exciting in the world today doesn't come near DoD. Therefore, we are going to have to reach out and find it. I used to be in the software business. Software kids today are going to work for Disney; they are not going to be as attracted to Lockheed Martin as was true before. And we better find out what is going on at Disney and at CNN and Biogen and so on. Because people at places like Microsoft are on the cutting edge of what is going on technologically in this society and they are not thinking about DoD problems. If you are going to adapt what they are doing, we are going to have to reach. By the same token, you need to find ways to reach the young people in the universities and industry who are leading the way through these kinds of changes. The people managing change in DoD are too old. I used to complain when I was in the Department when I asked a technical question they would send somebody with the answer that looked like me. I said I want somebody who looks like my kid! I want someone who is thirty or thirty-five years old. We tend not to do that in the Department—for all kinds of civil service reasons, or other habits. Even our advisory boards tend to be too old. They are very good people but we are missing a significant bet in terms of the future. We have to do civil service reform that will allow us the flexibility to hire and retain people who can provide us the skills we need in this new environment.

Finally, and most importantly, we need to reengineer a whole set of functions in the Department. Not agencies—functions. The agencies are not important. There is a lot of discussion, and was during the QDR (particularly from the military services), that we ought to go find the money in the defense agencies. My answer was, "That's terrific, I would love to do that. But the problem, General, is that money is your money. They are buying stuff for you. So would you mind telling me which one you want cut and by how much. Because I'm more than happy to do it, because I agree, they are too big." But I never got any answers to questions like that because people aren't serious. The issue is not defense agencies—the issue is functions. And we have lots of functions in the Department that we ought to reengineer. We ought to reengineer MIS, which is a disgrace. We ought to reengineer logistics. We ought to reengineer basic administration and base support, and spares, depots, health, housing, energy, and on and on. There is no end to the opportunities. The problem is that if it is to be done in a way that will create fundamental change, you must take some major risks and you must have a long-term program. In general, you will have to have a program that will last from three to five years. We need to set up a blue ribbon panel of experts who understand how to do this and have done it in industry and of academics who have studied it and have pointed the way in terms of what has happened in the private sector. We know who those people are and we can attract them to help in this enormously fascinating and important enterprise to do exactly the same things that have been done in the private sector. Obviously we need mechanisms that create extensive "buy-in" from the people who perform these functions, since they are the only ones who know how to reengineer their roles—those of us at the top or on the outside can't do that. We have to have incentives for people to do that. I think all of this is perfectly doable. It will take a major effort that will have to be put together in a way that it can be marketed and articulated and committed to such as has been done with acquisition reform. The Department has to buy into the fundamental arguments and the private sector has to buy in to the stake that it has in this opportunity. We, in fact, are leaving $10–15 billion of investment funds a year on the table because we are wasting it on other activities that are highly inefficient simply because we do not have the wherewithal or the political momentum and persuasiveness to make the changes that are necessary to have the Department operate in ways consistent with modern management practices.

We must accept our responsibility to make these changes. If we fail to do so, we will under-fund the investment programs and this could have truly dire consequences for our military capabilities in the first quarter of the next century.

NOTES

1. Keynote address at Bantle Symposium on Industrial Diversification: The Public Policy Perspective, Maxwell School of Citizenship and Public Affairs, Syracuse University, October 28, 1997.

PART TWO:

CORPORATE AND GOVERNMENT INTERESTS

4

SHAPING THE STRUCTURE OF THE AMERICAN DEFENSE INDUSTRY

DAVID S. C. CHU AND MATTHEW C. WAXMAN

Concern with the structure and focus of the defense industry is a recurring theme in American public policy debates. Its corollary is typically a call to intervene, or to change the rules of the marketplace in a manner deemed favorable to the desired result. At the height of the Cold War, some complained that the self-interest of the defense industry perpetuated the conflict, and they urged the "conversion" of the defense firms to civil pursuits (Congressional Budget Office, 1980; Gottlieb, 1997). In the late 1970s and the early 1980s, worry grew about the adequacy of the "lower tiers" of the defense industry—that is, the suppliers of materials and parts to major defense contractors (cf. Baumbusch and Harman, 1977). Concerned observers proposed steps to encourage more firms to consider defense business. During the Bush administration, critics complained that American competitiveness depended on the Department of Defense (DoD) using its funds to encourage specific technological directions; over the President's objections, they succeeded in funding an experiment in industrial policy.[1]

The Clinton administration brought its own set of proposals to restructure the defense industry. Soon after taking office, defense leaders summoned the chief executive officers of major prime contractors and overtly encouraged the merger of competitors. They were concerned about the prospective effect of limited defense procurement on the rate of capacity utilization (Korb, 1996: 22). At the same time, policy makers expressed considerable enthusiasm for opening up the technological treasure chest guarded by defense contractors, encouraging them to apply their military technological expertise to civil problems.[2] The chief result of the administration's industrial policy has been the rapid consolidation of the industry, and the concentration of certain supplier markets through mega-mergers (Donnelly and Clark, 1997).

With the exception of President Bush's objections, it is remarkable how seldom the call for government intervention in the defense sector has been challenged. This nearly uninterrupted advocacy seems curious when considered in light of the now widely recognized principle that governments intervene in market outcomes at their own peril, lest they produce unintended consequences worse than the condition they sought to correct. Some would reconcile this conflict by arguing the uniqueness of the defense market. As Gansler (1995) put it: "It would be proper to simply allow the market to operate to achieve the required efficiency and effectiveness if the defense marketplace was truly a free market, with numerous suppliers and buyers free to act in their own best self-interests. This is not an arena, however, in which it is even relevant to ask whether the government should be involved (p. 24)." But such a view, while logical, only begs the question of where the defense sector ends and other markets begin.

This chapter contends that before the debate on defense diversification can move forward, it must take a step back. Some of the assumptions that underlie it require re-examination; with simultaneous changes in defense resource levels, the nature of warfare, and U.S. strategic priorities, such re-examination is especially important. Designing the specific architecture of any government intervention in the defense industry can only proceed productively once basic questions have been re-explored, and the accepted answers either reaffirmed or revised. This chapter seeks to address three of those questions: What is a defense firm? Why should we care about diversification? And what kind of diversification is desired?

WHAT IS A "DEFENSE FIRM"?

It is astonishing that the debates over the structure and focus of the defense industry have advanced so far despite a significant hole in the literature, and that is a widely accepted and well-defined definition of a defense firm. The answer to the question "What is a defense firm?" is not merely of academic concern; the way we choose to define the scope of the industry depends on the policy objectives we have in mind, and it also reinforces those objectives by identifying the universe with which policy is concerned.

One way of defining defense firms would be based on the nature of the product the firm produces. That is, defense firms produce "arms" or "weapons." The Stockholm International Peace Research Institute defines defense firms this way, based on the volume of arms sales (*SIPRI Yearbook*, 1996). While most would certainly agree that weapons producers are defense firms, measures of arms sales do not mark a very clear boundary of the defense industry, nor do they account for the non-weapon technologies upon which the military depends. Can a computer be a weapon? Computer exports were controlled for military reasons in the Cold War—but much less so now, despite the advent of information warfare. Are they now less weapons? Is a militarized GPS receiver a defense item? Would our answer change if we acknowledged that the innards of the receiver are essentially commercial and that just the box is militarized? And what of such items as Navy support ships? It would seem that weapon producers constitute just a subset of defense firms.

Perhaps the most common way of defining defense firms is by volume of defense contracts. This can be thought of as a DoD's-eye view of the problem: who are DoD's most important suppliers? But some of these would not think of themselves as defense firms. In the 1996 *Defense News* Top 100 List, for example, AT&T is listed as the company least reliant on defense, with only 0.5 percent of its revenue derived from defense contracts. Yet it still ranked number 73 in total defense contracts. Similarly, General Electric derived only 3.1 percent of its revenue in defense, while ranking 21 on the total revenue list (*Defense News*, 1996). Are these companies defense firms? The military could surely not function without them, yet their corporate leadership would probably not view themselves as defense industrialists.

A variant, therefore, would define defense firms as those with a high fraction of their business composed of defense contracts. This reflects an industry's-eye view: which firms rely most heavily on DoD demand. Interestingly, this method and the preceding produce little overlap. For 1995, the two resulting top-10 lists looked as follows (*Defense News*, 1996):[3]

Total Defense Revenue	*Percentage of Revenue from Defense*
(1) Lockheed Martin	Dir. des Const. Nav. (100)
(2) McDonnell Douglas	Logicon (99)
(3) British Aerospace	General Dynamics (97)
(4) Hughes Electronics	Hindustan Aeronautics (92)
(5) **Northrop Grumman**	Tracor (91)
(6) Boeing	Oertikon-Contraves AG (90)
(7) **Loral**	Alliant Techsystems (83)
(8) Thomson Group	**Northrop Grumman** (83)
(9) General Electric	**Loral** (82)
(10) Raytheon	Australian Def. Ind (80)
	( ) = Percent of revenue from defense

Note that only two companies—Northrop Grumman and Loral—appear on both lists.

The primary deficiency of these metrics is their failure to capture the unique attributes of the defense industry as an economic sector, and the special set of relationships that exist between its members and DoD. As Tom McNaugher (1989) observed:

These firms do not face a market. They are private firms in the sense that they sell stock, borrow money in commercial markets, and seek an adequate return on investment. But the similarity to private firms stops there. Defense firms sell unique products to what has always been a single buyer, the U.S. government. The defense sector thus involves a monopsony in which prices are negotiated rather than set by market forces. Defense firms also face qualitatively different risks than most private firms. They engage in "winner take all" competitions rather than a fight for market share. Their sales are politically determined. And they face a constant demand for very high levels of technological advance. Thus risks are heavily subsidized if not covered entirely by the government. In short, defense contractors, especially those at the top of the defense industrial hierarchy, constitute a unique quasi-private, quasi-public sector of the nation's economy (pp. 150–151).

"Defense-ness" might therefore be thought of as the degree to which a firm has built the capability and competency to engage in business with agencies like DoD. Some of this involves understanding defense needs and how to meet them, including the particular bureaucratic politics that operate in any market. But any firm wishing to sell to DoD must also invest capital to develop the necessary expertise in, and continually accept additional overhead costs to ensure compliance with, the myriad accounting requirements found in the Federal Acquisition Regulations (FAR) and the specifications and standards imposed by DoD. A defense firm can thus be thought of as one that has the ability and the willingness to accept the inefficiencies and transaction costs associated with DoD contracts.

The validity of this approach is supported by the fact that a firm producing goods for DoD must often restructure entire portions of its organization and operations in order to do so. A number of recent industry studies have found, for example, that firms are likely to segregate their commercial and military production processes in order to comply with federal regulations and DoD-specific requirements (cf. Office of Technology Assessment, 1994; Van Opstal, 1991).

Anecdotal evidence also suggests that federal regulations and contracting requirements exert strong forces on firms' decisions to enter or exit the DoD supplier market:

> [M]any top U.S. companies, such as Hewlett-Packard, have concluded that DoD's business is not worth the intrusive government oversight that comes with it and refuse to do defense business on anything but a commercial basis. For example, managers at Intel Corporation recently tried to set up a government cost-accounting system needed before they could acquire DoD business, but after spending more than $2 million they decided the business wasn't worth the effort (Gansler, 1992: 51).

If investment requirements to do business with DoD affect not only the internal processes of firms but their strategic decisions about whether or not to become a defense supplier, then looking to such investment tells us much about the degree to which a company is a defense firm, from both the DoD's and the firm's perspective.

Not everyone agrees there are special costs in doing business with DoD, however. Indeed, some observers might go so far as to argue that it is actually easier to do business with DoD:[4] The rules of the game are published in the form of federal regulations (and thus easily knowable), and the business process is orderly and well structured. Moreover, one might ask if every market does not require investment in some specialized knowledge in order to be a competitor. The issue, then, becomes whether the investment costs of this kind for defense are substantially different from those of other sectors. Even if they are similar, provided they involve different capabilities, this definition would still yield a useful way to test which firms (in part or in whole) should be thought of as defense firms.

One might also point out that even if a defense firm is defined by its specialization in processes that allow it to do business in the defense sector, then perhaps we are talking less about "defense" firms and more about "government" firms, since the necessary specialization stems largely from generic federal regulations and practices. Perhaps it is better to think about defense firms as a subset of the larger category of federal government firms.

WHY SHOULD WE CARE ABOUT DIVERSIFICATION? WHAT KIND OF DIVERSIFICATION IS DESIRED?

As the preceding analysis reveals, defining the defense industry is not straightforward. Given the many possible approaches, which should we choose? The answer, of course, depends on the policy objectives sought. Since each definitional approach generates a different set of "defense firms," the approach we choose must reflect a reasoned choice about which attributes concern us most.

A major concern during the early part of the 1990s defense drawdown was the danger that, as defense procurement dollars dried up, the defense industry would dry up, too.[5] This threatened to leave the military without the necessary base to supply its needs, especially in the event of a crisis-induced surge in demand. As Gansler remarked (1992), "[T]he critical question is whether the defense industrial base that remains will be capable of producing next-generation equipment when it is needed and at a reasonable price" (p. 50).

With supply concerns in mind, it is natural to think about the defense industry in those very terms: how much does a firm contribute to DoD? If providing the military with its hardware and services needs is the issue, volume of contracts would be the definition of interest. The goal of diversification would be to ensure that there is a set of good future potential bidders. Such a goal implies that DoD would want firms to add to their defense business a set of activities that would keep them positioned as potential defense suppliers. (Given the record of some companies that have ventured far afield from their "core competencies," this implies diversification into closely related business lines.) Diversification policy would be seen as successful if it succeeds in reducing the fraction of the firms' business that is directly linked to DoD, while keeping them in business areas of interest to DoD, and responsive to the Department.

A different public policy interest in defense industry diversification revolves around the perceived overdependence of defense firms on DoD, causing perverse societal outcomes. These concerns were vocalized most loudly during the first few decades of the Cold War by critics of the "military-industrial complex" like Seymour Melman (1971):

> For the business units of industrial capitalism, the development of military industry has meant a transformation from the autonomous entrepreneurial firm to the military-industrial enterprise functioning under a state management. For the economy as a whole, the formalization of Pentagon capitalism and the outlays on its behalf have involved parasitic growth on a large scale and at a large opportunity cost. The economy and society as a whole bear the unknown cost of an array of depleted industries, occupations, and the industrial areas, and the cost of sustaining an economically underdeveloped population of 30 million among 200 million Americans (p. 7).

Aside from these economic distortions, Melman and others feared that the entrenched DoD-defense industry bloc would guide foreign policy to serve its own interests (Melman, 1988: 56–58), for example through its advocacy of foreign military sales (Gansler, 1995).

Ann Markusen (1997) argues that the defense industry continues to wield great influence over defense policy, despite the post-Cold War drawdown. She contends that this influence strongly pervades the industry restructuring debate itself: "The biggest impediment to conversion is the 'Iron Triangle,' that triumvirate of large defense contractors, the Pentagon, and Congress that has actively resisted even modest conversion initiatives over the last decade" (p. 87).

Concerns about the dependent relationship lead to a different definition of "defense industry," one cast in terms of industry reliance. Looking to the percentage of a firm's revenue derived from DoD contracts would reflect some of these broad policy issues. If reducing this dependence on DoD business is the public policy objective, then the goal of diversification would be to distance defense firms from DoD. Diversification policy would be seen as successful to the extent it succeeds in reducing the fraction of firm business linked directly to DoD (as above), and especially for firms with currently high dependence. But in contrast to the prior objective (where presumably it is better if diversification keeps the firm in business areas closer to those of interest to DoD), in this case the public policy interest would be seen as more successful the further away the new areas might be. In short, dilution of defense interest is the objective, where before concentration of defense interests was the objective.

A third strand of the diversification dialogue posits that defense research spawns technological developments that currently fail to reach the commercial economy. "Surely, the argument goes, the same minds that produced stealth technology and smart bombs can now help the United States win the trade wars in the international marketplace" (Berkowitz, 1992–93: 73). Encouraging or facilitating diversification might thus allow advanced technology to diffuse into the civil marketplace.

If military-to-civil technology transfer is sought, then a proxy for which firms are "defense," and thus which firms diversification policy should target, would look to the level of defense R&D conducted by the firm, especially such R&D devoted to basic science and technology. A combined measure would unite what the firm presumably has invested in military R&D with an attempt to assess how "closed" it is. The goal of diversification policy in this instance is to open the firm up, so that the civil sector can benefit from its technology. Diversification would emphasize how the defense firms' technologies might best help the civil sector.

Yet another strand in the diversification dialogue argues that the technological problem is the reverse of the one just portrayed—that DoD is not taking sufficient advantage of available commercial activity (Bingaman and Inman, 1992).[6] It is in dealing with this concern that the definition of a defense enterprise as one specialized in the arcana of government procurement may be most useful. The goal of diversification in this case is to bring the knowledge and practices of other markets to bear on defense, for which recasting government practices to parallel those of the commercial sector could be one of the instruments employed.[7]

ARE THESE APPROACHES RECONCILABLE?

Despite the differences among them, there is at least one outcome that would be consistent with all four views of the contemporary defense industrial sector and the objectives of diversification. If the future defense industrial base were indeed the broad array of enterprises that comprise the American economy—rather than being a subset of relatively specialized firms, as is now typical—then advocates of any of the preceding views regarding what is now the defense industry would be satisfied that their diversification objectives had been achieved. Those concerned with a set of ready suppliers to meet future defense needs would be delighted at so broad a spectrum of choices. Those fearing that a handful of powerful defense firms exert too much influence on American policy would applaud the potential dissipation of defense procurement among so many firms, few of which presumably would see defense so important to their business that they would engage in the conduct that present critics assert occurs. Those demanding the flow of military R&D to the civil sector, and vice versa, would take satisfaction in this new relationship between American industry and defense needs.

This outcome would only be readily achievable if the specialized knowledge necessary to do business with the Department of Defense were either easily obtained, or no longer differentiated the defense and non-defense sectors.

Reaching an outcome like this, that might satisfy the differing views of what constitutes the defense industry, and the differing diversification goals associated with those views, depends critically on minimizing the differences between defense business and other commercial activity. The principal target of policy, then, becomes not the firms themselves, or their particular behaviors or structures, but those actions of the government itself that separate defense firms from others. This, of course, constitutes a very different view of diversification policy, which up to now has largely focused on the firms and tried to encourage them to behave in different ways, thereby ignoring the incentives that cause them to behave as they now do–incentives that are often rooted in the government's own policies.

For those who might doubt the power of government incentives, and how they can restructure the defense industry, one need look no further than the aerospace industry. With recent mergers, encouraged by government overtures, the military aircraft supplier club has rapidly concentrated itself in only two firms: Lockheed Martin and Boeing. Moreover, as this example highlights, a firm-focused diversification policy would be challenging to craft for this industry, given the differences between the two principal firms. The former is a quintessential defense firm by any of the definitions offered earlier, while the latter is the world's leading producer of *commercial* aircraft but with a sizable defense business as well.

In contrast, concentrating on the government's practices would transcend such "one size fits all" difficulties. By opening the defense marketplace to new players, it might arguably help achieve the goal enunciated by the National Defense Panel (1997)—transforming the American military for the 21st century—by encouraging firms to compete for the defense procurement dollar with very different ideas about how to supply needed capabilities (albeit at the expense of current defense firms like Lockheed Martin and Boeing). And for those concerned with government meddling in the marketplace, such a policy would successfully answer worries about government intervention, for now the government would be intervening in its own bureaucracy so that marketplace forces can better serve the future American military.

NOTES

1. SEMATECH, a consortium of 14 U.S. semiconductor manufacturers and DoD, aimed at strengthening the competitiveness of the U.S. industry, especially relative to the Japanese. It involved initial DoD subsidies of $100 million per year. An evaluation published by the National Bureau of Economic Research

raises questions about SEMATECH's efficacy; its authors conclude SEMATECH led to a reduction of semiconductor firm R&D of $300 million per year, while the U.S. share of world semiconductor production rose only moderately through 1992 (see Irwin and Klenow, 1994, pp. 20 and 22). For more sanguine views of SEMATECH, see GAO, 1992 and Advisory Council on Federal Participation in SEMATECH, *SEMATECH: Progress and Prospects*, 1989.

2. In 1993, the Clinton administration established the Technology Reinvestment Project (TRP). Managed by the Defense Advanced Research Projects Agency (DARPA), the program sought to encourage the development of both militarily and commercially viable advanced technology.

3. Note that the Percentage of Defense Revenue list comprises only firms in the Top 100 Total Defense Revenue list.

4. Confidential interview with a senior defense procurement official.

5. See, for example, Office of Technology Assessment (1992). The Defense Conversion Commission (1992: 18), however, concluded the financial viability of the 25 largest defense contractors is not at risk and they will probably manage the drawdown successfully.

6. For a summary of the latter two strands of the technological problem, see Alic, *et al.*, (1992), pp. 3–27.

7. The Defense Conversion Commission hinted at this conclusion. See Transforming Defense: National Security in the 21st Century; A Report of the National Defense Panel, December 1997.

REFERENCES

Alic, J. A., Branscomb, L. M. Brooks, H., Carter, A. and Epstein, G. *Beyond Spinoff*, Boston: Harvard Business School Press, 1992.

Baumbusch, G. C. and Harman, A. J. *Peacetime Adequacy of the Lower Tiers of the Defense Industrial Base*. RAND: R-2184/1-AF, Santa Monica, CA: November 1977.

Berkowitz, B. D. "Can Defense Research Revive U.S. Industry?" *Issues in Science and Technology*, Vol. IX, No. 2 (Winter 1992–93), p. 73).

Bingaman, J. and Inman, B. R. "Broadening Horizons for Defense R&D." *Issues in Science and Technology*, Vol. IX(1), pp. 80–85, Fall 1992.

Congressional Budget Office. *Economic Conversion: What should Be the Government's Role?* Washington, DC, January 1980.

Defense Conversion Commission. *Adjusting to the Drawdown*, Washington, 1992.

Defense News, "Top 100 Worldwide Defense Firms", July 29, 1996.

Donnelly, J. and Clark, C. "Merger Mania Hits $53 Billion This Year—So Far." *Defense Week*, Vol. 18(27), July 7, 1997.

Gansler, J. S. "Restructuring the Defense Industrial Base." *Issues in Science and Technology*, Vol. VIII(3), p. 51, Spring 1992.

Gansler, J. S. *Defense Conversion*, Cambridge: MIT Press, 1995, p. 24.

Government Accounting Office. "Lessons Learned from SEMATECH", B-250107, September 1992.

Gottlieb, S. *Defense Addiction: Can America Kick the Habit*, Westview, 1997.

Irwin, D. A. and Klenow, P. J. "High Tech R&D Subsidies: Estimating the Effects of SEMATECH," NBER Working paper No. 4974, pp. 20 and 22, December 1994.

Korb, L.J. "Merger Mania". *Brookings Review*, Vol. 14(3), p. 22, Summer 1996.

McNaugher, T.L. *New Weapons, Old Politics*. Washington, DC: Brookings, 1989, pp. 150–51.

Melman, S. *The War Economy of the United States*. New York: St. Martin's Press, 1971, p. 7.

Melman, S. *The Demilitarized Society*. Montreal, 1988, pp. 56–58.

Markusen, A. "How We Lost the Peace Dividend". *The American Prospect*, No. 33, p. 87, July-August 1997.

National Defense Panel. *Transforming Defense: National Security in the 21st Century,* December 1997.

Office of Technology Assessment, *1992 Building Future Security:Strategies for Restructuring the Defense Technology and Industrial Base*, Washington, DC: GPO, June 1992, p. 79.

Office of Technology Assessment. *Assessing the Potential for Civil-Military Integration,* Washington, DC: GPO, September 1994.

SIPRI Yearbook 1996. Oxford University Press, 1966, Appendix 10A.

Van Opstal, D. *Integrating Commercial and Military Technologies for National Strength*. Washington, DC: Center for Strategic & International Studies, March 1991.

5

U.S. DEFENSE INDUSTRY CONSOLIDATION IN THE 1990s

KENNETH FLAMM

INTRODUCTION

The usual picture painted of the U.S. defense industry in the aftermath of the Cold War is one of firms under enormous pressure to restructure themselves, in order to face a fairly radical change in their economic environment—a sharply downsized defense budget, and much smaller market for their products. This chapter surveys available statistical evidence on changes in the size and organization of the core U.S. industries primarily selling to military customers, and in doing so challenges a number of widely held beliefs about the U.S. defense industrial base.

Contrary to a widespread perception, key pieces of the U.S. defense industry in the mid-1990s were about the size they were in the early 1980s, when the Reagan-era defense buildup began. As U.S. defense procurement outlays doubled, the data suggest that capacity indeed increased. But there appear to have been few major new companies entering into the industry, and therefore no significant declines in measures of concentration (across companies) in core defense-oriented sectors. In fact, to the contrary—it appears that a considerable **increase** in industrial concentration occurred in the aircraft, aircraft engines, and munitions industries in the 1980s even as defense procurement in these sectors soared. These changes occurred long before the dollars dried up and talk of the need to restructure the industry became common in the early 1990s. Given that there was little or no talk of too many companies chasing too few defense dollars back in 1982, why then was there so much discussion of the need for mergers a decade later, in 1992?

To the extent that inadequate data can be mustered to address this debate, they lean against the notion that the wave of mergers and acquisitions that took place in the U.S. defense industry in the early 1990s was somehow a "natural" outcome of a shrinking defense budget. Capacity increases in the 1980s were distributed among stable or even decreasing numbers of firms. The issue of how many **companies** are found within the industry was clearly a different question from that of how much **capacity** existed within the industry.

While we shall speculate about different possible answers to this question in the concluding section of this chapter, we mainly will be concerned with establishing basic facts about the configuration of the U.S. defense industry, and changes in its structure over the last fifteen years.

THE MYSTERIES OF THE DEFENSE PROCUREMENT BUDGET

The logical place to start is with the defense budget, the mighty river from which most defense industry revenues flow. Fig. 5.1 shows a couple of different ways of parsing the historical procurement figures in the defense budget, which are converted into estimated FY1998 dollars. Maximum dramatic effect is obtained if one examines budget authority (BA), or its

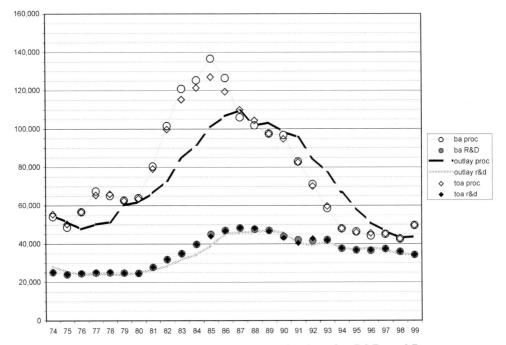

Fig. 5.1. U.S. Defense Budget Authorities and Outlays for R&D and Procurement, 1974–99 in FY 98 $, Based on FY 98 Budget

close relative, total obligation authority (TOA). BA is basically what Congress appropriates in any given fiscal year, while TOA is BA adjusted for shifting of unobligated funds from one fiscal year to the next (when permitted), as well as later rescissions and other fiddling at the margins by the Congress and the DoD. The historical data show procurement TOA dropping by two-thirds from a 1985 peak of about $135 billion, to about $45 billion during FY 1995–97. This "two-thirds drop in procurement" was and is often referred to by senior DoD officials when describing the painful adjustments needed in the defense industrial base.

Focusing on a peak year exaggerates the degree of fluctuation, however. For one thing, actual spending by DoD—outlays—adjusts only gradually to changes in appropriations, since it takes up to a decade for a system funded in one year to actually roll off a production line in finished form. Partial payments are made to contractors along the way, so any year's procurement spending is a sum of lagged payments derived from appropriations made up to a decade earlier. Thus, a defense industry would never have reasonably expanded the scale of its operations to the level needed to immediately produce a peak year's appropriated budget, or shrunk itself to a size just able to produce the single year output corresponding to the trough of defense appropriation cycle.

A more reasonable measure of the impact of defense procurement on industry is DoD's outlays, real spending in any given year. Fig. 5.1 shows a considerable decline—from a peak of about $110 billion FY1998 dollars in 1987, to about $45 billion in 1997–98—just under a 60 percent cut. As we shall see shortly, this somewhat reduced, but still very large decline also appears to overstate the real impact that budget declines had on output levels in key defense industries.

DEFENSE EQUIPMENT INVESTMENT

The defense department procures all sorts of goods and services, from paper clips and toilet seats, to fighter jets and tanks. Arguably, when discussing the condition of the defense industrial base, we are less concerned about the paper clips, and more concerned about the fighter jets—and other specialized, defense-unique products and systems that cannot be quickly and easily procured from commercial suppliers. DoD investment in military equipment—aircraft, ships, vehicles, missiles, electronics, and other defense equipment—might be considered a better measure of trends in government purchases from the industries that we tend to think of as making up the "defense industrial base."

Fig. 5.2 displays an index of real spending on defense equipment investment measured in 1992 dollars, as compiled by the national accountants at the Commerce Department's Bureau of Economic Analysis (BEA). Again, we see a substantial drop from a calendar year 1987 peak—about a 40 percent decline. But this is considerably reduced from

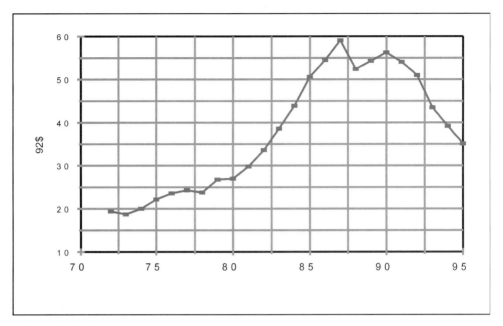

Fig. 5.2. BEA estimates of defense equipment investment, in constant 1992$.

the two-thirds drop in procurement budgets with which we started. And while total procurement outlays in 1995 had sunk well below a previous 1976 low, real 1995 equipment deliveries remained at a much higher level— above 1982 spending.

The considerably less steep decline in equipment investment spending may in part reflect the fact that sophisticated defense systems take much longer to produce and deliver than the more ordinary goods and services that DoD also procures. Cutbacks would therefore be stretched out in outlays over a longer time period, and the industrial impacts felt in a more gradual fashion. But it is also true that in relative terms, our military has spared the specialized systems and capabilities it is most concerned about protecting and maintaining—sophisticated, high tech equipment—from the sharpest relative cuts in the procurement budget.

The BEA breaks DoD equipment investment down by major systems, shown (as indexes with 1992 = 100) in Fig. 5.3. Expenditure is divided by price indexes for distinct types of systems, producing estimates of real purchases of that type of system.[1] As can be seen, there were vastly different trajectories for procurement outlays on different types of systems. Missile purchases tripled in the 1980s, soaring to a 1987 peak, then fell by two-thirds, to about where they were in 1980. Electronic equipment purchases more than doubled from 1980 to 1987 peak, then fell by about 85 percent through 1995. Ship outlays quadrupled from 1980 through a 1991 peak, then were halved by 1995. Vehicle purchases doubled from 1980 through a 1990 peak,

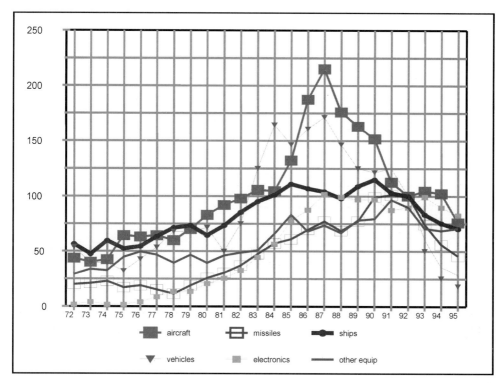

Fig. 5.3. BEA constant dollar equipment investment by type of defense capital good.

then fell by 40 percent through 1995. Other equipment outlays roughly quadrupled by 1987, then registered only a slight decline in the 1990s.

Thus far we have focused on the demand side of the U.S. defense industry, examining trends in spending by its primary customer, America's military services. The defense investments just charted miss a number of smaller, but increasingly important dimensions of U.S. defense industry's sales base. Spare and replacement parts, as well as products sold by our Defense Department to foreign militaries, are classified as consumption by the BEA, and buried in another set of accounts. Commercial sales of U.S.-produced military equipment to other nations do not show up in these figures at all.

DEFENSE AND SPACE INDUSTRIAL PRODUCTION

An alternative approach to measuring the defense industry starts with the supply side. Fig. 5.4 displays a monthly index of defense and space industry production maintained by the Federal Reserve.[2] This index peaks in early 1988, then drops by about 40 percent to a relatively stable level by 1996. Note that these figures are based on Census Bureau and industry association statistics that mix together military and nonmilitary production in industries

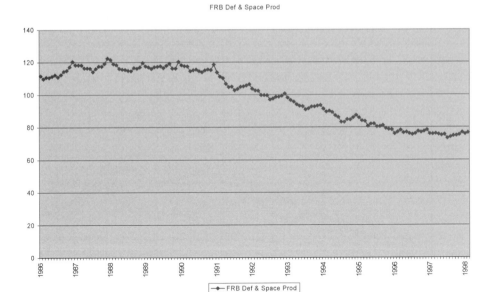

Fig. 5.4. Federal Reserve Board index of defense and space industrial production.

that ship much of their output to the military, so changes in civilian output
in industries producing military systems affects these trends.

CENSUS DATA ON DEFENSE MANUFACTURING

The closest thing to "real" data on the defense manufacturing base is found
in statistics collected by the Census Bureau. The most detailed data are
collected in the Census of Manufactures, which is undertaken every five
years. Less detailed estimates of manufacturing activity are available in
the Annual Survey of Manufactures (ASM), which uses a scientifically
selected sample of all U.S. manufacturing firms. Another useful survey is
one of the Census Bureau's Current Industrial Reports series, the MA-37D
Aerospace Industry (Orders, Sales, and Backlogs) report, which contains
a different (and non-comparable) itemization of aerospace industry companies'
shipments of selected products.

The Census of Manufactures and ASM collect data from individual
manufacturing establishments, which roughly correspond to factories or
plants. Establishments are assigned to an industrial sector according to
the classification of the principal products they produce. Certain census
data are also tabulated on a company basis, while detailed data on shipments
of products are also broken out in a separate analysis that aggregates all
like products across the sectors in which they are manufactured.

The lowest level of detail at which these Census Bureau sources report
useful data is the 4-digit SIC industry classification. There are twelve
4-digit SIC industries in which defense shipments have historically been

a driving factor and distinctive military product lines can be readily identified: Ammunition, except for small arms (SIC 3483), ordnance and accessories (SIC 3489), aircraft (SIC 3721), aircraft engines (3724), aircraft parts and equipment (3728), ship building and repairing (3731), guided missiles and space vehicles (3761), space propulsion units and parts (3764), space vehicle equipment (3769), and tanks and tank components (3795). In addition, there are two SIC industries in which more detailed data on mainly military electronic and electro-optical product lines are sometimes available: search and navigation equipment (3812), which contains search, detection, navigation, and guidance systems and equipment (SIC 38122), and optical instruments and lenses (3827), which contains sighting, tracking, and fire control equipment, optical type (38271).

Table 5.1 estimates the share of shipments from establishments in each of these 12 industries that are military or mainly military products in total industry output. For most of these industries, the shares have been estimated by taking the share of products that can be classified (at the five or seven digit level) as military in total products that are clearly classifiable as either military or commercial produced by **all** U.S. manufacturing establishments (including establishments not classified in that industry). In search and navigation equipment, however, that share was estimated by classifying products as military (reconnaissance and surveillance electronics, interrogate friend or foe (IFF) equipment, proximity fuses, sonar and antisubmarine warfare gear, electronic warfare equipment, and underwater navigational systems), dual-use but mainly military (search and acquisition, tracking, and instrumentation radars), and dual-use (navigational systems and instruments).[3] The military and mainly military categories produced the share estimates shown in Table 5.1. In optical instruments and lenses, all optical fire control equipment was classified as military products. The total value of military products output in 1987 and 1992 was then estimated by multiplying total industry output by the estimated military share.

Not surprisingly, for most of these industries the share of military output in the totals rises in the 1980s, then declines in the early 1990s. There are three interesting exceptions to this pattern. In shipbuilding, the military share of output in 1992 actually rises slightly above 1987 levels, reflecting the continued withering away of a commercial shipbuilding industry in the United States. In missiles and space vehicles, the military share of shipments continues to rise in the early 1990s, despite a continuing expansion in the commercial space launch market. Most likely, this reflects a dramatic uptick in the fielding of relatively expensive space-based sensor and command and control systems by the U.S. military, which dwarfs the continued growth in launches of relatively inexpensive commercial communications satellites. Finally, the share of specialized defense electronics in output of search and navigation equipment undoubtedly reflects the continually growing share of electronics in virtually every class of major defense system.

Table 5.1. Military output in U.S. defense-related industries, 1982–92

SIC		Military products % of sales			Mil product value $		Notes:
		1982	1987	1992	1987	1992	
3483	Heavy ammo	100.0	100.0	100.0	3983	3137	1
3489	Other ordnance	100.0	100.0	100.0	1678	1386	1
3731	Ship bldg & repair	62.1	83.7	84.0	6985	8732	2
3721	Aircraft	57.9	61.5	38.8	22138	21969	3
3724	Aircraft engines	NA	46.8	36.6	8812	7541	4
3728	Aircraft parts	49.8	60.5	29.6	11796	6552	5
3761	Missiles & space veh	71.2	76.5	85.0	12255	11754	6
3764	Space propulsion	74.1	76.5	58.4	2652	3162	7
3769	Space veh Equipment	70.1	70.7	68.3	2372	3086	8
3795	Tanks & tank comp	100.0	100.0	100.0	3017	2503	1
3812	Search & nav equip	79.0	85.0	86.3	28898	29708	9
3827	Optical instr & lens	35.5	37.7	34.4	750	787	10

Notes:
1. All products military assumed.
2. Self-propelled ship building and ship repair for military as share of total for all customers.
   Share of covered products in total industry sales: 85.7   95.3   91.6
3. Aircraft sold to or designed for military customers, R&D for military customers or designs, conversion or overhaul of military aircraft, other services for military aircraft as share of totals for conversion or overhaul of military aircraft, other services for military aircraft as share of totals for both military & commercial.
   Share of covered products in total industry sales: 99.9   98.9   96.6
4. Military engines (built for military aircraft or to military specs), R&D on military engines, other services for military engines.
   Share of covered products in total industry sales: NA   97.0   91.7
5. Mechanical power transmission, hydraulic subassemblies, pneumatic subassemblies, landing gear, R&D on aircraft parts, other subassemblies for military aircraft as share of totals for military and commercial.
   Share of covered products in total industry sales: 94.5   91.5   87.3
6. All guided missiles and services assumed military. Guided missiles plus space vehicles, R&D on space vehicle other services for space vehicles for U.S. government military customers as share of totals for all customers.
   (1987 only: R&D on complete space vehicles allocated to U.S. military and all other customers on basis of 1982 ratios.)
   Share of covered products in total industry sales: 100.0   99.8   89.9
7. Complete missile & space vehicle propulsion units, R&D on propulsion units, other services on propulsion and other propulsion unit parts for U.S. military customers as share of totals for all customers.
   Share of covered products in total industry sales: 100.0   98.8   74.8
8. Airframe, space capsule and other space vehicle, and other parts and accessories, and R&D, for U.S. military customers as share of totals for all customers.
   Share of covered products in total industry sales: 71.9   98.1   88.9
9. Search, detection, navigation & guidance equipment (SIC 38122) share of all search, detection, navigation, & guidance equipment & aeronautical, nautical, and navigational instruments × share "military or mainly military" in 38122.
   Share of separately classified products in total industry sales NA   97.5   97.5
   Share of "military and mainly military" in SIC 38122.   85.4   91.2   86.3
10. Sighting, tracking, and fire-control equipment share of all optical instruments and lenses.
   Share of classified products in total industry sales: NA   95.3   96.4

More surprising—for at least a quick moment—is the fact that the largest defense industry in 1992 and 1987 was defense electronics. Military electronics products shipped by search and navigation equipment producers were about 50 percent greater than the value of military aircraft shipped in the U.S.!

The primacy of defense electronics is confirmed by Table 5.2, which displays the number of manufacturing establishments **primarily** specializing in the production of military products (classified at the five-digit level), and the number of employees and production workers in these establishments. (Prior to 1987, it is sometimes also possible to count establishments for which these products are 75 percent or more of output, also displayed in Table 5.2.) Note that employment in SIC 38122 (search, detection, navigation, and guidance, which the notes to Table 5.1 estimate to be over 90 percent military output in 1987 and 1992), was 228,000 in 1992, compared with 122,000 employed in making military aircraft.

The data in Table 5.2 also tell us something about the number of factories (that is, manufacturing establishments) primarily producing military goods. Note that while data are available on the number of **companies** producing over $100,000 worth of detailed military products (whether or not it is their primary product), the low output threshold makes these numbers of limited value.

Military aircraft. The number of factories primarily manufacturing military aircraft dropped from 26 in 1982, to 13 in 1987, then rose to 21 in 1992. (In this particular case, changes in the number of **companies** making over $100,000 in military aircraft resembles the pattern of change in factories: from 29 in 1982, to 15 in 1987, then up to 27 again in 1992. The number of companies doing R&D on complete military aircraft dropped sharply, from 11–12 in the 1980s, to 5 in 1992.)

Military aircraft engines and parts. The number of military engine factories rose in the 1980s, then dropped back somewhat in 1992. (The number of companies shipping over $100,000 in complete military engines and doing R&D on engines generally rose over this entire period.)

Military ship building. The number of military shipyards rose in the 1980s, then fell somewhat in the early 1990s.

Guided missiles. The number of guided missile factories rose from 1982 to 1987, then dropped back a bit by 1992. (Roughly the same pattern prevailed with the number of companies.) The number of missile R&D establishments increased slightly (from 5 to 6) with the defense procurement expansion of the 1980s, then dropped to 3 by 1992.

Tanks and tank parts. There was an increase in the number of tank part factories in the mid 1980s, falling back to the numbers of the early 1980s by 1992.

Search, detection, navigation, and guidance systems. There was little change in the number of plants producing these systems from 1987 to 1992.

Table 5.2. Profile of U.S. Defense Manufacturing Establishments, 1977–92

SIC 92	Product	Number mfg establishments with primary product:						Employees (1000s)						Product Workers (1000s)					
		1977	1977 >75%	1982	1982 >75%	1987	1992	1977	1977 >75%	1982	1982 >75%	1987	1992	1977	1977 >75%	1982	1982 >75%	1987	1992
37211	Military aircraft (including all aircraft for U.S. military and any other aircraft built to military specifications)	13	4	26	17	13	21	100.6		118.6		144.2	122.4	48.0		56.6		71.4	46.8
37241	Military engines (for U.S. military aircraft and any other aircraft built to military specifications)	8	1	9	2	13	11												
37312	Military, self-propelled ships, including combat ships, etc., new construction	12	5	12	6	17	15	81.9	50.2	84.0	29.4	75.9	79.6	61.5	33.6	63.4	19.6	55.1	55.8
37314	Ship repair, military	34	20	51	37	60	52	11.0	3.5	12.2	6.8	19.9	14.6	9.6	3.1	10.1	5.6	16.0	12.0
37611	Complete guided missiles	11	3	10	2	16	15	41.6		30.7		35.0	24.9	15.4		13.2		16.6	11.1
37613	Research and development on complete guided missiles	7	3	5	3	6	3	18.3		14.3		25.5		5.9				8.8	11.3
37616	Other services on complete guided missiles			7	6		5												
3795	Tanks and tank components	24	19	43	36	56	42	12.4	10.6	18.1	15.9	16.4	10.0		8.7	12.5	10.6	11.0	9.8
38122	Search, detection, navigation, and guidance systems and equipment					261	273					335.7	228.0					140.9	92.0
38271	Sighting, tracking, and fire-control equipment, optical-type	39	23	34	23	28	32	7.2	2.4	7.1	3.0	6.4	4.4	4.3	1.7	4.7	2.0	3.6	1.7
	Total	148	78	197	132	470	469	273	67	285	55	659	484	158	47	166	38	326	225

Optical sights and fire control. The number of factories was fairly stable over the 1980s and 1990s.

Note that in 1982, at least, many of these plants were producing significant amounts of other products (which might be either military or commercial). In only 17 out of 26 plants producing military aircraft in 1982, did these aircraft account for three-fourths or more of shipments. Analogous numbers for military engines were 2 out of 9 plants, 6 out of 12 military shipyards, 2 of 10 guided missile factories, 36 out of 43 tank component factories, and 23 of 34 sighting and fire control plants. If not dual-use, a significant number of military system factories were clearly "multi-use."

Generally, as perhaps might have been predicted, the number of plants in all these sectors tended to increase somewhat over the 1980s, then fall in the 1990s. What is most notable, perhaps, is the contrarian decline in 1987, in both factories (and companies doing R&D) that was evident in military aircraft. This is consistent with other data suggesting a trend toward greater concentration in aircraft in the 1980s. In military aeroengines, shipbuilding, and guided missiles, numbers of factories in 1992 remained somewhat greater than those producing in 1982; in tanks and optical fire control, the number of plants in 1992 stood roughly at 1982 levels.

INDUSTRIAL CONCENTRATION IN DEFENSE INDUSTRIES

The standard measure of industrial concentration is the so-called Hirschman-Herfindahl Index (HHI), defined as the sum of the squared market shares of all companies in an industry. The HHI will approach zero in a perfectly competitive industry (where each firm's market share is infinitesimal), 10,000 (i.e., 100 percent squared) with a monopoly. To give these numbers a more intuitive dimension, in an industry made up of two identical (symmetric) firms, a duopoly will have an HHI of 5000, an industry with three firms an HHI of 3333, four firms 2500, five firms 2000, and a ten-firm industry an HHI of 1000. Historically (based on its 1984 Merger Guidelines), the U.S. Justice Department scrutinized mergers in which significant (i.e., greater than 50 point) increases in HHI occurred, in highly concentrated industries (where postmerger HHIs would lie above 1800). This legal "red zone" for levels of concentration is represented by the rectangle drawn in the upper right quadrant of Fig. 5.5.

By matching establishment data on shipments from the manufacturing censuses of 1982 through 1992 with the identities of the corporate owners of those factories, the Census Bureau calculated HHIs for shipments by U.S. companies in these years, for 4-digit SIC industries. While this level of detail is somewhat more aggregate than that shown in Table 5.1, the results are revealing. Fig. 5.5 plots changes in HHIs over the 1982–92 period on the Y-axis, for all 4-digit U.S. manufacturing industries, against the HHI for each industry in 1992 (on the X-axis, with a logarithmic scale).

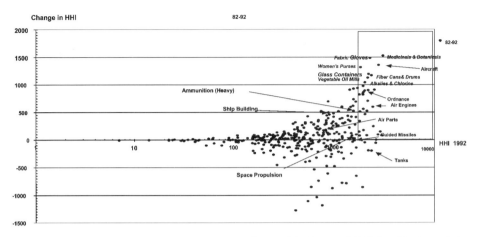

Fig. 5.5. Changes in Hirschmann-Herfindahl indexes (HHIs) of industrial production concentration from 1982–92 vs. levels of HHIs in 1992.

Overall, there was relatively little apparent change in the distribution of HHIs among U.S. manufacturing industries. Tanks were among the most concentrated of U.S. industries in all years (HHIs near 2500), but actually became a little less concentrated over this decade. Space propulsion and guided missiles tended to have HHIs near 1500 in all years—a significant degree of concentration, but not unusual given the small number of producers in these sectors. Two notable changes really leap out: huge jumps in concentration within aircraft (1400 points) and aircraft engines (600+ points), sectors in which output in general, and military shipments in particular, greatly increased through the 1980s, and a similarly great leap in HHI (about 900 points) in the ordnance industry. It is striking that this occurred long before any budget downsizing or the mergers and acquisitions of the 1990s, and in an overall market that was growing rapidly.

If we look at other, non-defense industries where the HHIs for U.S. production increased by very large amounts over this decade (such industries with increases in HHI exceeding 1000 points are identified in italics in Fig. 5.5), it is likely that something very different was going on. These are for the most part apparel items, containers, and specialized chemicals, where an increase in imports may have left a very small number of U.S. producers accounting for the bulk of American production, but not necessarily the bulk of the U.S. market after imports are factored in. In defense markets, however, imports have been negligible, so increases in HHIs track real increases in market concentration.

DISAGGREGATED DATA ON MILITARY PLATFORM PRODUCTION

A final, useful pass at sizing defense output changes over the last two decades can be had by examining the most detailed data on military products shipments. Fig. 5.6 compares shipments of complete military

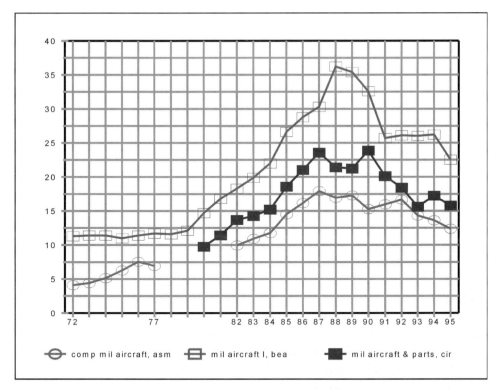

Fig. 5.6. Diverse estimates of military aircraft output.

aircraft, as reported in the census and ASM, with shipments of military aircraft and parts reported in the CIR, and DoD outlays on aircraft investment as reported in the national income accounts by the BEA. All current dollar values have been converted to 1992 dollars using the BEA military aircraft investment deflator.

The ASM and CIR shipments data are in rough agreement. Shipments of complete aircraft increased by 75 percent from 1982 through a 1987 peak, then fell by about 30 percent to a 1995 level that was still 25 percent above 1982 production. The data seem to be telling us that the production of finished military aircraft from the core facilities of the systems integrators making up the heart of this defense industry fell by a lot less than the DoD outlay data would have led us to believe. How can this be?

One factor could be that shipments of military aircraft to non-DoD customers increased in relative terms. In fact, we know this to be the case. Exports of military aircraft did in fact rise in relative terms.

But it is puzzling that DoD's military aircraft outlays are roughly double the output of complete military aircraft from U.S. factories, particularly since the factory data include exports, and the outlay data do not. Also, the BEA outlay data are not supposed to include R&D, repairs and parts, and services, which are classified as consumption in BEA's national accounts.

There are three factors that seem to provide plausible explanations for the huge discrepancies. The first explanation is that outside contractors with subcontracts are supplying subsystems and equipment that are being integrated into aircraft as contractor or government furnished equipment, and not included in the billing for work done at the aircraft plant. A typical aircraft program, for example, has separate contracts for the engines, avionics, fire control, airframe and assembly, etc. The shipments for complete military aircraft recorded in the manufacturing census, therefore, should probably be viewed as the value of the airframe, assembly, and systems integration work done at the aircraft factory. The values reported for military aircraft and parts sales in the CIR, then, add to airframe and assembly revenues the value of all subcontracts for equipment, parts, and subsystems provided from **within** the aerospace industry. The large discrepancy, therefore, between DoD aircraft investment, and U.S. industry shipments of aircraft and parts, would then be interpreted as the value of subsystems and services supplied under separate contract to military aircraft and part establishments by producers of products classified in other industrial groups.

A second factor relates to timing issues. Many subsystems and parts are delivered to aircraft producers well in advance of delivery of a finished aircraft, since they do not require as lengthy a production and assembly process. The value of these subsystems shows up in equipment investment at the time of delivery of the finished subsystem, not when the completed aircraft is delivered. Thus, we would expect subsystems and parts costs to be front-loaded over time, creating "bulges" in the difference between total aircraft investment and assembled aircraft deliveries towards the beginning of major military aircraft programs. Discrepancies between equipment investment and aircraft and parts sales may also exist for definitional reasons (because progress payments on cost-plus contracts are recorded as sales as they occur, while investment is recorded only on final delivery of a finished product).

Over time, however, the cumulative differences between complete aircraft shipments, aircraft and parts shipments, and total investment should correctly reflect the relative shares of airframe and assembly producers, subsystems suppliers from within the aerospace industry, and subsystems and parts suppliers classified in industries outside the aerospace sector.[4] In both cases, the missing revenues should be showing up some place else at some time: if it is a defense unique product or service, in some other defense sector; if it is a dual-use product, in a commercial industry.

A third possible source of discrepancies is that in the real world, procurement contracts not infrequently may require additional development work that is in effect funded by the procurement contract (as opposed to a separate R&D contract). Such additional development work would end up being imbedded in BEA estimates of equipment investment as a non-recurring cost, despite the BEA accounting rule which excludes R&D

as consumption, and therefore increasing the discrepancy between aircraft investment and aircraft and parts shipments.

Further light is shed on this issue if we look at analogous data for real spending on (in 1992 dollars) guided missiles, shown in Fig. 5.7. Guided missiles have the great advantage of being almost exclusively a military product, so "guided missile" and "military guided missile" are virtually synonymous—obviously not the case in aircraft.

Fig. 5.7 shows that complete missile factory shipments account for a stunningly low fraction (on the order of a third) of the value of missile investments by DoD. If we add on missile R&D (using census/ASM data) we get a total that is at least 50 percent higher, and if we also add on services on complete guided missiles, we get a figure that is virtually identical to "missile systems and parts" as reported in the CIR, though still substantially short of the BEA investment figure for most years. Because CIR has a separate item for "R&D (under contract)," we infer that the R&D reported in the Census/ASM figures is mainly R&D undertaken in the course of working on procurement contracts, not on separate R&D contracts. "Services" on complete missiles presumably include the systems test and evaluation, and systems project management that are typically included in a procurement contract. There is a much reduced discrepancy between BEA missile investment and guided missiles, R&D, and services

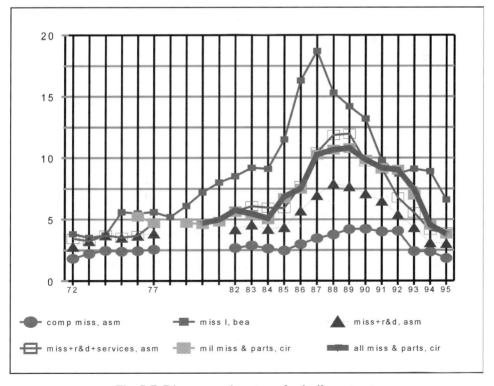

Fig. 5.7. Diverse estimates of missile output.

(as measured by the Census of Manufactures and ASM), or military "missile systems and parts" (as measured by the CIR). Much of the remaining difference is likely to be propulsion units for missiles (which unfortunately cannot be broken out separately from propulsion units for space vehicles), and R&D subcontracted out to firms in other industrial sectors (this is suggested by the coincidence of R&D spending increases on missiles in the ASM with a similar pattern for total BEA missile investment).

What substantive conclusions seem to follow from attempts to deconstruct the data shown in Figs 5.6 and 5.7? The most important one seems to be that in both military aircraft and missiles, there was considerably less volatility in shipments of completed systems from the factories of the systems integrators than there was in shipments of parts and subsystems from subcontractors. If we picture outlays as an offset to the right, with a lower peak and higher trough than appropriations, then deliveries of finished platforms by systems integrators are shifted still further to the right, and have a still lower peak. In military aircraft, 1995 deliveries of finished planes and subsystems dropped by about 37 percent from a 1988 peak. For finished aircraft alone, the decline was under 30 percent from a 1990 high. In guided missiles, platforms and parts, deliveries fell by about two-thirds from a 1987 peak, while sales by systems integrators' assembly plants fell a "mere" 50 percent. The potential problems of cyclical volatility and excess capacity would seem to be greater for the parts and subsystems contractors than for the systems integrator primes.

We conclude this discussion by briefly examining military ship construction, which looks rather different from aircraft or missiles. Fig. 5.8 shows that

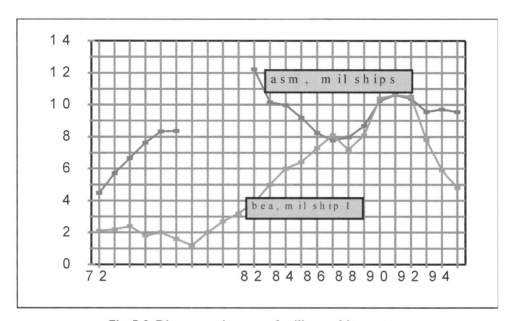

Fig. 5.8. Diverse estimates of military ship output.

from 1986 to 1992, military shipbuilding investment as reported by the BEA was roughly the same as military ship output as reported by the Bureau of the Census. Prior to 1982, exports may account for this discrepancy. After 1992, exports also may account for some of the discrepancy, though it is hard to see how they could account for all of the discrepancy. The census/ASM numbers suggest shipbuilding was down by about a quarter from a 1982 peak. The BEA investment numbers tell a different story: 1995 DoD ship outlays down by about half from a 1991 peak. Which number tells a truer story about the condition of U.S. ship constructors? As we shall see next, the census/ASM portrait seems closer to the reality of capacity utilization.

CAPACITY UTILIZATION

Another approach to understanding how serious an issue underutilization of capacity might be for the defense industrial base is to look at direct measures of capacity utilization. Unfortunately, the only available data are for 4-digit industrial sectors, which—except in a couple of cases—mixes commercial activities with defense shipments. The data are nonetheless suggestive.

Until 1989, the Census Bureau estimated capacity utilization in U.S. manufacturing industries using two concepts: "preferred" capacity was the level of utilization a manufacturer would desire to achieve on purely economic grounds, while "practical" capacity was the maximum output a plant could reasonably attain with realistic work schedules even if it was not as economically attractive. From 1989 on, Census instead measured "full" capacity—the maximum level of output under normal operating conditions—and "national emergency" capacity—the maximum output an establishment could expect to sustain for a year or more under national emergency conditions. Though not strictly comparable, we have matched "full" with "preferred" capacity, and "practical" with "national emergency" capacity in constructing Table 5.3.

Table 5.3 shows capacity utilization rates in 1994 (the most recent available year) as a percentage of capacity utilization in 1989 (the oldest strictly comparable data) and the maximum value for capacity utilization during the 1980s (which may be a year that is not strictly comparable to 1994). In making these comparisons, the first set of rows (full, rather than national emergency capacity utilization) is the one most relevant to the economics of peacetime capacity utilization in the U.S. defense industrial base. We are defining our own concept of "peak" capacity utilization relative to a 1980s peak utilization rate, since no industry has ever attained 100 percent full capacity as defined in the Census data, historically.

Two of these sectors—aircraft and aircraft engines—are mainly commercial products these days (55 percent commercial for aircraft in 1995, 75 percent for engines). Aircraft engines and aircraft parts seemed to have considerable

Table 5.3. Capacity utilization in defense systems industries

		94/80s peak	94/89
Full			
3721	Aircraft	68.8	88.7
3724	Aircraft engines & engine parts	70.2	82.5
3728	Aircraft parts & equipment, n.e.c.	78.8	82.7
3731	Ship building & repairing	85.9	85.9
3761	Guided missiles & space vehicles	112.2	118.6
3764	Space propulsion units & parts	61.4	71.8
3769	Space vehicle equipment, n.e.c.	78.3	108.3
3795	Tanks & tank components	58.9	82.8
3812	Search & navigation equipment	82.4	82.4
National emergency			
3721	Aircrft	60.3	84.6
3724	Aircraft engines & engine parts	62.3	62.3
3728	Aircraft parts & equipment, n.e.c.	55.4	59.4
3731	Ship building & repairing	67.6	67.6
3761	Guided missiles & space vehicles	105.8	105.8
3764	Space propulsion units & parts	62.3	69.6
3769	Space vehicle equipment, n.e.c.	53.2	85.4
3795	Tanks & tank components	43.1	72.1
3812	Search & navigation equipment	71.2	71.2

excess capacity in 1994 (40 percent below peak), while aircraft was only 15 percent below peak.

One sector—tanks—was purely military, and seemed to have the most excess capacity. Utilization was almost 60 percent below its 1980s peak. Even when compared with 1989, utilization rates were down by almost 30 percent. Shipbuilding is mainly military in the U.S. today (roughly 80 percent of sales), and capacity utilization was down by almost one-third from its 1989 peak.

In missile and space vehicle propulsion, which mixes commercial space engines with military and missile engines, capacity utilization was down 30 percent from 1989, closer to 40 percent when compared with a 1980s peak. In space vehicle equipment, utilization was almost 50 percent down from its 1980s peak, but only 15 percent off from 1989 levels. Search and navigation equipment, which also mixes military with commercial products, was down 30 percent from its utilization peak.

In missiles and space vehicles (of which 41 to 56 percent of sales were missiles in the 1990s) capacity utilization was actually 6 percent above its 1980s peak! This is extremely interesting for a number of reasons. Missiles had the sharpest declines in investment relative to their 1980s peak of virtually any defense sector we looked at. Missiles producers also experienced relatively little consolidation prior to the mid-1990s. In fact, the most distinctive institutional factor that may have been different in missile systems compared with some of these other defense sectors was that the Department of Defense embarked on a number of experiments with competition in the 1980s, paying the overhead required to support dual sourcing of several missiles. The impression that missiles were a relatively lean and mean sector going into the 1990s is supported by other data.

WARNING SIGNS OF EXCESS CAPACITY IN DEFENSE MANUFACTURING ESTABLISHMENTS

One final source of relatively disaggregated information on the cost structure for defense manufacturing can be derived from the 5-digit level data for manufacturing establishments **primarily** producing military products shown in Table 5.2. If we examine only these establishments, we can compile the figures shown in Table 5.4. Prior to 1987, some data are also available for establishments for which these products are 75 percent or more of output.

In particular, if overcapacity exists, we would expect output to be small relative to sunk investments in plant and equipment. In a cost-plus contracting environment, with all other things held equal, we would also expect to see the share of total costs attributable to capital—depreciation and profits—increase. Deducting off employment costs from value added produces an estimate of the share of capital and profits in the value of output, and it is straightforward to calculate this ratio.

(Note that we also can think of a more expansive notion of capacity—one that includes a minimum overhead of technical, scientific, and professional manpower. One might deduct off the cost of production labor from value added and interpret the resulting share in shipments as the share of capital and professional overhead in output. But this would be much dicier, since no one could reasonably argue that the entire professional labor force in, say, 1987, constituted "overhead"!)

In military aircraft, Table 5.2 showed overall employment and the production workforce increased through 1987, then falling by 1992. Table 5.4 shows the share of capital in output was increasing in 1982, falling in 1987 back to 1977 levels, then rising again in 1992. One can only speculate about what this means, but one consistent story would be of military aircraft producers laying on new capacity in 1982 in anticipation of budget increases, hiring workers, more fully utilizing capacity, and building more airplanes by 1987, then laying off workers, producing less, and utilizing their plant less fully again in 1992. A vaguely similar pattern seems to hold in tanks and tank components, but the share of capital there rises astronomically in 1992, supporting the impression of truly massive overcapacity formed in our review of capacity utilization data.

In guided missiles, the work force drops from 1977 to 1982, then rises in 1987, then drops again by 1992. From 1977 to 1982, the story looks

Table 5.4. Proxies for overhead in defense manufacturing establishments, 1977–92

SIC92		Share value added—payroll						Share value added—wages					
		1977	1977 >75%	1982	1982 >75%	1987	1992	1977	1977 >75%	1982	1982 >75%	1987	1992
	Radar												
37211	Military aircraft (including all aircraft for U.S. military and any other aircraft built to military specifications)	25		39		25	34	43		56		41	51
37241	Military engines (for U.S. military aircraft and any other aircraft built to military specifications)												
37312	Military, self-propelled ships, including combat ships, etc., new construction	17	17	26	33	27	29	31	36	36	43	42	44
37314	Ship repair, military	24	24	24	24	17	24	32	32	32	31	26	31
37611	Complete guided missiles	30		43		24	12	53		63		40	24
37613	Research and development on complete guided missiles	41		63		46		68		83		67	
37616	Other services on complete guided missiles												
3795	Tanks and tank components	25	23	29	29	19	39	30	28	37	37	26	48
38122	Search, detection, navigation, and guidance systems and equipment					34	38					56	60
38271	Sighting, tracking, and fire-control equipment, optical-type	28		30		39	30	47		42		52	47
	Total	26	20	37	30	30	34	43	34	52	41	48	52

similar to that in military aircraft—the share of capital increases sharply, perhaps as new capacity is added in anticipation of a defense boom. As in aircraft, that share drops in 1987, as output increases. But unlike aircraft, both the workforce and the share of capital in output drop in 1992, confirming our earlier hint that capacity utilization appears to have actually improved in the early 1990s in missiles.

CONCLUSION

DoD policy toward the defense industrial base in the 1990s had many dimensions. We have not considered several of the more important threads— acquisition reform, and a shift toward commercial, non-defense-unique products and suppliers, wherever possible, in order to foster greater competition even if the numbers of traditional defense suppliers were to decline. Instead, we have focused in this chapter on understanding and interpreting existing data on industries which primarily serve military customers, and contrasting what the data seems to say with the logic of recent DoD industrial policy toward its traditional supply base. The policy has at least implicitly followed the following chain of reasoning:

- massive, if not unprecedented, overcapacity is or will soon be present among Defense's core suppliers;
- in a cost-plus contracting environment, overcapacity will translate into higher costs for new systems;
- by encouraging and providing incentives for consolidation among large defense primes, Defense can provide the primes incentives to retire unneeded capacity.

The data we have used as a lens through which to view these propositions are generally unsatisfactory. Poor as these data are, however, it is basically what is out there. I do not believe there are any secret data bases in the Pentagon—or in the Aerospace Industry Association, for that matter— which contain clear answers to the questions I am raising.

Some basic points have emerged from our analysis:

- In general, the "right" number describing the overall relative size of the defense industry's contraction from its 1980s' production peaks seems to be nearer to 40 percent than to the 60–70 percent decline that is frequently cited.
- Reduction in the number of companies is not the same as reduction in the amount of capacity. An important topic for further study would be the measurement of what actual write-downs in capacity have taken place within the prime contractor supplier base, as the result of the policy experiments of 1993–97.
- There was a significant increase in industrial concentration in the aircraft sector in the 1980s, despite a large increase in the scale of demand.

Further encouragement of consolidation in aircraft manufacturing in the 1990s seems guaranteed to have ratcheted the degree of concentration up even further, and the time may be ripe to reconsider the virtues of continued mergers in aircraft and aircraft engines. The government's apparent opposition in early 1998 to the proposed merger of Lockheed Martin and Northrop Grumman would seem to indicate that a new skepticism has now crept into policy.

– A number of pieces of data suggest that the steepest relative declines in output may have been in shipments of parts and components by subcontractors, with relatively lower peaks and higher valleys experienced by the systems integrator primes. One wonders whether the real issue for defense industrial policy should be to identify and worry about defense-unique industrial "choke points," where small markets, high fixed costs, and inherently small numbers of suppliers guarantee tough tradeoffs between economies of scale and competition. How many such specialized subcontractors and component suppliers really exist is an interesting question, and a useful subject for further research.

– In at least one case—missiles—the industry seems to have improved capacity utilization despite a declining market, and without undergoing an extensive merger and acquisition wave (though that has since changed). This was a sector in which DoD conducted some interesting experiments designed to increase competition in the procurement process. The Clinton administration put a significant effort into reducing generic entry barriers into defense procurement as a procompetitive defense industrial policy. The time may be ripe to think about novel ways to encourage competition in defense industries with sector-specific entry barriers.

A FINAL QUESTION

Pressures to consolidate in the U.S. defense industry might be interpreted as related to one of two alternative dynamics. One is based on a vision of competitive pressures forcing change. Given significant economies of scale in defense systems (based on the economics of fixed investments in R&D and plant, as well as learning economies), fewer competitors will be able to "fit" into a smaller market in the long-run and maintain a market rate of return on their capital investments, as they compete with one another. Some will therefore be forced out by the "natural" pressures of competition as the market shrinks.

An alternative view is that the absence of true price competition in the real, cost-plus world of defense production contracting means that Defense will simply pay for whatever overhead costs exist within those defense-oriented firms it chooses to maintain through the awarding of contracts. Therefore, it is Defense itself, in its quest to achieve greater bang per buck of procurement spending, which must create pressure on its contractor base to reconfigure itself to reduce overhead. The pressure can be in the form of positive carrots (like agreements to share overhead savings with contractors when they take steps to reduce them through mergers and acquisition) or negative sticks (like letting companies go bankrupt when they don't receive sufficient contract awards to keep themselves profitable). In both views of the world, government will play a key role in determining defense industry configuration through both its policies toward mergers and acquisitions, and its contracting practices.

But did novel economic pressures really make a new policy toward the defense industry inevitable in 1992? The data suggest that in 1992 we were at about the same level of real defense output as in the early 1980s. Since the number of companies producing this output, and available measures of industrial concentration, were mainly about the same in 1992 as in 1982, this raises an important question. If the amount of output, and the number of producers involved in manufacturing it, were about the same in 1992 as in 1982, why was a reduction in the number of companies producing this output now desirable? Or to put it another way, why was there no pressure to undertake (or at least discuss) consolidation and restructuring of the defense industry back in 1982, if the problem was so apparent in 1992?

One logically consistent answer would be to argue that the industry was facing an imminent crisis back in 1982. In fact, this was precisely the answer that Bernard Schwartz, former CEO of defense giant Loral (now for the most part absorbed into Lockheed Martin) gave when asked this question in October 1997, at a presentation at the Johns Hopkins University's School of Advanced International Studies. Schwartz's prepared remarks, calling for prophylactic policies to discourage further vertical integration within the defense industry, have been published as *Defense Industry Consolidation: Where Do We Go From Here*. Only the knowledge that an incoming Reagan administration was likely to vastly increase defense procurement, it might be argued, kept restructuring of the defense industry off the policy agenda.

The problem with this argument is that it flies in the face of the facts of profitability in the defense industry. Simple calculations of returns on equity or investment for aerospace firms[5] or more complex methodologies designed to estimate economic profits in defense firms (Field, 1993) show that the early 1980s were a period of high levels of profitability for defense companies. Paradoxically, defense contractors' rates of return and profitability fell steadily after 1983, through the remainder of the decade, even as defense procurement soared.

A more compelling argument hinges on an increase in the soaring costs of R&D for new, higher-tech military platforms. A typical fighter aircraft R&D program, for example, ran about $6 to $8 billion in 1990s vintage dollars in the late 1970s and early 1980s. Today, the price tag on R&D for a new stealth fighter has roughly tripled, to $20 billion or more (current estimates for the F-22 and Joint Strike Fighter, for example). This makes it much more expensive to have a larger number of programs. With a decline in the number of new defense programs also comes a reduction in the number of producers that can be sustained on an ongoing basis. Exploding price tags for investments in new technology needed to develop advanced platforms—not a reduction in the overall size of the procurement pie, but an increase in the size of the minimum slice needed for a new system—seems a more persuasive argument about why 1992 was so different from 1982.

NOTES

1. The BEA price indexes used to produce real output estimates for major defense systems, described in Ziemer, R. C. and Kelley, P. A., "The Deflation of Military Aircraft," have some weaknesses that are discussed in Alexander, A. J., "Comment," both of which may be found in Foss, M. F., Manser, M. E. and Young, A. H., eds., *Price Measurements and Their Uses*. Chicago: NBER and University of Chicago, 1993.
2. This index has **not** been adjusted for seasonal factors.
3. 1992 data from the 1993 Current Industrial Reports, rather than the Census of Manufactures, was used to get a product breakdown for SIC 3812 for 1992. In all other cases, Census of Manufactures data were used.
4. Another timing difference might tend to reduce the discrepancy between shipments and outlays shown in Fig. 5.8 in the early years of a program, then increase it at the end. BEA excludes from its equipment investment figures progress payments made to contractors on work-in-progress except in shipbuilding), since investment in the national accounts is recorded on a delivery basis, and outlays in the federal budget are recorded on a cash basis. Work-in-progress in the national accounts is recorded as a change in business inventories, and inventory investment is classified as government consumption. In FY 1996 through FY 1998 budgets, estimated DoD procurement outlays exceeded BEA national defense gross investment by 13 to 18 percent of investment (LaBella and Webb, 1998).

   Since contractors are instructed to record progress payments plus estimated share of contract profit on cost-plus contracts as sales revenues in the CIR when they are received, not when the finished system is delivered, the value recorded in BEA investment would tend to be less than shipments at the start of a program, when there were no deliveries, and greater than shipments at the very end of the program, since the total value of delivered systems would exceed payments on work-in-progress on unfinished systems. On fixed-price contracts, the CIR instructs contractors not to report sales until the finished product is delivered, and we would expect reported sales to coincide with the relevant value added into BEA equipment investment. In both cases, the missing revenues should be showing up some place else at some time: if it is a defense

unique product or service, in some other defense sector; if it is a dual-use product, in a commercial industry.
5. For example, using data from the Census Bureau's Quarterly Financial Survey of Manufacturing.

REFERENCES

Alexander, A. J. "Comment." In Foss, M. F., Manser, M. E. and Young, A. H., eds., *Price Measurements and Their Uses*. Chicago: NBER and University of Chicago, 1993.

Field, J. L. "Economic Profit in United States Government Defense Contracting: Theory and Practice of a Fair and Reasonable Return." Doctoral dissertation, Graduate School of Business Administration, Harvard University, pp. 264–268, 1993.

LaBella, R. G. and Webb, M. W. "Federal Budget Estimates, Fiscal Year 1998." *Survey of Current Business*, pp. 11–12, March 1997.

Schwartz, B. "Defense Industry Consolidation: Where Do We Go From Here?" Talk given at Johns Hopkins School of Advanced International Studies, Washington, DC, October 6, 1997.

Ziemer, R. C. and Kelly, P. A. "The Deflation of Military Aircraft." In Foss, M. F., Manser, M. E. and Young, A. H., eds., *Price Measurements and Their Uses*. Chicago: NBER and University of Chicago, 1993.

6

THE NEW INDUSTRIAL POLICY?

SEAN O'KEEFE

Diversification is a natural corporate strategy option anytime the market changes. It is an inevitable economic motivation that will drive corporations to look at such strategy options. In considering what public policy alternatives ought to be considered in dealing with diversification questions, I've come to the view that the arguments usually center on a variant of a long-standing debate over the propriety of setting industrial policy.

Advocates of defense industrial policy are driven by three primary factors. The first is industrial base concerns—the concern that somehow if we do not continue to demonstrate our capacity to produce something we will lose that capability forever. This argument assumes there is superior knowledge, that somehow that capacity, that capability, that prowess will need to be demonstrated again. Each time our nation has adopted this policy we have been sorely disappointed. In 1954, the Congress passed the National Defense Stockpile Act. It was designed specifically to put aside the kinds of natural resources—or resources that were considered necessary—to deal with wartime circumstances to guarantee that there would not be shortages. This was a direct outcome of the World War II era experience. Congress sincerely believed that there were certain commodities required in a wartime situation. But as a consequence of this forecasted judgment, today we maintain the world's most impressive stockpile of asbestos, which at that time was considered to be an absolute critical commodity and today we can't get rid of the liability.

The second factor in this drive for advocacy of defense industrial policy is that economic markets produce winners and losers. As a matter of course, the defense acquisition process seeks to soften this effect. The essential premise of defense acquisition policy is that there is going to be a winner, there is going to be a loser, then there must be a determination of how much each ought to get. There are a few "winner take all" competitions.

The third factor is that DoD is accustomed to being the principal, if not primary and exclusive, driver of the customer requirements. This assumes a level of control and influence over supply as well as demand.

All three of these factors will be changing as a consequence of recent public policies and corporate strategies, which are variants of this industrial policy debate.

The current policies and strategies can be categorized into six major approaches. The first is the defense conversion argument that emerged in the late 1980s and the early 1990s, which suggested that industrial capacity needed to "convert" from military to commercial focus. This argument has largely succumbed to a particularly strong economy, which has absorbed the jobless rate for defense layoffs as well as economic expansion that we have enjoyed in the United States. These factors have made the question of whether we should convert from making missiles to making baby carriages on defense production lines a moot point.

The second major initiative is publicly financed mergers and acquisitions. The investment bankers have been the chief beneficiaries of this particular public policy. The initiative assumed that it is in the taxpayer's best interest to see these mergers and acquisitions occur. We have motivated these industrial actions by public financing, but given the market conditions, the mergers would probably have occurred in some variation without public prompting.

The third factor is the policy for dual-use technology. Initially, several billion dollars were planned for programs which advanced the policy. The flaw in this approach is that it assumes that a market will emerge once we have fueled and provided the necessary subsidies in order to yield the technology that has applications in not only defense but also in commercial form. The unstated assumption is that if the new demand doesn't emerge or support that market, then the public will continue to subsidize. In part, this is the industrial policy argument perpetuating itself.

The fourth factor is the advocacy for technology demonstration efforts. Frankly, I think these have been among the more significant accomplishments to demonstrate the edges of technology and its application in a wide range of different categories and sectors. That has worked out as well as we could have expected and probably has been one of the more significant contributions of the Perry-White legacy.

The fifth factor is what the Clinton administration liked to refer to as the "procurement holiday". The argument was that we could just stop modernizing for the time being because we are enjoying production deliveries from a variety of different acquisition programs initiated in the Reagan-Bush years. The argument follows that as those weapon systems are delivered, we don't need to replace at similar rates since the inventory will exceed the now smaller force structure requirements—so we can take a "procurement holiday." That was the initial strategy of the Clinton administration in 1993, but that strategy became a "permanent vacation." There is no plan to support replacement of the systems that are currently operating even at lower rates of production. The future years defense program (FYDP) simply doesn't support such a program.

The last factor that has become a major public policy initiative that incorporates this new variant of industrial policy is an export promotion policy. The policy has been to reverse the Carter era impediments which required the defense industry to recoup a portion of research and development costs associated with foreign sales as well as marketing expenses. This administration has not only removed the requirement for recoupment but also actively encouraged marketing activities to be underwritten as a cost of doing business overall. We have indirectly and by public subsidy encouraged an activist kind of export policy in the hopes that it would fill the market void created by reduced U.S. demand. But we're finding the foreign demand simply isn't there, and for the limited demand that is there, the competition is fierce.

Let me conclude with three observations. First, these policies are further evidence of the enduring wisdom of the law of unintended consequences. These latest variants of industrial policy do create a new environment. To be sure, this is a very different environment from the one in which we began a few years back. The Department of Defense is no longer the only, and in some cases, not even the primary, customer of these new mega-merger companies we see today. In fact, I think we are trending towards minority market status for the range of product and services, which is really going to change the nature of the government–industry relationship. Until there is a change in the mindset of business–government relationships in this case, the Defense Department is going to find itself in a distinctly disadvantaged position of trying to get the attention of an industry that is otherwise occupied. The industry is occupied by what Gerry Susman has defined as a natural tendency to explore diversification alternatives. Being one of a range of customers, as opposed to being the principal if not only customer, is going to change the nature of the government–industry relationship and make contract administration very different from what we have seen in the past.

The second observation is that the incentives for outsourcing DoD services will be overwhelming. Absent an unlikely ground swell for increasing the total defense budget, reducing the cost of operations is the only other source of funds for any significant procurement and modernization. This is the next best opportunity that the private sector has to change in the way the federal government does business. In this area there is going to be a very strong push from the Defense Department, not necessarily for product, because there isn't much left, but for services.

Third, the calls for a new round of base closures will further diminish the organic, government industrial base. This is an irony, given that this administration has advocated efforts to "privatize in place," a notion I've never quite understood, for the Kelly and Sacramento Air Logistic Centers. This approach doesn't get rid of unneeded capacity, it simply changes the management of the excess capacity. On the one hand, this makes some constituencies very pleased because it tempers the need for a new round of base closures—particularly gratifying to state and local communities. On the other hand, ironically, the same people who were bitterly opposed to base closures in the late 1980s—the Defense Department, the military—are now strong advocates of removing infrastructure and reducing the cost of carrying around this extended burden of excess capacity.

The challenge now is to access technology with less than a commanding market position. That's the real problem and what's really unique about the way these diversification strategies have evolved. The industrial base now is more properly defined as the technology base, because the capacity to produce things is something that we've seen diminish and for those who advocate more base closures as well other adjustments in the size of the defense budget for procurement, you're going to see that capacity in the commercial sector dwindling. That means that the industrial base—in looking at the technology base—requires the manufacturing fabrication sector to "unlearn," and disregard the old ways of doing business. They're reversing the ways they looked at the process in the past. That is an element of good news. We are no longer in the same mode, interestingly, of perpetuating the view that the industrial base is there to demonstrate our capacity to do something precisely the way that we did it in the past, in case we need it in the future. Interestingly, all this means, I think, that we are ultimately driven to the Admiral Bill Owens' definition of the "revolution in military affairs." The "system of systems" approach, as he defined it, may be the only remaining answer because we won't be able to buy many new systems. As a consequence we better get used to the systems we've got and figure out how those systems can best be put to bear and employed. The budget simply isn't going to support an adequate level of modernization as it is right now. By the most conservative calculation we're at least $20 billion off on the modernization side of the equation and nobody is predicting big increases. If anything, the alternative is an ever-shrinking force. In that case, and if Ann Markusen's view is representative of the majority opinion, we may need to get used to the idea of Bill Owens' prediction of a military force of four divisions, 170 ships, and twelve wings, absent a change in policy. That argues that we adopt the "systems of systems" approach to employ what we have now in new and different ways to deal with what we see as emerging mission requirements.

There is a huge cost of tinkering with industrial policy and what we have seen in the last five years is yet another manifestation of that cost.

Nevertheless, those results, in my view, may yet be beneficial. They may yet bring us to a different position in which we are employing the assets we have today in a more creative way to deal with the challenges that we see in the future. But, I'd much rather it be a conscious policy—not one we simply back into, which is precisely where we are at this point.

7

NATIONAL SECURITY IN AN ERA OF GLOBAL TECHNOLOGY MARKETS: DoD'S DUAL-USE DILEMMA

THEODORE M. HAGELIN[1]

INTRODUCTION

Throughout history, technical superiority has been the prime determinant of military and industrial advantage. The race to achieve technical supremacy is more challenging today than ever before. Powerful private market forces are converging with national government policies to produce an accelerating pace of technical advance, an expanding scope of technical availability and an increasing merger of civilian and defense technical systems. Never before have more people had access to more sophisticated technologies capable of more positive and negative uses. This confluence of events has important implications for U.S. defense and economic policies.

In this chapter I will first review the private market factors shaping today's global economy and how they have contributed to the progress and proliferation of advanced technology. Next I will discuss DoD's dual-use strategy and how it is likely to operate within the context of global technology markets. I will then consider the major domestic and international regimes under which the U.S. attempts to control the transfer of critical military technology and the dilemma between the promotion of U.S. high-technology industries and the protection of U.S. national security interests. Finally, I will briefly outline some of the policy options available to reconcile the goals of economic growth, defense efficiency and national security. I conclude that there are basic conflicts between U.S. civilian and defense policies which soon must be debated if we are to preserve U.S. technical advantage in both sectors.

PRIVATE MARKET FACTORS

The link between technical capability and industrial competitiveness has produced a strong world demand for advanced technologies. This global demand has, in turn, given rise to multinational corporations and business alliances, and to international pools of technical and managerial workers. U.S. firms are highly competitive in the global market for advanced technology, which is one of the few areas in which the U.S. economy enjoys a sizeable trade surplus.

Manufactured products that incorporate advanced technologies accounted for nearly 20 percent of all U.S. trade in goods and services in the 1990s.[2] U.S. technology trade is especially strong in information technologies, aerospace and electronics which accounted for almost 85 percent of total U.S. technology product exports in 1994.[3] The U.S. is also the major net exporter of technological know-how and intellectual property in which it enjoys a 3-to-1 trade advantage compared to a 1.3-to-1 trade advantage in advanced technology products.

The U.S. advanced technology trade surplus reflects both the superiority of U.S. technical know-how and the business incentives for U.S. firms to transfer technology to foreign markets in both developed and developing countries. Developing countries offer U.S. firms an opportunity for increased sales, greater market share and higher return on investment because they require an abundance of new technology to build their economic infrastructures. The GDP growth rate in developing countries has been considerably greater than in Europe and North America for more than a decade, while wages, operating costs and taxes in developing countries have often been considerably less. This combination has proven to be a powerful lure for U.S. high-technology firms. In the future, it is likely that U.S. high-technology firms will continue to seek market opportunities in developing countries in order to grow. It is also likely that this growth will result in developing countries gaining expanded access to increasingly sophisticated technologies.

In pursuing international market opportunities, U.S. firms have also forged a host of strategic business alliances with foreign firms in developed countries. Today, the economics of technology markets forces each firm to concentrate on its core skills and knowledge, on doing what it does best.[4] U.S. firms are often rich in research, intellectual property, technical know-how and management, but short on competence in manufacturing, marketing, sales and support systems in local foreign markets. Strategic business alliances with local firms are often critical to the successful entry of U.S. firms into foreign markets. These business alliances, however, also serve as conduits for the transfer of advanced technical know-how among individuals, corporations and, in some cases, governments.

One form of business alliance, which contributes directly to the global spread of advanced technical know-how, is the research and development joint venture. Since the 1970s, there has been a sharp increase in international research and development partnerships. The known number of international multi-firm R&D alliances grew from 86 in 1973–76 to 988 in 1985–88 and this growth is estimated to have continued through the 1990s.[5] There has also been a marked trend by U.S. companies to make substantial R&D investments overseas. From 1985 to 1993, the overseas R&D investment of U.S. firms increased three times faster than did the R&D performed domestically. Overseas R&D investment now constitutes more than 10 percent of industry's domestic R&D spending.[6] Most of the overseas R&D investment has occurred in Germany and the United Kingdom, but considerable recent growth has taken place in Asian countries, especially Japan, Singapore and Indonesia.

It is also worth noting that there has been a dramatic increase in R&D investment by foreign firms in the U.S. In 1993, foreign companies accounted for about 15 percent of all industrial R&D funding in the U.S. compared to a 9 percent share in 1985.[7] Altogether there were about 635 foreign-owned freestanding R&D facilities in the U.S. in 1994 of which roughly one-third were owned by Japanese companies.[8]

The growth of multi-national R&D alliances in the private sector is being supported by a growing number of international science and engineering students studying in foreign and U.S. universities. From 1975 to 1992, higher education in science and engineering expanded rapidly in several regions of the world. The total number of science and engineering degrees awarded in six Asian countries (China, India, Japan, Singapore, South Korea and Taiwan) almost equaled the total number of science and engineering degrees awarded in Europe and North America in 1992.[9] Foreign students have accounted for a steadily increasing proportion of doctoral degrees awarded in science and engineering fields in U.S. graduate schools. In 1993, doctorate degrees earned by non-U.S. citizens reached 57 percent in engineering fields and 47 percent in mathematics and computer science.[10] Of the approximately 8000 foreign doctoral recipients in 1993, only about 30 percent remained in the U.S. after graduation.[11]

The rapid rise in science and engineering education in many areas of the world, coupled with the continuing increase in foreign science and engineering students in U.S. graduate schools, will result in an expanding global technical knowledge base, and a growing number of science, engineering and managerial workers capable of supporting many different companies, countries and research missions.

THE DOD'S DUAL-USE STRATEGY

Recent reforms instituted by the Department of Defense will accelerate the pace and diffusion of technical advances creating new challenges to U.S. national security. Under pressure to cut costs and to operate more efficiently, the DoD has adopted a "dual-use" strategy for the development of new technologies.[12] Dual-use technologies are technologies which have both military and civilian applications. The DoD's dual-use strategy attempts to leverage the strength of private sector technology markets in three ways. First, dual-use technologies allow DoD to benefit from the production economies of scale generated by civilian demand, thus lowering unit acquisition costs. Second, the incorporation of civilian technology into defense systems allows DoD to avoid the costs required to independently develop state-of-the-art military technical components. Third, the use of already developed commercial technologies allows DoD to reduce the cycle time required to upgrade existing systems and implement new systems. The progressive merger of civilian and military technologies will result in the increased global availability of advanced technology having potential military application. Whatever is available to the DoD in the commercial market place is, of course, also available to other nations and groups.

DoD's dual-use strategy is also intended to benefit defense prime contractors and subcontractors. The development of dual-use technologies allows defense contractors to diversify into commercial markets and thus hedge against down-turns in defense spending. The financial strength of defense contractors is important to DoD and the dual-use strategy allows DoD to support defense contractors without expending actual funds. From the perspective of the defense contractor, a successful dual-use strategy depends upon an expanding merger of military and civilian technologies and an accelerating conversion of applications from one field to the other. The natural business incentive of defense contractors operating in a dual-use environment will provide a strong impetus to the diffusion of advanced technology with direct military potential.

The essence of DoD's dual-use strategy is the leveraging of commercial sector technology in order to reduce system acquisition costs and increase the availability of state-of-the-art technology. Paul Kaminski, former Secretary of Defense for Acquisition and Technology and one of the chief architects of the dual-use reforms, has described three pillars of the dual-use strategy:

- Invest in dual-use technologies critical to military applications.
- Integrate military and commercial production.
- Insert commercial components into military systems.[13]

As this list suggests, the implications of DoD's dual-use strategy extend beyond the utilization of commercial technologies for military applications. DoD's dual-use strategy also seeks to integrate military and commercial production systems. This integration would require the increasing combination of defense and civilian lines of business within single firms. The ultimate goal of DoD's dual-use strategy in the words of Dr. Kaminski is "to move from separate industrial sectors for defense needs and commercial markets to an integrated national industrial base."[14]

In support of its dual-use strategy, DoD has instituted a number of new programs. The *Dual-use Applications Program* is intended to fund service managers in pursuing the development of dual-use technologies through government–industry R&D partnerships. The *Commercial Technology Insertion Program* is intended to accelerate the inclusion of commercial technologies into defense systems.[15] Other DoD programs which also support the dual-use strategy include the *Small Business Innovation Research Program, Small Business Technology Transfer Program,* and the *Government–Industry–University Research Initiative.*[16]

DoD's dual-use strategy will likely result in ever more sophisticated military grade technology becoming available in commercial markets. Just as DoD seeks to leverage the civilian market, private firms will seek to leverage the defense market to underwrite the development of increasingly advanced technology for sale in commercial markets. The problem is most acute in the growing number of technical areas where civilian and military applications are essentially indivisible—what could be called "pure" dual-use technologies. Computers, radar, satellite systems, sensors, lasers and aerospace are just a few examples of technologies which have inseparable military and commercial applications and can be used for either purpose with minimum modification.

These "pure" dual-use technologies provide the core of the military's most advanced fighting systems such as the Army's Force XXI, the Navy's 20/20 Vision and the Air Force's New World Vistas.[17] All of these systems require near real-time information gathering, processing, decision-making and communication. The technologies necessary to support these military needs are, of course, equally valuable in banking, finance, manufacturing and a host of other commercial applications. The transfer of these technologies for commercial use cannot be accomplished without also making these technologies available for military application.[18]

Finally, it should be noted that there are a number of other federal programs which also contribute to the pace and diffusion of technical advancement. These programs, collectively known as *Federal Cooperative Technology Programs,*[19] are intended to facilitate the commercialization of federally supported research by providing the federal funding recipient with the right to elect title to any patents resulting from the research project.[20] The federal government is the largest funder of basic and applied research and universities are the largest recipients of these research funds. These federal programs have resulted in dramatic increases in technology transfers from universities to private industry. Although these programs are designed to benefit U.S. firms and the U.S. economy, there is little capability of controlling the ultimate destination, or application, of these technologies once a transfer has been completed. Congress has also sought to encourage the transfer of technology developed at federal laboratories through the allowance of exclusive licensing agreements, and cooperative research and development agreements, between federal laboratories and private firms.[21] Giving away federally-owned technology to promote economic competitiveness and growth is a popular notion within Congress. However, federal laboratory technology transfers also cannot be effectively restricted from international markets.

DoD is a major provider of federal research funding in universities, private industry and federal laboratories. DoD research performed in these institutional settings is often freely available for commercial exploitation by private firms. DoD funded research cannot be exploited for commercial purposes, however, without also running the risk of exploitation of military purposes.

TECHNOLOGY TRANSFER CONTROL REGIMES

In the face of the powerful forces encouraging technology advancement and diffusion, U.S. efforts to control the transfer of critical military technology have had limited success. The U.S. attempts to regulate transfers of sensitive technology through a complex set of domestic and international control regimes. Serious, and possibly worsening, problems have surfaced in each of these control regimes.

Domestic Export Controls

Domestically, the U.S. export control system is divided into two sub-regimes: munitions items are regulated pursuant to the Arms Export Control Act[22] and dual-use items are regulated pursuant to the Export Administration Act.[23] The Department of State has responsibility for munitions exports and the Department of Commerce has responsibility for dual-use exports. In general, the ability to deny export of dual-use items listed on the Commerce Department's Commodity Control List[24] is more limited than the broad denial authority afforded the State Department in administering the U.S. Munitions List.[25] DoD has no formal responsibility for technology export control, but generally provides consultative input to the Departments of State and Commerce. Recent studies have shown serious deficiencies in the administration of the U.S. export control system which could result in critical military technologies being licensed for lawful export.

One of these studies, which focused on stealth technology, found that (i) the definitions of stealth-related technologies contained in the Commodity Control List and the Munitions List failed to clearly assign individual technologies to one list or the other; (ii) the jurisdictional responsibilities of State and Commerce were unclear in a number of areas; (iii) DoD input into technology export decisions was often limited; and (iv) a large number of applications for export of stealth-related dual-use technology reviewed by the Commerce Department were never seen by the State Department or DoD.[26] The report found that between 1991 and 1994 only 15 out of 166 applications were referred to either State or DoD for review.[27] If this report is accurate, it suggests that DoD not only has limited opportunity to protest dual-use exports, but that DoD is often unaware of dual-use technology exports and hence denied the opportunity to monitor the end-users and end-uses of this technology.

The conflict between the Commerce Department and State Department over technology export control is, of course, part of a larger contest between U.S. economic and national defense goals. In this contest, which has now been going on for over a decade, domestic economic growth and foreign trade have clearly prevailed, and U.S. export control of advanced technology in all fields has been substantially liberalized. Practical pressure to liberalize U.S. technology export controls has also come from other nations that supply the same or similar technologies to the world market which the U.S. refuses to export, and from U.S. technology firms anxious to increase their share of foreign markets.

As global economic competition intensifies and technology becomes increasingly dual-use, the strain on the U.S. technology export control system will grow greater. Balancing the costs and benefits of U.S. export controls is extremely difficult, but inherently biased in favor of economic interests. The cost of export controls can be measured by lost jobs, lost contracts, lower profits, higher trade deficits and many other objective standards. The benefit of export controls cannot be quantified. The loss of military advantage, the lessening of national security, diminished deterrence and strategic vulnerability are far less concrete than economic indicia. In the battle between a dollar in hand and a military threat in the future, the dollar always seems to win.

International Technology Export Controls

Although the U.S. has struggled gallantly to bring nations together to control the proliferation of nuclear weapons, missile technology and weapons of mass destruction, it has confronted numerous obstacles. The demise of COCOM (a partnership between the U.S. and its NATO allies to control the transfer of sensitive technology to the former U.S.S.R. and Eastern Block nations) illustrates the problems with international export control regimes.[28] COCOM functioned relatively well during the height of the Cold War, but as the Cold War wound down the COCOM member nations became increasingly divided over export controls and increasingly willing to take unilateral action to promote domestic economic interests. COCOM was completely dismantled in March 1994, but as yet no successor regime has been put into place.[29] Potential members of a new control regime disagree strongly over what procedures should apply to the review of technology exports, over what technologies should be subject to export controls, over how technologies should be classified, and over what nations should have access to various classes of technologies.[30] The treatment of Russia in a successor control regime to COCOM has been a particularly divisive question. These tough issues, which always reflect differing domestic economic interests and foreign policy priorities, will not be any easier to solve in the future than they have been in the past.

Other multilateral technology control regimes confront similar problems. The Missile Technology Control Regime, the Chemical Weapons Convention, the Biological Weapons Convention and the Nuclear Suppliers Group are all fraught with difficulties regarding (i) the list of items to be controlled; (ii) the countries included as members or associate members in the control regime; (iii) the benefits offered for membership or adherence to the control regimes; and (iv) the list of countries of greatest export concern.[31] These multilateral control regimes are voluntary, unverified agreements which in most cases have little or no institutional presence, no independent budget and no permanent staff. Enforcement is at the discretion of individual members. Although these multinational regimes have been useful in building some consensus for restricting certain items from export to certain countries, they have not prevented members from contributing to weapons of mass destruction programs in countries such as Iraq, Iran and Libya.[32] For example, Russia, which is a member of the Nuclear Suppliers Group and of the Missile Technology Control Regime, has sold reactors and possibly uranium enrichment technology to Iran as well as missile technology to India. And China, which is a member of the Missile Technology Control Regime, is alleged to have sold missile technology and provided nuclear assistance to Pakistan.

A final disheartening point must be noted regarding the proliferation of weapons of mass destruction. The technology export control regimes discussed above are only effective, of course, where the potential technology transfer is known to exist through some formal notification requirement or application procedure. Underground technology transfers are not subject to regulation by any entity whether domestic or international. In an appendix to his testimony before the Senate Armed Services Committee on the DoD Fiscal Year 1997 Budget, Dr. Gordon Oehler, Director of the Non-Proliferation Center at the Central Intelligence Agency, listed more than 80 separate nuclear smuggling incidents around the world in a two-and-a-half year period, spanning from November 1993 to March 1996.[33] It is reasonable to suspect that the extent of smuggling of chemical and biological weapons of mass destruction is at least equal to nuclear smuggling and quite possibly greater. Chemical and biological weapon technologies have become increasingly popular with terrorists because they are easier to use than nuclear technology and because the component parts are easier to obtain on the open market. It is estimated that more than 25 countries have chemical weapons systems and that more than 10 have biological weapons systems.[34]

Clearly, the U.S. must continually strive to strengthen both domestic and international technology control regimes. Equally clearly, U.S. military security cannot be wholly dependent upon these regimes. The power of economics is too great and the power of consensus is too limited. And even if these obstacles could be overcome, there would remain the ever present problem of underground technology transfers.

POLICY OPTIONS

The U.S. has different policy options available to reconcile the goals of high-technology industrial growth, defense efficiency and national security. Selecting the optimum combination of policies will require a careful assessment of the positive and negative consequences, and a careful balance of U.S. domestic and international interests. The following are some of the policy options which could be pursued.

- The U.S. could expand its efforts to strengthen both domestic and international technology transfer control regimes. Stricter control regimes, however, could disadvantage U.S. high-technology companies in global markets and could be resisted by other nations seeking to promote their own industrial and foreign policy self interests.
- The U.S. could reverse its policies of funding advanced research and development and of transferring federally funded technology to the private sector for commercial development. Withdrawal of federal support for R&D and technology transfer programs, however, could disserve the progress of science and engineering, damage U.S. universities and ultimately retard U.S. global competitiveness.
- The DoD could moderate its dual-use reforms and accept a greater separation of civilian and military technologies. Any retrenchment in the dual-use initiatives, however, could come at the expense of diminished defense efficiency in terms of higher development and procurement costs, and longer cycle times to implement new systems.
- The U.S. could increase investment in advanced defense research projects to maintain critical technical force advantages. Increased spending on defense research, however, could be perceived to come at the expense of civilian research and could be opposed by groups supporting additional research in such fields as health and the environment.
- Finally, the U.S. could seek to monitor cutting-edge R&D in critical technical fields throughout the world, tracking the places, persons and projects of greatest strategic importance. Such a monitoring scheme, however, could threaten proprietary technical and economic information

in the hands of private firms, and could be challenged by foreign governments as a plan for economic, or military, espionage.

CONCLUSION

There is a growing conflict between U.S. economic and defense policies.
Current policies in both sectors have been formulated to advance specific
objectives without sufficient regard for broader, unintended, consequences.
U.S. economic policy is being driven by global technology competition,
in many cases coming from countries which actively subsidize and
coordinate domestic industries. The U.S. has sought to support its
own domestic high-technology industries by encouraging free trade,
liberalizing export controls, transferring federal-owned technology and
funding advanced science and engineering research. U.S. defense policy
is being driven by demands for reduced spending and greater operational
efficiency. DoD's dual-use initiatives have sought to respond to these
post Cold War imperatives by utilizing commercial components in military
systems, encouraging the integration of commercial and military production
lines, and investing in dual-use technology research. Although each of
these policies seeks to promote an important individual national interest,
in combination they are contributing to an explosion of advanced
technology around the world which poses new threats to U.S. national
security.

The U.S. confronts a difficult set of policy issues today. A coherent
resolution of these issues will require wide public debate with input from
industry, government and universities. At the end of a millennium in which
technology life-cycles have gone from centuries to months, this debate can
no longer be postponed.

NOTES

1. I would like to thank the Klein and Bantle Symposia participants for the
   stimulating discussions which contributed to the ideas in this chapter and
   especially Sean O'Keefe for his helpful comments on a first draft of this chapter.
   I also want to thank Mr Michael Herrman, a third year student at Syracuse
   University College of Law, for his assistance in preparing this chapter for
   publication.
2. National Science Board, *Science & Engineering Indicators – 1996*, U.S.
   Government Printing Office, 1996 at 6-6. (Hereafter "Science & Engineering
   Indicators.")
3. Ibid at 6-7.
4. See generally, Dratler, J., Jr. *Licensing of Intellectual Property*, Law Journal
   Seminars-Press, pp. 1–2 and 1–38, 1994; Porter, M. *Competitive Strategy:
   Techniques for Analyzing Industries and Competitors*. Free Press, 1980 and
   "What Is Strategy." *Harvard Business Review*, Nov.-Dec., 1996.

5. Science & Engineering Indicators, supra note 1 at 4-43. See also, The Maastricht Economic Research Institute on Innovation Technology's Co-Operative Agreements and Technology Indicators Data Base (Merit-Cati). From The Evolution Of Technology And Markets And The Management Of Intellectual Property Rights, 72 *Chi.-Kent L. Rev.*, p. 369, 1996.
6. Science & Engineering Indicators, supra note 1 at 4-44.
7. Ibid at 4-45.
8. Ibid at 4-45.
9. Science & Engineering Indicators, supra note 1 at 2-2.
10. Ibid at 2-23, 2-24.
11. Ibid at 2-28.
12. See 10 U.S.C. S2511 which has been updated as PL 105-85 (HR 1119) November 1997, "National Defense Authorization Act for Fiscal Year 1998."
13. Hearings Before Senate Armed Services Committee (testimony of Paul Kaminski), March 20, 1996, Hearing Report at 98.
14. Ibid at 98.
15. Ibid at 84, 85.
16. Science & Engineering Indicators, supra note 1 at 4-18.
17. Hearings Before Senate Armed Services Committee (testimony of Gen. Charles Krulak, USMC, Adm. Jay Johnson, USN. Gen. Thomas Moorman, USAF, Lt. Gen. Ronald Hite, USA), March 5, 1996, Hearing Report at 5-73.
18. The six high tech industries identified on the basis of R&D intensity are (from highest to lowest R&D intensity) aerospace; office and computing equipment; communications equipment; drugs and medicines; scientific instruments; and electrical machinery.
19. See Berglund, D. and Coburn, C. "Partnerships: A Compendium of State and Federal Cooperative Technology Programs, 1995."
20. Pub. L. No. 96-517, 94 Stat. 3020, 1980 (codified at 35 U.S.C.SS.202, 1980); see also 35 U.S.C.S.203, 1980 (patent rights in inventions made with federal assistance).
21. Congress is constantly tinkering with the laws governing technology transfer. In the 103rd Congress, 243 bills were introduced that dealt with technology transfer and 80 of these bills referenced, or sought to amend, the Stevenson-Wydler Act. See generally Rudolph, L. Overview of Federal Technology Transfer, 5 Risk: Health Safety & Env't 133, 134, 1994. (Addresses all federally supported R&D and specifically discusses H.R. 820, H.R. 1432, and H.R. 523, which are identified as "noteworthy".) H.R. 820 contains the National Competitiveness Act of 1993, the Manufacturing Technology and Extension Act of 1993, and the Civilian Technology Development Act of 1993. "These proposals seek to boost the nation's international competitiveness by strengthening our technology base and fostering the development of advanced products, particularly in manufacturing". Ibid.
22. U.S.C. 2778 et seq.
23. U.S.C. App. 2403(c) and 2405(h)(2)–(4), as amended and is continued in effect by Executive Order 12924 of August 19, 1994, and by notices of August 15, 1995, and August 14, 1996.
24. The Commerce Department's Commodity Control List (CCL) specifies the type of export license required for a specific export to a given destination. The CCL

divides goods and technologies into export categories and indicates the country group level of control. See, 50 U.S.C. app. SS 2403(b), 2404(c)(1).

25. For a good discussion, see Coping With Export Controls 1997, DEPARTMENT OF COMMERCE EXPORT CONTROLS. Hunt, C. 760 PLI/COMM 37, 1997. See also, 22 C.F.R. §121.1 International Traffic in Arms, The United States Munitions List, 1998.

26. Hearings Before Senate Armed Services Subcommittee (testimony of David Cooper), May 11, 1995, Hearing Report at 8.

27. Ibid at 9.

28. COCOM was a body which coordinated national controls on technology and strategic exports. Cecil Hunt, Multilateral Cooperation in Export Controls—*The Role of COCOM*, 14 TOL.L.REV. 1285, Summer 1983. Its membership was comprised of the NATO states (except Iceland) and Japan. COCOM maintained three lists of controlled goods: the International Atomic Energy List, the International Munitions List and the International List (containing dual-use items not on the first two lists).

29. However some have argued that "Australia Group and the Wassenaar Arrangement on Export Controls for Conventional Arms and Dual-Use Goods and Technologies" is sometimes characterized as the successor regime to COCOM. See, Coping with U.S. Export Controls, EXPORTING TO SPECIAL DESTINATIONS; TERRORIST-supporting and embargoed countries, Connaughton, A.Q., 760 PLI/Comm 317, 321, 1997.

30. Hearings Before Senate Armed Services Committee (testimony of Dr. Zachary Davis), May 11, 1995, Hearing Report at 21-22.

31. Ibid.

32. Hearings Before Senate Armed Services Committee (statement of Sen. Bob Smith), March 27, 1996, Hearing Report at 203-204. The countries that the Secretary of State has determined, under §6 of the 1979 Export Administration Act (50 U.S.C. app. 2401 et seq.), to have repeatedly provided support for acts of international terrorism are: Cuba, Libya, Iran, Iraq, North Korea, Syria and Sudan. Embargoed destinations are Cuba, Libya, Iran, Iraq, and North Korea. In addition, a limited export embargo applies to Angola (UNITA).

33. Hearings Before Senate Armed Services Subcommittee (testimony of Dr. Gordon Oehler), March 27, 1996, Hearing at 221-227.

34. Ibid at 227-228.

8

PUBLIC POLICIES AND CORPORATE STRATEGIES: A POST-COLD WAR VIEW FROM EUROPE

TERRENCE GUAY

INTRODUCTION

One of the main purposes of this volume is to identify the changing corporate strategies and public policies that are shaping the development of the defense industry since the end of the Cold War. While the focus of this collection is on the United States and the U.S. defense industry, it is instructive to compare the situation in the United States with Europe, specifically Western Europe. The purpose is not to diminish the tremendous difficulties of determining the appropriate corporate strategies and public policy responses in the United States, but to highlight the problems that other countries with significant defense industrial sectors are also facing. The comparative approach is a useful analytical tool that can aid us in identifying opportunities from the experiences of others, while also being mindful of the differences that exist in other political and economic systems.

PUBLIC MANAGEMENT CHALLENGES: REGULATORY, LEGAL, AND POLITICAL CONSIDERATIONS

Perhaps the greatest problem we encounter when comparing the U.S. and European defense industry is the fact that there is no truly "European" defense industry. Europe is divided at both the buyer and supplier levels. Weapons systems are procured at the national level with relatively little coordination between national governments. While Europe exists as a geographic region, and institutions like the European Union (EU) have developed policies that affect each of its 15 members, Europe is still configured as a group of nation-states. At the same time, there are no truly "European" (or transnational) defense firms because the owners, workers, and production facilities (but not sales) of companies are almost entirely concentrated within particular countries. While this is also true in the United States, U.S. defense firms, especially after the recent flurry of mergers and acquisitions, have the benefit of economies of scale and scope that smaller European companies lack. Thus, such definitional and structural problems make it somewhat misleading to refer to "European" defense industrial policies and strategies, and to compare them to any other national situation.

A second challenge for Europe is that military equipment spending has decreased even more rapidly than it has in the United States. Equipment spending by EU countries[1] declined from $28.6 billion in 1988 to $17.6 billion in 1996—a 38 percent cut (SIPRI, 1997). In the United States, military equipment spending dropped from $80.3 billion to $58.6 billion (or 27 percent) over the same period. While the end of the Cold War and changes in threat perception were the primary justifications for reducing military spending in both Europe and the United States, an additional reason for sharp defense cuts in Europe during the 1990s has been the push toward economic and monetary union (EMU). The problem is compounded by the fact that Europe as a whole spends much less on defense in general ($160.1 billion for the 15 EU countries versus $226.4 billion by the United States in 1996), and military equipment in particular, than the United States does. A smaller market that is shrinking more quickly exacerbates the difficulties in determining the appropriate public policies and corporate strategies. The desire by each of the large European countries (Britain, France, and Germany) to be self-sufficient in arms production breeds tremendous inefficiencies. Although the U.S. armaments market is about twice as large as the EU market, the U.S. industry has been rationalized from 20 companies in 1980 to just three main suppliers today (Tucker, 1997). By contrast, in the fall of 1997 Europe still had six companies making civilian aircraft, six making fighters, three helicopter producers, 12 missile producers, six big defense electronics companies, and five manufacturers in the field of satellites.

The severity of these problems could perhaps be lessened if they were tackled in a more coordinated manner. But this, too, poses a problem since Europe cannot decide which political level is the most appropriate for addressing these issues. There are several institutions that have the potential to respond to defense industry restructuring. The North Atlantic Treaty Organization (NATO) has a poor record in promoting transatlantic armaments collaboration, and European members (particularly France) are suspicious of U.S. intentions when such proposals are floated in this forum. The Western European Union (WEU), created in 1948 and "reborn" in the 1980s, is still in embryonic form. It lacks the infrastructure to coordinate pan-European industrial restructuring, and has not been successful in developing a joint armaments procurement agency as was proposed in the 1991 Maastricht Treaty. The EU perhaps holds the most promise for an institutional solution. The Commission is experienced in deciding corporate mergers and acquisitions, supporting cooperative research and development (R&D) projects, funding conversion programs, working to open public procurement, and developing external trade policies (essential for a common armaments export policy). However, the EU's common foreign and security policy (CFSP) is weak, its defense dimension is non-existent, and the British are opposed to an EU defense industrial policy. Complicating the matter further is Article 223 of the 1957 Rome Treaty, which provides the legal foundation for states to argue that the production and trade of armaments is a national—and not an EU—policy area. Thus, it is at the national level that most European governments prefer to deal with defense industry restructuring. With each major country seeking to remain self-sufficient in armaments production, it remains difficult for cross-border restructuring to make significant strides, and for a regional public policy approach to be implemented.

A fourth major problem for Europe is that defense industry restructuring is not a high priority for most governments. The two primary policy concerns in Europe today are achieving EMU and reducing the region's high unemployment rates, which have been stuck in double digits in many EU member states for most of the 1990s. Interestingly, these issues are having paradoxical effects on Europe's defense industry. EMU is forcing governments to make drastic budget cuts, and defense is a popular target. New projects like the Future Large Aircraft transport plane are on hold, while current weapons systems are being refurbished rather than replaced. At the same time, governments are unwilling to force companies to rationalize, reduce capacity, and cut workforces at a time when European unemployment is so high. But beyond the economic repercussions of EMU is the political will to see the common currency project succeed as the next grand step in the European integration process. Unfortunately for Europe's defense industry, that same political will is lacking in a policy area where vision and cooperation are sorely needed. In sum, policymakers, for the most part, are addressing defense industry restructuring and diversification only indirectly.

ACQUISITION REFORM

The political, regulatory, and legal considerations mentioned above shape the acquisition process for military equipment. Since the governments of nation-states, and not institutions, have the predominant influence over

defense firms, each country has its own procurement policy. The production
of military equipment in Europe is concentrated in France, Britain, and
Germany, while Sweden, Italy, and Spain are distant "second-tier" producers.
It is France, Britain, and Germany that have the largest military forces,
defense budgets, and armaments industries in the EU. However, the
acquisition policies of these three countries differ considerably, as do current
discussions of reform.

France has been the slowest of the three major European states to cut
defense spending in the post-Cold War era. In 1987, Britain spent $42.6
billion on military expenditures, Germany spent $40.6 billion, and France
spent $42.3 billion (SIPRI, 1997). By 1996, Britain had reduced defense
spending by 26 percent to $31.5 billion and Germany had made cuts of 25
percent to $30.5 billion. France, however, had cut its military spending by
only 9 percent to $38.4 billion. Consequently, higher French defense budgets
have been accepted by the public because, with more than 90 percent of
the funds spent domestically, it has served as a form of industrial policy
(Ridding, 1994). Cuts in defense and military R&D spending would eliminate
high-technology jobs and put further pressure on the country's persist-
ently high unemployment rate. Defense spending thus represents a very
explicit industrial policy. However, this policy has become very costly. Many
observers regard much of the French defense industry to be inefficient and
unnecessary—particularly the firms that are mostly state-owned (such as
Aérospatiale, the tank- and munitions-maker GIAT, and the shipbuilder
DCN).

However, French governments on both the right and left have been
unwilling to force domestic firms to take the drastic actions (including
closing plants and laying off workers) needed to be more internationally
competitive. It is no coincidence that, of the three major European countries,
France is the most ardent supporter of both nationalist industrial policies
and a "European preference" for collaborative projects (meaning the exclusion
of U.S. defense firms). Yet France's grand military designs now clash with
the reality of the country's domestic economic and financial situation and
its desire to see EMU succeed. In early 1996, Paris announced a restructuring
program designed in part to improve the efficiency of its defense industrial
sector (*Economist*, 1996b; *Economist*, 1996c). Thomson-CSF, the defense
electronics company, was to be privatized, while the government would try
to merge aircraft-makers Aérospatiale and Dassault. The intention is to
concentrate the country's defense electronics and military aircraft into two
separate groups. Privatization of these companies, as well as aircraft
engine-maker SNECMA, would provide some of the capital injection needed
to rationalize excess capacity, layoff workers, and increase R&D. The
Délégation Générale pour l'Armament, the industrial arm of the defense
ministry, was to be reorganized with an emphasis put on improving
productivity by 30 percent. Further proposals in September 1996 called

for the proportion of France's arms budget spent in cooperation with its European partners to more than double over the next six years from 15 percent to 34 percent (Owen, 1996). However, by late 1997 France had made little progress in privatizing parts of its defense industry, or in forcing French companies to merge (privately-owned and profitable Dassault is strongly opposed to pooling its business with state-owned and loss-making Aérospatiale). The future of acquisition reform in France is still very uncertain.

The British approach to acquisition contrasts with the French experience. Britain's armaments procurement policy has been described as "value for money" since the mid-1980s. This means that orders for weapons are open to bidding by any defense firm—domestic or foreign. In theory, decisions are made based on cost and quality. Many European defense firms and governments (particularly the French) criticize the UK because the value for money principle often results in U.S. companies winning arms contracts.[2] Thus, the British government proclaims to have a "hands-off" policy toward its domestic arms industry. Roger Freeman, former Minister of State for Defence Procurement, said that "[h]ow [the defense] industry is structured and what it sells, must principally be determined by companies operating within the market and not by Government" (Freeman, 1995). Despite Britain's professed openness to foreign arms producers, the bulk of its defense money stays in the country. In the government's 1995 defense white paper, a breakdown of British arms purchasing over the previous five years revealed that 79 percent of procurement spending went to UK companies, 9 percent to companies elsewhere, and 12 percent to joint ventures (Gray and Clark, 1995). Nonetheless, the British government has been more willing than its continental counterparts to let market forces lead the reorganization of its armaments industry. In the 1980s, the Thatcher Governments privatized British Aerospace (BAe), Rolls Royce, and a number of naval yards. In the 1990s, the Conservatives let BAe and GEC compete in the stock market for ship-maker VSEL, and allowed GKN, a military vehicles company, to succeed in making a hostile bid for Westland. While the British defense ministry feels that such corporate acquisitions lead to greater efficiency, and therefore lower procurement costs, the country's armaments industry has probably reached the point where further consolidation (such as a merger of BAe and GEC) could lead to a monopoly situation.

Germany's acquisition philosophy falls somewhere in between the French and British approach. On the one hand, Germany supports French initiatives to foster cooperative weapons projects in Europe. The political integration that such programs imply is important to Germany and its general European policy. On the other hand, Germany, like Britain, has become more cost-conscious. Unification was more expensive than originally expected, a sluggish economy has hampered tax revenues and forced an increase in social spending, and the drive to achieve EMU has led to severe budget cuts, particularly in defense. The effects of Germany's steep defense cuts and focus on equipment cost have been evident in the national debate over participation in the Eurofighter program. Until the fall of 1997, when Germany's parliament gave final approval to the project, the country's involvement was uncertain because many politicians, especially from the Social Democratic party, did not think that it was a high budgetary priority—even though many aerospace workers in Bavaria would benefit (Atkins, 1997). The German government is also becoming increasingly wary of French initiatives that support European weapons systems, but minimize the extent and severity of industrial inefficiencies (*Economist*, 1997). The French are upset that DASA decided not to merge its missiles and satellites businesses with parts of Aérospatiale, on account of the state-owned French firm's unprofitability. Germany, consequently, has the potential to play a key role in nudging Europe's defense industry toward the Anglo-Saxon model.

At first glance, it would seem that smaller European countries can avoid the difficult policy choices that the British, French, and Germans face. With modest defense industrial sectors and few workers employed in arms production, small countries would appear to have the flexibility to acquire weapons systems based on cost, quality, and defense needs. But even here, smaller countries are often pressured by the largest three (particularly France) to "buy European" rather than U.S.—even if that means purchasing more expensive and lower quality products. If small countries become tied into a European arms procurement agency (discussed below) the pressure to buy European will only intensify.

An alternative to the purely nationalist approach to military equipment acquisition is the formation of multinational consortia. Such projects enable governments to undertake projects that would simply be too expensive, technologically difficult, or risky to do alone. While European governments have worked on collaborative projects for decades (including the Tornado and Eurofighter aircraft, various missile programs, and the current Horizon frigate), the success of these efforts has been mixed. The practice of *juste retour*, whereby production is allocated to companies in proportion to the quantity of weapons systems that will be purchased by participating countries, is widely seen as inefficient, but somewhat better than having each country produce the system on its own (Moravcsik, 1990). The four-nation Eurofighter is behind schedule in large part because Germany has been indecisive in its commitment to the program. The Horizon frigate has faced obstacles because the participating countries (France, Italy, and Britain) each envision different roles for the ship. National variations, delays, and the need to incorporate ever-changing technology into weapons systems increase their cost. Nonetheless, as will be discussed further below, collaborative programs are still popular in Europe.

The acquisition reform that would have the most significant impact on Europe's defense industry is the formation of a joint armaments procurement agency. The Maastricht Treaty called on the WEU to explore the possibility of setting up a European Armaments Agency. The purpose of the program is to coordinate weapons needs and harmonize technical specifications, thereby reducing costs and enabling the interoperability of equipment in military missions. To date, the initiative has gotten nowhere as member states feel that "conditions do not currently exist for the creation of an agency conducting the full range of procurement activities on behalf of member nations" (SIPRI, 1995). However, a bilateral program formed by France and Germany in January 1996 has developed into a multilateral project including Britain, Italy, the Netherlands, Belgium, and Spain. The major substantive difference between this initiative (known as the Joint Armaments Cooperation Organization) and the agency proposed in the Maastricht Treaty is that the former is a loose, ad-hoc arrangement, while the latter would institutionalize the process within the WEU (or, perhaps someday, in the EU).

It is not yet clear how successful this approach to acquisition will be, and whether it will assist defense industrial restructuring. Some French and German defense officials are concerned that too many participants may reduce the efficiency and effectiveness of the arms agency, bringing their bilateral initiative to a virtual stop.[3] In other words, a multilateral approach to arms procurement would risk the bureaucratic problems and national differences that the bilateral agency had hoped to avoid. A second problem is the different philosophical approaches to weapons procurement that different countries follow. The agency will have to determine whether procurement decisions will be based on value, preferences for European defense firms, wider industrial or foreign policy considerations, or some combination of these concerns. It may not be a good omen for the long-term health of Europe's defense industry that, in its eagerness to join the Franco-German agency, Britain was forced to give ground on its value for money philosophy, and accept the notion that some preference would be given to European defense firms in the awarding of contracts.

CORPORATE STRATEGIES

The defense industry is unlike virtually any other economic sector. Its close relationship with government has fostered the long-entrenched notion that countries need to be self-sufficient in the production of arms. The justification for national arms industries today has moved beyond the simple national defense explanation. The provision of technological "spin-offs" to civilian applications and good-paying manufacturing jobs are economic arguments used by business executives and government officials to preserve the status quo. Governmental control over weapons exports also makes arms production a valuable foreign policy tool and a symbol of national prestige. It is not surprising then that the organization of defense industries along national lines remains the dominant corporate structure.

In Britain, BAe dominates the production of aircraft and missiles, and GEC is the major military electronics supplier. Both companies have grown through acquisitions of smaller, primarily British, companies involved in defense electronics, shipbuilding, and munitions. In Germany, Daimler Benz's DASA became a conglomerate involved in aircraft and missiles production after a series of acquisitions in the late 1980s and early 1990s. Siemens is the primary electronics firm (although in October 1997 the company announced that it was selling its defense electronics business to BAe and DASA). Aérospatiale and Dassault are the French aircraft companies, Aérospatiale and Matra specialize in missiles, and Thomson-CSF and Matra produce military electronics. GIAT makes tanks, artillery, and munitions, and, while small relative to the other major French defense firms, has been active in acquiring even smaller European arms producers. Celsius is Sweden's major defense firm, and Saab-Scania specializes in fighter aircraft. The most prominent defense resources in Spain and Italy are concentrated in the aerospace sector: in Spain it is CASA; in Italy it is Alenia and Agusta (a helicopter manufacturer). The national champion approach has accelerated over the past ten years. Because of the political sensitivities of cross-border mergers and acquisitions, most deals have involved large companies buying up smaller firms—either within their borders or in countries with small arms industries (like Belguim).

Despite the influence of the national champion model, a second corporate strategy has begun to emerge. While collaboration on weapons systems has existed since the 1960s, there has been an increased emphasis on such efforts in recent years (Skons, 1993; Anthony and Wulf, 1992). Economics and, to a lesser extent, politics are the driving forces. The expense of increasingly sophisticated weapons technology has been a primary impetus for cooperation. Defense firms realize that the breadth of technology needed to develop and manufacture state-of-the-art weapons systems is beyond the capabilities of individual companies. Cross-equity participation and joint ventures allow for the promotion of both high-level consultation on collaborative ventures and some technology sharing. Some firms have created a new company together, such as the decision by Aérospatiale and DASA to merge their military and civilian helicopter businesses in Eurocopter. About 70 percent of DASA's defense business is tied up in joint projects (*Economist*, 1996a).

Strategic alliances also allow for inter-firm cooperation. In the case of BAe, for example, five of its last six military aircraft have been the result of collaboration. In February 1995, Saab-Scania reached an agreement with BAe in which the British firm will handle the international marketing of the Swedish aircraft manufacturer's latest generation JAS 39 Gripen fighter aircraft (Verchère, 1995). Initially the agreement is to market the aircraft, but if significant export sales are won, manufacturing work will be shared between the two firms. The Swedish firm benefits from the international network and contacts that the UK has from selling defense equipment for decades around the world. Prior to joining the EU in 1995, Swedish arms companies were hampered by relatively strict national export laws.[4]

Another form of cooperation is teaming arrangements (SIPRI, 1993). These are joint bids between two or more companies for contracts to develop and produce major weapons as a means of sharing the risks involved. The companies within a team retain their individual autonomy, although one company assumes the lead as prime contractor. This is a particularly attractive option for defense firms trying to break into foreign markets. In these cases, foreign companies team up with a respected domestic firm in making a bid for an armaments project. Such relationships provide participating players mutual access to each other's home market.

A variation of this strategy is to team up with U.S. defense firms. BAe, for example, has joined Lockheed Martin in a competition to build America's next generation fighter plane—the Joint Strike Fighter. While the transatlantic option is attractive in that European companies gain access to more advanced technology, it has major drawbacks. One problem is that the U.S. government maintains tight control over the re-export of technologies to third countries, which could dampen the likelihood of significant international sales by European producers. However, the larger difficulty is political in nature. Many European firms and governments fear that an increase in cooperative relationships with U.S. weapons makers will simply turn Europeans into subcontractors for the United States. This, in turn, would make Europe highly dependent on the United States, and affect Europe's ability to conduct an independent foreign policy.

With the armaments market shrinking at both the national and global levels, a conversion from defense to civilian production is a strategy that is being erratically pursued. Perhaps because large European companies such as DASA, Aérospatiale, and BAe are already fairly diversified between military and civilian products, European governments do not appear to be offering much help to defense firms to convert to a higher proportion of civilian work. The EU has a defense conversion policy, blanketed within the regional development policy area, but little money has been allocated. The EU budgeted only ECU500 million[5] for the KONVER program during the 1994—99 time period (*Bulletin of the European Union*, 1994). But beyond this limited program, firms are on their own if they choose to pursue a conversion or diversification strategy.

A more promising strategy for European defense firms would be to restructure along sectoral lines. A merger of BAe and DASA, for example, would create a "Eurochampion" in military (and civilian) aerospace. A similar outcome in defense electronics would occur if GEC were to fully link up with Matra or Thomson-CSF. Such horizontal integration in Europe would create global companies that could better challenge U.S. defense giants Lockheed Martin and Raytheon. Yet the political obstacles again come into play, making it difficult to achieve what may be most practical on purely economic grounds.

While it may appear that European defense industry rationalization and restructuring is floundering with no particular direction, corporate executives put much of the blame on governments. Many defense firms feel constrained because their governments have not made their industrial restructuring policies clear. A study by Ernst & Young (1994) concluded that the British government, as well as other European governments, need to take the cross-border rationalization of Europe's defense industry more seriously before economic events overtake the government's ability to influence this sector. But the blame goes both ways. At an aerospace industrial meeting in 1995, European ministers told industry executives that they should expect no government help unless firms took further steps to rationalize and become internationally competitive (Skapinker, 1995). The chairman of DASA countered that the industry would find it easier to put aside its different national interests in a more politically integrated Europe. Although Europe's defense firms may be able to reach some agreement on what should be done to address the industry's problems, it is questionable whether these companies are yet willing to accept some of the prescriptions, particularly those which will be hardest hit by rationalization. Recent controversies[6] over the allocation of Eurofighter production highlight the contradiction between the rationalization that firms say is necessary to strengthen Europe's armaments industry, and the actions individual companies are willing to take (Gray, 1996). However, the pressure from governments is increasing. In December 1997, Britain, France, and Germany told European aerospace and defense electronics companies to present a detailed consolidation plan within four months. "The message is to rationalize or die. It's as blunt at that," said Britain's Defense Secretary George Robertson (Goldsmith, 1997).

A notion that industrial activity leads government policy suggests that these are independent actions, and underestimates the intimate relationship between politics and economics in any discussion of defense industries. Consequently, corporate interests differ somewhat by country of origin.[7] The Germans appear the most interested in a more active EU role in defense industry matters. Increased aid for, and coordination of, defense R&D programs is highly desired, as are proposals to improve the competitiveness of European industry *vis-à-vis* America's aerospace and defense sector and harmonize European arms export regulations. French industrialists seem to find greater European cooperation in armaments production as a nice ideal, but are realistic enough to know that states are reluctant to cede this area of national security to a supranational institution like the EU or WEU. British firms are the strongest advocates of free market solutions, and BAe and GEC seem to see institutional involvement in defense industry matters as almost irrelevant to their corporate strategies. While trade associations like the European Defense Industries Group will continue to push for government coordination of funding decisions, in-service dates, technical harmonization, and critical technologies, the interests and strategic decisions of Europe's defense firms will always be shaped by their specific national context.

PUBLIC POLICY OPTIONS

Given the situation and constraints described above, Europe presently has three public policy options. First, European governments could decide to reduce national control over defense industrial policy, and transfer more authority to an institution. The EU, rather than the WEU, is best placed to handle such a responsibility. The EU currently has policies that: regulate the trade of dual-use goods; decide corporate mergers and acquisitions; monitor state aid to companies; subsidize cooperative R&D programs; force national governments to open public contracts to competition; and assist the conversion of regions that have been economically dependent on military bases and defense firms (Guay, 1998). While these policies have much less effect on Europe's defense industry than national policies do, it is significant that virtually all of them have been developed only since the late 1980s. The Single European Act spurred Europe to reduce internal trade barriers and create the "Eurochampions" in non-defense sectors that would be necessary to compete in the global economy. The policies had a spillover effect, and have pushed the EU, particularly the European Commission and Parliament, to begin to address the specific circumstances of Europe's defense industry.

Yet, for the EU to be a major defense industrial policymaker, much more needs to be done. The EU currently has no real armaments export policy, and the political will to strengthen the institution's CFSP, a necessary prerequisite for such a policy, remains lacking. While support for cooperative R&D programs in the civilian sector remains fairly strong, there is reluctance to create a policy that explicitly funds defense R&D, although a significant portion of civilian R&D has applications for military equipment (Walker and Reppy, 1987). Even the EU's competition powers have limits because national governments can prohibit the Commission from reviewing defense industry mergers and acquisitions if they so choose. Finally, the unwillingness of member states to fully commit themselves to a joint procurement agency only prolongs the region-wide rationalization that is necessary if Europe hopes to increase its competitiveness *vis-à-vis* U.S. defense firms. At present, EU members appear very reluctant to take the political steps (such as removal of Article 223, strengthening CFSP, and developing an EU defense role) needed to bring the institution to the next level as a defense industry policymaker.

A second outcome is the formation of "Eurochampions"—defense industry firms that restructure at the continental (as opposed to national) level. This would take the form, for example, of a European aerospace company in which BAe, DASA, and perhaps Saab-Scania and/or Dassault merge into a single entity. But, as stated above, there are major political obstacles to this approach, too. Governments in Britain, France, and Germany would be reluctant to see "foreign" companies acquire domestic defense firms, close plants, lay off workers, and take "home-grown" technology outside of their territorial borders.

A third outcome is the continued reliance on national champions as the primary source of military equipment. Much of the recent evidence suggests that governments in the three major countries have encouraged domestic companies to integrate horizontally first. This has been achieved in Germany through the consolidation of the country's aerospace industry around DASA and defense electronics in Siemens, and in Britain by BAe in aerospace and GEC in electronics. Proposals from Paris over the past two years suggest that this should soon happen in France, as Dassault and Aérospatiale try to merge, and defense electronics and telecommunications reorganize around Thomson-CSF and Alcatel-Alsthom. The next step in this scenario would be vertical integration, whereby most of a country's defense industrial base is regrouped around a single company, rather than a continuation of horizontal integration through cross-border deals. This has recently happened in Germany, as Siemens has sold off its defense electronics business to DASA and BAe. Executives at GEC and BAe have conducted "on again–off again" negotiations throughout the 1990s on a possible merger between Britain's two defense giants. If the French government succeeds in its defense industry consolidation efforts, it will be left with one large aerospace company and two major defense electronics firms (the Thomson-CSF/Alcatel-Alsthom entity and Matra) which may find themselves under government pressure to take a final national step. Smaller countries like Italy and Spain have already consolidated most of their defense sectors. While this option preserves the autarkic status quo, it limits the extent to which Europe can create defense firms that have the economies of scale and scope that are necessary to compete with the U.S. defense giants. European governments and defense firms will have to decide whether the economic necessities of further industrial restructuring should overcome the political realities that have hindered progress to date.

NOTES

1. While the EU currently has 15 member states, the figures cited here do not include Austria, Finland, France, Ireland, and Sweden since SIPRI does not report comparable figures for these five countries.

2. For example, the British government agreed to buy 25 Lockheed Hercules transport aircraft in 1994 and 14 Boeing Chinook transport helicopters in 1995.
3. Interview with French Defense Ministry officials, June 5, 1995 and German Defense Military officials, June 8, 1995.
4. Over the 1991–95 period, Sweden was the world's 17th largest supplier of major conventional weapons. However, Swedish arms exports were only about 10 percent of either German, British, or French sales.
5. Approximately $550 million at early 1998 exchange rates.
6. In 1994, Germany suggested that it wanted to reduce its order of Eurofighters from 250 to 140 aircraft. However, it wanted to retain at least 30 percent of the work to ensure political support for the program. Since most of the domestic work would be performed by DASA, the German aerospace company lobbied hard for this workshare. This was a much higher percentage than German orders justified, and was unacceptable to the Eurofighter partners, especially Britain (and BAe). A compromise was eventually worked out whereby Britain will reduce its order as well, but receive 38 percent of the work. Germany would get 30 percent.
7. Interviews conducted of European defense industry executives in Brussels, May-June 1994.

REFERENCES

Anthony, I. and Wulf, H. "The Economics of the West European Arms Industry." In Brzoska, M. and Lock, P., eds., *Restructuring of Arms Production in Western Europe*. New York: Oxford University Press, pp. 18–35, 1992.

Atkins, R. "Eurofigher Set for Lift Off as Germany Gives the Go-ahead." *Financial Times*, p. 2, November 27, 1997.

Bulletin of the European Union, "Community Initiatives," p. 72, December 1994.

Economist, "A Farewell to Arms Makers," pp. 69–75, November 22, 1997.

Economist, "Building Eurospace Corp," pp. 59–61, September 7, 1996a.

Economist, "France's New Global Bid," pp. 45–46, March 2, 1996b.

Economist, "France Opens up its Arms," p. 70, February 24, 1996c.

Ernst & Young. *The UK Defence Industry: Securing Its Future*, May 1994.

Freeman, R. "HCDC/TISC Inquiry into Defence Procurement and Industrial Policy." Opening Statement by Rt. Hon. Roger Freeman, MP, Minister of State for Defence Procurement, May 23, 1995.

Goldsmith, C. "Consolidation of European Aerospace and Defense Firms Urged by Three Nations." *Wall Street Journal*, p. A18, December 10, 1997.

Gray, B. "Dispute over Fighter Project Resolved." *Financial Times*, p. 2, January 19, 1996.

Gray, B. and Clark, B. "Forces Told to be Ready to Meet Unexpected Threats." *Financial Times*, p. 9, May 4, 1995.

Guay, T. *At Arm's Length: The European Union and Europe's Defense Industry.* London: Macmillan, 1998.

Moravcsik, A. "The European Armaments Industry at the Crossroads." *Survival*, Vol. 32 (1), pp. 65–85, 1990.

Owen, D. "Paris Deepens Defence Reform." *Financial Times*, p. 3, September 12, 1996.

Ridding, J. "France to Increase Defence Spending." *Financial Times*, p. 3, April 20, 1994.

Skapinker, M. "Aerospace Industry Urged to Rationalise." *Financial Times*, p. 2, September 30, 1995.

Skons, E. "Western Europe: Internationalization of the Arms Industry." In Wulf, H. ed., *Arms Industry Limited*. New York: Oxford University Press, pp. 160–190, 1993.

Stockholm International Peace Research Institute (SIPRI). *SIPRI Yearbook 1993: World Armaments and Disarmament*. New York: Oxford University Press, 1993.

Stockholm International Peace Research Institute (SIPRI). *SIPRI Yearbook 1995: World Armaments, Disarmament and International Security*. New York: Oxford University Press, p. 459, 1995.

Stockholm International Peace Research Institute (SIPRI). *SIPRI Yearbook 1997: World Armaments, Disarmament and International Security*. New York: Oxford University Press, 1997.

Tucker, E. "Bangemann Call for Aerospace Restructuring." *Financial Times*, p. 3, September 25, 1997.

Verchère, I. "BAe Boost for Saab-Scania Exports." *The European*, p. 15, February 17, 1995.

Walker, P. and Reppy, J. *The Relations Between Defence and Civil Technologies*. Dordrecht: Kluwer, 1987.

9

DOGFIGHT: EXPORTING SUPERSONIC COMBAT AIRCRAFT IN THE POST-COLD WAR ERA

ROBERT CALLUM

INTRODUCTION

> Cry "Havoc", and let slip the dogs of war.
> *Julius Caesar, Act III, Scene One.*

The decisive dogfights of the post-Cold War era are taking place among marketers, not pilots. Decreased demand and the spiraling costs of supersonic combat aircraft (fighters and bombers) have conspired to form a world where production runs are growing shorter. Yet, increased research and development (R&D) expenses mean that production runs must be longer to diminish per unit fixed costs. The need to create longer runs in the wake of falling demand seems somewhat paradoxical. Not surprisingly, the United States and Europe have dealt with that mandate in different ways.

U.S. industry has downsized and rationalized, becoming leaner and discarding an insular focus. The remaining key producers have entered the supersonic combat aircraft export market as serious players, attempting to recoup lost domestic profits via international expansion. Through aggressive marketing campaigns and partnerships between government and industry, the United States has emerged as the dominant supplier of combat aircraft (indeed, of arms in general) to the world. Visiting the historical trends of U.S. defense firms, we will explain why changes in the post-Cold War environment have altered business as usual.

The Europeans have tried to perpetuate continental overcapacity by artificially lengthening production runs, either through a dependency on the export market or through the mechanism of collaborative development. French defense industrial policy and the consortium Eurofighter project will illustrate European desire to avoid rationalization. Their maneuvers were successful throughout the Cold War when competition in the export market was fairly tame. As competition mounts, illustrated in part by the efforts of the Swedes and the Russians, the Europeans will be under growing pressure to rationalize and combine. Decreased demand and an oversupply of fighters may well lead to the fiscal failure of France's *Rafale* combat jet and the Eurofighter. Such an outcome could either catalyze a long-delayed rationalization of the European defense industry, or else lead to the demise of this industry as an independent entity.

THREATS AND TECHNOLOGY CONSPIRE TO REDUCE DEMAND

If the Cold War had an indisputable winner, it was the worldwide defense industry. In 1987, world spending on weapons reached a peak of $1.36 trillion (all figures given in constant 1995 dollars, U.S. Arms Control and Disarmament Agency, 1997: 1). The implosion of the Soviet Union, however, marked a dramatic decrease in arms expenditures. In 1995, defense spending had declined by 34 percent, to $864 billion (ibid). Export/import figures have fallen as well. Whereas in 1987, $82.4 billion in military hardware changed hands among countries, arms sales were only $31.9 billion in 1995 (ibid: 100). The downturn for the industry as a whole is mirrored in the numbers for supersonic combat aircraft. In the period 1984–1986, 1602 were bought and sold on the world market, but that number plummeted to 692 for the period 1993–95 (ibid: 168).

While the changed nature of the post-Cold War threat environment plays a key role in diminished demand, there is also a second factor: cost. The fighters of today are much more expensive than their ancestors. The French Mirage F1, which entered service in the 1970s, cost five times as much (in constant 1975 dollars) as the *Ouragan*, the first indigenously produced French fighter of World War II (Kolodziej, 1987: 141). The average price of fighters worldwide increased 10,000 percent in constant U.S. dollars from 1945 to 1985 (ibid). This rate of increase shows no sign of slowing down. The unit price of the U.S. F-16, which entered production in the 1970s, is about $23 million (Goldsmith, 1997), while the sticker for the new U.S. F-22, set to enter production around the turn of the century, will be approximately $100 million (GAO, 1997: 10). The exponential growth in cost is due to the increasing complexity of R&D. The French estimate that the fixed costs for their new *Rafale* fighter will be between 30–40 percent of total costs (Moravcsik, 1990: 67) and the British assume fixed costs of 50 percent for many of their new weapons systems (Taylor, 1994: 103).

For those R&D investments, today's fighters are far more advanced than their predecessors. However, even if the threat environment were to stay constant, the increased capabilities of each successive aircraft generation mean that less of them are required to counter the same threat. As the fixed costs of R&D increase, and the production runs of aircraft decrease, unit prices march ever upward. Norman Augustine took this trend out to its logical extreme when he remarked that "in the year 2054 the entire U.S. defense budget will purchase just one tactical aircraft" (Augustine, 1982: 55). Imagining the technological capabilities and immense cost of this aircraft . . . one shudders to think what might happen if it were lost during an air show.

THE U.S. RESPONSE

As the costs of combat aircraft soared and capabilities improved, it is likely that even without the compounding factor of the Cold War's resolution, some defense industry restructuring would have occurred in the United States. But as the defense budget was slashed in search of a "peace dividend," the U.S. defense industry realized that the halcyon days of the Reagan buildup were over. Prodded by then-Secretary of Defense Les Aspin's "last supper"[1] in 1993, the industry hastened to adjust. Layoffs by firms such as Northrop, Hughes, Lockheed, General Dynamics, Litton Industries, and TRW marked a spate of "downsizings" (Matthews, 1992: 99) and acquisitions, culminating in the mergers of Lockheed and Martin Marietta; Boeing and McDonnell Douglas; Raytheon and Hughes; and most recently, Lockheed Martin and Northrop Grumman. There is now some question as to whether the Lockheed Martin/Northrop Grumman merger will be approved (Wilke et al., 1998; Biddle et al., 1998). Nowhere was this industry rationalization more apparent than in the military aerospace sector. Whereas in 1987, the United States had seven major producers of military fighters or bombers (Lockheed, Martin Marietta, General Dynamics, Boeing, McDonell Douglas, Northrop and Grumman) pending the approval of the Lockheed Martin/Northrop Grumman merger it will have two: Lockheed Martin and Boeing.

Along with an industry rationalization came an industry realization: the surviving U.S. military aerospace concerns would be forced to export. During the height of the Cold War, seemingly unending domestic consumption meant that U.S. firms did comparatively little exporting. In the period 1984–86, the United States exported only 13 percent of the 1602 supersonic combat aircraft sold worldwide (U.S. Arms Control and Disarmament Agency, 1997: 168). Moreover it was politics, not economics, which drove the few sales that did occur. Since the mid-1960s, the United States has used arms transfers as tools of statecraft. The U.S. strategy of flexible response, which was first articulated by then-Secretary of Defense Robert McNamara in 1961 and became official NATO policy in 1967, put more emphasis on conventional forces in Europe. The United States, however, did not want to pay for the increase in domestic European capabilities necessary to make the strategy viable. Instead, it relied on arms sales to key allies, doubling U.S. arms exports in real terms from 1963 to 1968 (Krause, 1992: 101). The promulgation of the Nixon Doctrine in 1969 furthered the political imperatives of arms sales.[2] In essence, the United States promised the material, but not the soldiers, to allies who were unworthy of a direct security guarantee. This created additional markets for U.S. arms, but again, the United States was selling only to the chosen few who met political requirements.

Which is not to assert that economics never entered the equation. The so-called "Sale of the Century" in 1975 is a good example. By the mid-1970s, Denmark, Norway, Belgium and the Netherlands all needed to replace aging F-104 Starfighters. The two alternatives under discussion by the four countries were General Dynamics' F-16 and Dassault's Mirage F1 M53. The U.S. government lobbied on behalf of General Dynamics, extolling the virtues of the F-16's superior performance and cheaper price. General Dynamics also offered better "offset"[3] arrangements than Dassault. The French, relying on "European brotherhood" to clinch the sales, lost out to the United States in each country (Kolodziej, 1987: 335).

With the end of the Cold War, the marketing effort behind the "sale of the century" became the rule, not the exception. Starting in 1990, the U.S. Secretary of State has directed all overseas missions to support the marketing efforts of U.S. defense companies (GAO, 1995). In February 1995, this policy was expanded to include actively involving senior government officials in promoting sales of particular importance to the United States and supporting Department of Defense participation in international air and trade shows (ibid).

U.S. participation in the international fighter bazaar began long before 1995, however. Then-Secretary of Commerce Ron Brown spoke at the 1993 Paris Air Show. During the U.S. pavilion's opening ceremonies, he told industry representatives that "we will work with you to find buyers for your products in the world marketplace, and we will work with you to close the deal" (Hartung, 1993). Brown's words were not idle. U.S. firms are now aided implicitly and explicitly by their government in the quest to sell arms abroad. Implicitly, for those allies that the United States may be willing to defend directly, such as the Saudis, the purchase of U.S. arms is a tacit condition for such protection (Taylor, 1994: 104). Explicitly, changes in tax laws allow firms to write-off marketing expenses in foreign countries.[4] Further, for certain foreign sales the price of weapons no longer has to include a charge for R&D. In essence, this allows U.S. firms to sell weapons at below their true cost. In fiscal year 1994, the Defense Security Assistance Agency waived these R&D costs on sales to nine allied countries, saving those countries $273 million, and making U.S. products more competitive in the process (GAO, 1995).

U.S. marketing efforts seem to be paying off. The U.S. share in the supersonic combat aircraft export market has gone from 13 percent in 1984–1986 to 41 percent in 1993–1995 (U.S. Arms Control and Disarmament Agency, 1997: 168). In Central Europe, Poland and the Czech Republic are seriously considering the F-16 and the F/A-18 for their modernized air forces (*The Economist*, 1997a: 72–73). The development costs of these aircraft have been paid long ago, so for roughly $23 million (for an F-16), one can get a plane whose capabilities are not terribly below that of the new European fighters.[5]

The military aircraft export market may be shrinking, but U.S. firms hold a bigger share than ever before. This fact is cause for concern in London, Paris and Bonn, since the growth in U.S. share is at the expense of the Europeans. According to a recent survey, from 1991–95 U.S. defense firms increased their international sales by 60 percent, while the average European company suffered a 12 percent decline in those sales (Dowdy, 1997: 95). Europe finds itself holding a smaller piece of a dwindling pie. But U.S. entry into the global market has only served to highlight a long-standing problem: oversupply.

THE EUROPEAN REACTION

Even during the height of the Cold War, the European continent faced a glut of supply as each country sought military self-sufficiency. Since even the largest national markets in Europe are less than one-sixth the size of the United States, a policy of national production will place the efficiencies that come from economies of scale out of reach (Bangemann, 1996). While U.S. domestic demand is twice that of the European Union (EU), the EU countries have ten helicopter and warplane manufacturers to four for the United States (Grant, 1997: 11 and *The Economist*, 1996: 63). By some calculations, European industry has three times as much capacity as demand warrants (Steinberg, 1992: 41). John Weston, Group Managing Director of British Aerospace, is not alone in noting that U.S. rationalization means the U.S. defense budget will be supporting approximately one-third the number of contractors as Europe, on twice the amount of defense spending as the United Kingdom, France, Germany, Italy, Spain and Sweden combined (Weston, 1996). Such a rationalization, if unanswered, "is a recipe for making the European defense industry unable to compete" (ibid).

Yet, due to politics, economics and national pride, the Europeans have been unable to muster the will to downsize their defense industries to the extent necessary. With domestic markets too small to support indigenous production, the Europeans have been forced to rely on the worldwide export market and on collaborative projects to lengthen production runs and sop up excess supply. These strategies have historically served the Europeans well, but that was before the Cold War ended and the United States entered the export market in earnest.

Presently, there are three fighters in development on the continent: the French *Rafale*, the consortium Eurofighter (a collaborative project of the United Kingdom, Germany, Italy and Spain), and the Swedish *Gripen*. In addition, the Russians are entering the export market with great deals on late model MiGs. The reduction in world demand and the newfound pressure on the export market makes it likely that either the *Rafale* or Eurofighter (if not both) will fail to lengthen their production runs in any meaningful way. That eventuality would leave both the sponsoring countries and their respective aerospace industries with substantial debt. Such a crisis may be the only way to catalyze a long-delayed rationalization of the European market.

French Defense Industrial Policy

In an ironic twist, the existence of a French defense industry owes much to the United States. Between 1945 and 1955, France received twice as much U.S. military aid than any other NATO country (Krause, 1992: 128). They used that aid in part to re-create their indigenous arms industry. By 1958, that industry was a firmly established producer of military armaments, and had begun to export. Contemporaneously, the French made the decision to develop nuclear weapons. As they diverted money into the nuclear program, less funding was available for conventional procurement. France, however, was not developing nuclear weapons as a substitute for conventional capability. They wished to have both a nuclear deterrent, and high quality (albeit smaller) conventional forces. But the production of conventional arms in the small numbers necessary for domestic use would have resulted in exorbitant unit prices. Turning to the export market to lengthen production runs was the only way that France could afford a modern, indigenously built conventional military (Kolodziej, 1987: 83).

Since the early 1960s, the continued existence of an independent French defense sector has been predicated upon export success. For most French military aircraft, more than 60 percent of production has been sold abroad, and for several platforms, this number has been closer to 70 percent (ibid: 95). Not only has France sought to lengthen runs with export sales, but they have subsidized domestic procurement as well. The Mirage III fighter that sold abroad for 13 million French Francs could be had at home for only 4.66 million (Krause, 1992: 141). The exported Mirages may have cost more, but they were tailored to the buyer's needs. While the United States follows a policy of standardization, selling the same export model F-16 to all takers,[6] the French listen to their clients' needs and tweak their product accordingly (Simon, 1993: 77).

Yet, that very flexibility has created a troubling paradox for France. As the French have become more dependent on export sales, weapons have been designed with the needs of foreign markets, especially developing countries, in mind. Unfortunately, the strategic concerns of Third World nations are seldom congruent with those of fighting a conventional war in Central Europe (Kolodziej, 1987: 102).The exhortation of the Pompidou administration in the early 1970s, to produce "simple, exportable and less costly" (ibid) weapons, has at times fated the French Air Force to use sub-standard fighters.[7] After all, that is what the domestic industry is producing. Further, weapons do not sell well on the world market without the backing of the producer's armed forces. No one wants to buy a jet that was not good enough for the French Air Force.[8] So the French fly light fighters, because that is what sells in developing countries. Yet, one can rightly ask whether France has allowed the exigencies of the export market to adversely affect the quality of their domestic weaponry. If the whole point of export dependency was to produce high quality conventional arms for the domestic market, and that domestic industry is now designing weapons with foreign needs in mind, has the original goal of the program been lost?

The truth is that the national security goals have changed. It is now not the defense of French territory that is uppermost in the minds of policymakers, but rather the continued viability of the domestic defense industry. In the late 1980s, 50 percent of French aerospace workers derived their jobs from the export market (Bajusz and Louscher, 1988: 13). As one official from Thomson (a French defense electronics concern) said, "the day we stop exporting . . . that's it. We close up shop" (Krause, 1992: 141).The export market has become a lifeline for French industry, one that cannot be severed without placing continued domestic production in doubt.

Unfortunately, the decreased demand for fighters and the increased competition on the export market make France's aerospace sector increasingly vulnerable. During the 1984–86 period, France held a 6.9 percent share of the supersonic combat aircraft export market. In the period 1993–95, their market share declined to 1.4 percent (U.S. Arms Control and Disarmament Agency, 1997: 168). The success of 1992, when Taiwan purchased 60 Mirage 2000 jets for $6.5 billion after a fierce marketing effort by the French Defense Minister and other government officials, has not been repeated (Hartung, 1993).

Into this environment is born the *Rafale*, Dassault's replacement for the obsolete Mirage 2000. France was an original member of the Eurofighter consortium, but chose to leave that collaborative project in 1987 because the prototype designs did not suit their needs. The French, ever mindful of export sales, wanted a lighter, smaller plane that would increase marketing possibilities in the Third World (Matthews, 1992: 155). The French also needed a naval version of a fighter that could be launched off their new *Charles de Gaulle* aircraft carrier (Swardson, 1997: C01). The goal of the

Rafale program was to create a lighter plane at "about half of the estimated price" of the Eurofighter (Simon, 1993: 30).To that end, France has spent between $7.4 and $9 billion (40–49 billion French Francs) to develop the *Rafale* (Simon, 1993: 326; Lorell *et al.*, 1995: 56).

For that money, they have created a plane that equals the Eurofighter in weight, sells for approximately the same price as the Eurofighter, but is not nearly as effective. In combat simulations run by British Aerospace and Britain's Defence Research Agency, the *Rafale* trailed the Eurofighter badly, and even lost out to upgraded versions of the old U.S. F-15 (Lorell *et al.*, 1995: 27–30).[9] At a unit price of $49–55 million (250–300 million French Francs), Dassault must sell 800–950 planes, in addition to planned domestic purchases, to recoup their development investment. In the heady days of the Cold War, this may have been a reasonable quota, even with potential competition from other fighters in Europe. In this era, however, a recent dissertation, including country-by-country analysis of potential aircraft purchase patterns, estimates the upper limit on *Rafale* exports at 200 (Simon, 1993: 327). It is unclear how even the most determined marketing and politicking effort will close a 600-plane shortfall.

The Eurofighter

French industrial policy does not represent the only way to shelter overcapacity. A second model is that of the collaborative project, whereby several countries agree to produce a weapon jointly, sharing R&D costs along with procurement numbers via some "fair" strategy. The Tornado, Alpha Jet and Transall aircraft were all built collaboratively. The Eurofighter consortium, however, is the continent's most ambitious collaborative endeavor (Callum, 1998). In 1983, five countries, the United Kingdom, Germany, Italy, Spain and France (France was a founding member of the consortium but withdrew in 1987 to develop the *Rafale*) agreed to finance the development of a next-generation fighter, the Eurofighter. The consortium would be run on a *juste retour* model.

Juste retour (just return) consortia use an allocation mechanism that seems "fair." Each country assumes costs and receives benefits in proportion to their investment in R&D, along with the number of units they buy. What is fair politically unfortunately leads to gross inefficiencies economically. *Juste retour* consortia are often plagued by delays and cost overruns. The consortium must massage differing tactical requirements, translate technical documentation, agree on production facilities and workshares, ensure that production cycles harmonize with replacement cycles, and clearly define management responsibility, control, and coordination. The president of an Italian arms firm estimates that joint projects take 25 percent, and projects with three nations 50 percent, longer to complete because of these delays (Matthews, 1992: 152).

Those delays are just one *juste retour* consortia cost. There is a duplication of assets, as production lines are mirrored in several countries. Parts are shuttled between sites, increasing expenses by 15 to 20 percent (Valéry, 1994: M16). Along with production inefficiencies, there can be competency inefficiencies. A member country is entitled to a share of the work, regardless of whether its industry is suitably equipped. Some countries join *juste retour* consortia as a means of accessing technology not readily available in their own industrial base. Thus, consortia can lead to the increase of an already bloated industrial capacity.

Delays, redundancies and incompetence all affect the fighter's bottom line. In an age when the profitability of projects is dependent on export sales to lengthen production runs, the inherent costs of *juste retour* can muddle the entire logic of cooperation. The Eurofighter has only recently entered production, in part because of German funding delays (Callum, 1998: 30). When the fighter begins rolling off the line, the plane's unit price of approximately $58 million will be contingent on doubling the production run beyond the 620 planes that the four consortium countries have committed to buying. While the Eurofighter can claim to fly rings around the *Rafale*, the claim is moot while the Eurofighter is but a "paper airplane." The Eurofighter is in real danger of ceding orders to the *Rafale* simply because the French fighter has entered the production phase. More ominously, there are cheaper alternatives to both the *Rafale* and the Eurofighter that may appeal to many buyers.

This circumstance has not stopped the governments of the consortium countries, most notably the United Kingdom, from engaging in pre-production marketing. Early 1993 saw then-Prime Minister John Major on tour in the Persian Gulf, selling 48 Tornado combat aircraft to Saudi Arabia and 36 Challenger tanks to Oman. The two packages together were worth $8 billion, and, according to Major's calculation, 20,800 jobs in northwest England, Leeds and Newcastle (Hartung, 1993). The Saudis were impressed by Major's high-level salesmanship, but also by something more concrete: he promised the Kingdom access to the Eurofighter when the jet entered production (ibid). That promise enticed the Saudis back in 1993. Today, with alternatives galore, it is unclear if anything but political motivation would cause the Saudis to wait for the long-delayed Eurofighter.[10]

The Swedes and the Russians

Saab, the Swedish maker of the *Gripen*, appears poised to win a portion of the great Central European lottery. The Swedes created a cheap, modern fighter primarily for their own use, but NATO expansion has made the *Gripen* a hot commodity. The Swedes had the foresight to make their fighter interoperable with NATO standards.[11] The result is that the Czechs and the Hungarians, strapped for cash but needing to upgrade capability, are seriously considering *Gripens* for their air forces (*The Economist*, 1997a: 72–73). The Eurofighter and *Rafale* are not even in the running. The $25 million *Gripen* cannot outfly either of the pricier planes, but for NATOs-to-be on a budget, the *Gripen* offers the chance to buy European, for less.

For those with less lofty ambitions and not quite the largesse, there is always a slightly used MiG-29. The Russians are breaking with Soviet tradition and selling for economic, not political reasons. While the Soviets were much more active than the United States in the Cold War arms game, with a 59 percent share of the supersonic combat aircraft export market from 1984–86 (U.S. Arms Control and Disarmament Agency, 1997: 168), the impetus of their arms transfers was much the same. According to a Russian official, the 1980s saw the Soviets give away as much as $80 billion in military hardware, as "aid" (Taylor, 1994: 111). Moscow no longer gives "aid," but it does offer attractive deals. At $18 million, the MiG-29 is probably the best value in today's export market, especially for a "rogue" state that cannot pass the background check for an F-16.

The Russians, moreover, are quickly learning to grasp the intricacies of Western marketing in the newly competitive export world. At the 1993 Paris Air Show, the Moscow Aircraft Production Organization handed out free MiG-29 lapel pins, along with a slick information packet including a picture of a MiG-29 in a steep climb and a slogan, "The Guarantee for a Peaceful Sky" (Hartung, 1993). While their grasp of English may not be perfect, the Russians understood the success that the French had with the policy of customization. The Russians were ready to install a number of options to suit the MiG to the needs of the buyer, including "Western" navigation systems, updated radar and the capability to launch laser-guided munitions (ibid). While the MiGs have been plagued by poor reliability in the past, Mikoyan (the Russian producer of the MiG) is addressing that concern by creating a worldwide logistical network for spare parts (Simon, 1993: 61).

Recently, an unlikely buyer of MiG-29s has surfaced. The United States bought 21 MiG-29s from Moldova to prevent the planes, which have the capability to carry nuclear weapons, from being sold to Iran (Myers, 1997). While the deal served as a U.S. diplomatic and intelligence coup, one wonders at its long-term implications. The Russians, who were informed of the deal because of "diplomatic sensitivities," evidently did not try to stop the transfer (ibid). The reason may well be that the Russians like this U.S. precedent, since the value of the plane will certainly increase if the United States is recognized as a player with an abiding interest in acquiring excess supply.[12] Regardless of future U.S. purchases, however, MiGs will remain a low-cost option for many buyers in the fighter market, and will represent an alternative to those who might otherwise have been forced to turn to the *Rafale* or Eurofighter.

CONCLUSION: RATIONALIZATION OR RETREAT?

During the height of the Cold War, with the United States and the Soviets selling for political motives, the Europeans could shelter their overcapacity through collaborative efforts or by exporting in quantity abroad. But with U.S. industry primed and rationalized, and the Russians willing to trade almost anything for hard currency, the worldwide export market for supersonic combat aircraft has suddenly grown quite crowded. Excess supply and decreased demand produces intense competition, and such competition usually means that the inefficient producers are forced from the market.

The inefficiencies in the European market are intrinsic. An internal market, which is half as small as that of the United States, cannot support ten helicopter and aircraft manufacturers much longer. It is unclear whether anyone will pay inflated prices for mediocrity (the *Rafale*) or will be willing to wait and see if the Eurofighter ever leaves the ground. In the former case, it is simply impossible for a country the size of France to develop an affordable and competitive mid-tier fighter. In the latter case, the delays of consortium development will call into question the ultimate viability of the craft. Real collaboration, in the form of European cross-border mergers and acquisitions, must follow if the Europeans are to be competitive over the long term with the United States.

However, the impediments to that eventuality appear daunting. Job loss, the merging of corporate cultures and the loss of control over national armament production seem insurmountable barriers to the creation of an integrated "European defense industry." There is a real danger that if business as usual continues in Europe, the healthiest of the European defense players may end up as "subcontractors" to their American cousins. The F-22 is slated to be the last service-specific combat jet produced in the United States. The follow-on to that fighter, the Joint Strike Fighter (JSF), will be shared among the U.S. Air Force, Navy and Marines in different configurations. Compared to a unit price of $100 million for the F-22, it is estimated that the Air Force version of the more advanced JSF will sell for $28 million a copy (*The Economist*, 1997b: 70). This price will be made possible by adopting several production methods from civil aerospace, and an unusually long production run of 2900 (ibid).

The opportunity to develop and buy a technically advanced, inexpensive fighter has already attracted much attention in Europe. The U.K. has invested $200 million in R&D, and the Royal Navy has promised to buy 60 jump jet JSFs (ibid). Norway, Denmark and the Netherlands have invested smaller amounts and other countries are expected to follow (ibid). British companies such as British Aerospace (BAe), GEC-Marconi, Rolls-Royce and Messier-Dowty (an Anglo-French venture) are involved in both the Boeing and Lockheed Martin JSF development teams (ibid).

While BAe is working with the Lockheed Martin team, the thought that Britain will buy the JSF for the Royal Air Force, in addition to the Royal Navy, horrifies many BAe executives (*The Economist*, 1997c: 75). This eventuality may leave no indigenous European fighters in production after the *Rafale*, Eurofighter and *Gripen* close their lines. If the JSF becomes the only combat jet in development at some future time, then European aerospace firms will have no choice but to sign on as junior partners to Boeing and Lockheed Martin. Without the wherewithal of the U.S. titans, the Europeans are in real danger of becoming future subcontractors to U.S. firms (ibid). The French were therefore not necessarily engaging in hyperbole when they described British participation in the jump jet version of the JSF as a "betrayal of Europe" (ibid).

Yet, one can rightly ask what the alternative was for the British. Production of the Eurofighter may yet become a disaster, and the next, next-generation fighter promises to be even more expensive unless production runs the length of the JSFs can be created. Europe could, in theory, create runs of that magnitude, but only by engaging in true cross-border rationalization, where the main aerospace concerns of the United Kingdom, Germany and France combine to battle the size and might of the U.S. giants. As we have noted, however, the obstacles to that eventuality are quite formidable.[13]

Yet, the impending failures of either the Eurofighter or the *Rafale* may be the shock that the Europeans need. With world demand static and a glut of supply, there are too many fighters to go around. When the *Rafale* and/or the Eurofighter fail to sell, the continent will be faced with the better part of a $100 billion development bill (Grant, 1997: 4), and no way to recoup that investment. The resulting trauma and shock will not be easy, but in the end, a rapid and forced restructuring may be exactly the strong medicine the Europeans need to retain a place in the supersonic combat aircraft export market. To do nothing risks losing the export dogfight permanently, long before planes ever reach the air.

ACKNOLWEDGMENT

The author would like to thank Christine Meyers and Maura Jane Flood for their help in preparing this chapter.

NOTES

1. Secretary Aspin and Deputy Secretary William Perry invited a dozen defense industry executives to dinner in the Pentagon. Aspin told the assembled group that there was twice the number of people at dinner than the government wanted in five years' time, and warned that the Department of Defense was ready to see some firms exit the market. The implied threat, "combine or die," along with a policy of government subsidies covering some merger-related costs, have clearly helped to speed rationalization of the U.S. defense industry, see Augustine, 1997; Dowdy, 1997: 91.
2. The Nixon Doctrine stated that in the future, the United States would look "to the nation directly threatened to assume the primary responsibility of providing the manpower for its defense", see also Krause: 102.
3. "Offsets" are deals whereby some of the plane's purchase price is offset by agreements to produce components, manufacture, or assemble some or all of

the plane in the purchasing country. Import deals are politically "sweeter" if some domestic economic benefit can be created.

4. Presentation by former Secretary of the Navy Sean O'Keefe at the Louis A. Bantle Symposium on Defense Industry Diversification, October 28, 1997.

5. Lockheed Martin propaganda places the F-16 on par with the *Rafale* and Eurofighter. Analysis by RAND, which had the not-so-hidden agenda of furthering the case for F-22 and Joint Strike Fighter production, places the F-16 far below the European models. The truth is probably somewhere in the middle. Norway is presently comparing the F-16 to the Eurofighter for a forty-plane procurement. Whatever Norway's decision, we have good reason to believe that the two planes are fairly close on price/performance grounds since each of the four countries involved in the Eurofighter project contemplated, at various times, pulling out of the consortium and buying the F-16 instead, see Goldsmith, 1995; Lorell *et al.*, 1995: 27–30.

6. Note, however, that this does not mean the export version of the F-16 is the same as the domestic version. The United States typically exports its planes with under-powered engines and reduced electronic counter-measure capabilities, but the export version offered to Turkey would be the same as the one offered to Taiwan.

7. During the "sale of the century" in 1975, the Air Force was forced to fly the Mirage F1 and shelve plans for a more advanced fighter. The failure of the sale allowed the F1 to be retired and the Mirage 2000 to enter production. One French Air Force official called the sale's failure a "victory" for the French Air Force, see Kolodziej (1987: 103).

8. Note the failure of Northrop to sell the F-20 as a simple replacement for the F-5. Foreign buyers decided that any plane that was not good enough for the U.S. Air Force was not good enough for them, see Taylor (1994: 110).

9. Naturally, the French dispute the finding, but even the rosiest reading has the *Rafale* slightly under-performing the Eurofighter.

10. Political motivation, of course, is a key component in any arms transaction. Witness the 1994 decision by the Kuwaitis to divide future arms purchases among all five permanent members of the United Nations Security Council, see General Accounting Office (1995).

11. To that end, Saab converted the Heads-Up Display symbols and icons, replacing metric with imperial units and introducing other language changes with the help of an English dictionary, see Morrocco (1996: 21).

12. The author is indebted to Terrence Guay for sharing this theory.

13. The recent joint statement by British Prime Minister Tony Blair, German Chancellor Helmut Kohl, French President Jacques Chirac and French Prime Minister Lionel Jospin calling on their respective national defense industries to unify in the face of U.S. competition is an important first step. Still, there remains a large gap between political rhetoric and actual rationalization, see Nicoll *et al.*, 1997.

REFERENCES

Augustine, N. *Augustine's Laws and Major System Development Programs*. New York: American Institute of Aeronautics and Astronautics, p. 55, 1982.

Augustine, N. "Reshaping an Industry: Lockheed Martin's Survival Story." *Harvard Business Review*, May–June 1997.

Bajusz, W. D. and Louscher, D. *Arms Sales and the U.S. Economy: The Impact of Restricting Military Exports*. Boulder, CO: Westview Press, p. 13, 1988.

Bangemann, M. "The Future of Europe's Defense Industry." *The Future of Europe's Defense Identity*, Brussels, June 18, 1996, http://www.eurunion.org/news/speeches/960618mb.htm.

Biddle, F. M., Cole, J. and Ricks, T. E. "Lockheed, Northrop Grumman Plan Is Unlikely to Satisfy U.S. Demands." *Wall Street Journal*, March 11, 1998.

Callum, R. "The Eurofighter Consortium: A Harbinger of Rationalization for the European Defense Industry?" *National Security Studies Quarterly*, Vol. IV(1), pp. 21–40, Winter 1998.

Dowdy, J. J. "Winners and Losers in the Arms Industry Downturn." *Foreign Policy*, pp. 91, 95, Summer 1997.

Economist, "American Monsters, European Minnows." p. 63, January 13, 1996.

Economist, "Contortions Galore." pp. 72–73, November 8, 1997a.

Economist, "The Richest Plus in the Skies." p. 70, November 22, 1997b.

Economist, "A Farewell to Arms Makers." p. 75, November 22, 1997c.

General Accounting Office. "Military Exports: A Comparison of Government Support in the United States and Three Major Competitors" (GAO/NSIAD-95-86), May 1995.

General Accounting Office. "Tactical Aircraft: Restructuring of the Air Force F-22 Fighter Program" (GAO/NSIAD-97-156), p. 10, June 1997.

Goldsmith, C. "After Years, Germany Expected to Clear Way for Eurofighter." *Wall Street Journal*, July 11, 1997.

Grant, C. "Survey: Global Defence Industry." *Economist*, pp. 4, 11, June 14, 1997.

Hartung, W. D. "Welcome to the U.S. Arms Superstore." *The Bulletin of the Atomic Scientists*, September 1993, http://www.bullatomsci,org/issues/1993/s93/s93hartung.html.

Kolodziej, E. A. *Making and Marketing Arms: The French Experience and Its Implications for the International System*. Princeton, NJ: Princeton University Press, pp. 83, 95, 102, 103, 141, 335, 1987.

Krause, K. *Arms and the State: Patterns of Military Production and Trade*. Cambridge, UK: Cambridge University Press, pp. 101, 102, 128, 141, 1992.

Lorell, M., Raymer, D. R., Kennedy, M. and Levaux, H. *The Gray Threat: Assessing the Next Generation European Fighters*. Santa Monica, CA: RAND, pp. 27–30, 56, 1995.

Matthews, R. *European Armaments Collaboration: Policy, Problems and Prospects*. Chur, Switzerland: Harwood Academic Publishers, pp. 99, 152, 155, 1992.

Moravcsik, A. "The European Armaments Industry at the Crossroads." *Survival*, Vol. 32, p. 67, January–February 1990.

Morrocco, J. D. "UK Prods Partners On Eurofighter's Fate." *Aviation Week & Space Technology*, p. 21, September 9, 1996.

Myers, S. L. "U.S. Buying MiGs So Rogue Nations Will Not Get Them." *New York Times*, November 5, 1997.

Nicoll, A., Skapinker, M., Graham, R. and Atkins, R. "Defence: EU Leaders Call for Industry Shake-up." *Financial Times*, December 10, 1997.

Simon, Y. "Prospects for the French Fighter Industry in a Post-Cold War

Environment: Is the Future More than a Mirage?" Ph.D. diss. RGSD-106, RAND Graduate School, pp. 30, 61, 77, 326, 327, 1993.

Steinberg, J. *The Transformation of the European Defense Industry: Emerging Trends and Prospects for Future US-European Competition and Collaboration*, R-4141-ACQ, Santa Monica, CA: RAND, p. 41, 1992.

Swardson, A. "French Breakaway Leaves European Arms Industry Adrift." *Washington Post*, p. C01, August 19, 1997.

Taylor, T. "Conventional Arms: The Drives to Export." In Taylor, T. and Imai, R. eds., *The Defence Trade: Demand, Supply and Control*. London: Royal Institute of International Affairs, pp. 103, 104, 110, 111, 1994.

U.S. Arms Control and Disarmament Agency. *World Military Expenditures and Arms Transfers 1996*. Washington, DC: U.S. Government Printing Office, 1997.

Valery, N. "Survey: Military Aerospace." *Economist*, p. M16, September 3, 1994.

Weston, J. "The European Defence Industry in the Global Market." *XIIIth NATO Workshop on Political-Military Decision-Making*. Warsaw: Poland, June 19–23, 1996, http://www.csdr.org/Weston.html.

Wilke, J. R., Ricks, T. E. and Biddle, F. M. "Lockheed Martin Faces Hurdles In Deal for Northrop Grumman." *Wall Street Journal*, March 10, 1998.

PART THREE:
CORPORATE PERSPECTIVES

10

THE POST-COLD WAR PERSISTENCE OF DEFENSE SPECIALIZED FIRMS

ANN MARKUSEN

INTRODUCTION

Defense industry analysts, reflecting on the future of the American defense industry at the end of the Cold War, expected that defense firms would survive but become smaller and less defense dependent. It was widely believed that they would invest the considerable cash accumulating in the wake of the 1980s buildup, along with portions of their skilled workforce and proprietary technology, in product development for non-defense markets and that they would seek to integrate both R&D and production across a historic military and civilian divide (Miller, 1991; Morrocco, 1991; Schine 1991; Velocci, 1991). Government pronouncements favoring procurement reform and dual-use technology development encouraged this prognosis.

Instead, at the upper end of the size distribution, firms have become larger and remain relatively defense-specialized, almost entirely as a result of mergers among the largest, most defense dependent firms, divestiture of military divisions from other more diversified firms, and the absorption of divested defense units by other defense contractors (Table 10.1, Fig. 10.1). Much of the cash reserves of the largest contractors was expended in financing these mergers, and in some cases, considerable additional debt has been taken on.

Although there remain civilian projects to which these corporations are deeply committed, especially in aircraft and communications satellites, many conversion efforts were dismantled or shunted aside as a result of the predominant preoccupation with mating and consummating these

Table 10.1. Major defense contractors by nation, sales 1993, 1996

Company	Country	Defense revenues $bn
Lockheed Martin	United States	19.39
Boeing/McDonnell Douglas	United States	17.90
Raytheon/Hughes/Texas Instruments	United States	11.67
British Aerospace	Britain	6.47
Northrop Grumman	United States	5.70
Thomson	France	4.68
Aerospatiale/Dessault	France	4.15
GEC	Britain	4.12
United Technologies	United States	3.65
Lagardere Groupe	France	3.29
Daimler/Benz Aerospace	Germany	3.25
Direction des Constructions Navales	France	3.07
General Dynamics	United States	2.90
Finmeccanica	Italy	2.59
Litton Industries	United States	2.40
Mitsubishi Heavy Industries	Japan	2.22
General Electric	United States	2.15
Tenneco	United States	1.80
TRW	United Staes	1.71
ITT Industries	United States	1.56

Source: U.S. Department of Defense, *100 Companies Receiving the Largest Dollar Volume of Prime Contract Awards from the Fiscal Year 1993 and 1996*

mergers. In some cases, existing dual-use capacity was abandoned in favor of new, segregated facilities for military and civilian markets.[1] To the extent that they moved from defense markets, the largest firms concentrated on other government markets where their political know-how would hold them in good stead. These outcomes are documented elsewhere (Oden, forthcoming, 1998; Markusen, 1997c, 1997e).

The purpose of this chapter is to explore in greater depth why the mergers came to dominate the restructuring process. Advocates of dual-use and civil/military integration should understand the broader set of forces and active agents participating in the post-Cold War restructuring process if they hope to coax the industry toward lessened defense dependency and greater reliance on civilian markets. Those who would dismiss these developments as involving only the largest contractors should heed the merger dynamic for the simple reason that it threatens next to engulf smaller firms in the same defense-dedicated consolidation process (Office of the Secretary of Defense, 1997).

In what follows, I focus first on the choices facing the largest defense contractors, characterizing their motivations and enumerating their assets.

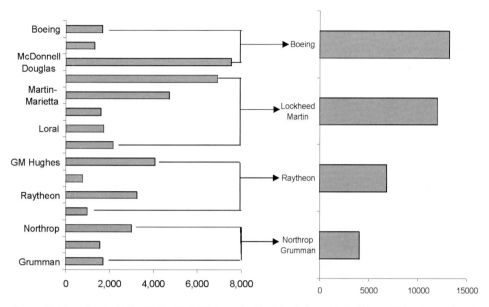

Source: All data are from the DoD publication (P01) 100 Companies Receiving the Largest Dollar Volume of Prime Contract Awards from the Fiscal Year of 1993 and 1996.

Fig. 10.1. U.S. Defense mergers in the 1990s

I then turn to a group whose leadership and activities have been poorly understood in this process—the business consultants and investment by "pure play" major players. I show that their initiatives have been rationalized as congruent with more general financial preferences for "focused" corporations, but are also driven by their expectations of returns from cost-cutting, real estate sell-offs, and enhanced market power and political clout. In the final part of this chapter, I review briefly the role of the Pentagon, itself a large and internally conflicted organization with top management turnover, in preparing itself for and responding to the defense mergers and, more generally, in managing defense industrial base restructuring.

My account relies upon deductive propositions about the behavior of these three sets of agents, an emphasis on the small numbers of major opinion makers and players and thus on gaming aspects of this process, and a marshaling of evidence in support of my interpretation, which stresses the dominant role of the financial/consulting sector placing pressure on a conflicted and diverse set of industry managers, abetted by an accommodating role, at least for a few key years, of top Pentagon managers.

DEFENSE CONTRACTOR CIRCUMSTANCES AND CHOICES, 1989–97

The challenges facing defense contractors, large and small, in the post-Cold War era are well known and well documented in other chapters in this volume. The industry had enjoyed an unprecedented real demand increase of 50 percent in budgeted procurement in the period 1978–87. With lags in outlays of about three years, this demand had carried them well into the early 1990s and generated large profits and cash reserves. Many had become more defense dependent over this period, as flush Pentagon demand compensated for stiffer competition in civilian markets (Markusen and Yudken, 1992: Ch. 3, 4).

The end of the Cold War, conjoined with the growing preoccupation with a budget deficit, created a crisis for firms. It was clear that there would be a long-term real decline in defense spending of rather major proportions. By 1997, procurement budgets had fallen by 70 percent from the Reagan era peak, a rather abrupt and steep decline: outlays, as Ken Flamm shows in Chapter 5, had fallen by at least 40 percent. There were, however, reasons to believe that defense firms would weather the reversal with grace—they possessed a formidable stock of proprietary leading-edge technologies, employed a highly skilled workforce, and enjoyed considerable cash reserves and ongoing income streams from past defense orders.

We can assume that the goal of defense firm managers is broadly to maximize profits, but that they may be somewhat more strategic and longer term in their outlook than their civilian sector counterparts. Large weapons systems generally span two decades or more from design to production, and long-term contracts from the buyer enable relatively long time horizons. Most defense firm managers, at least until recently, are engineers who have risen up through the ranks. Their priorities for corporate growth and investment generally stress the reinvestment of returns in R&D intensive, high tech market niches where innovative rents account for the lion's share of returns. What assets do defense firm managers possess or control, which of these are unique, and what are their options in a period of crisis of this sort?

Defense firms possess four distinct types of capital: fixed, political, cash and real estate. They own considerable fixed capital in the form of buildings and equipment. Some of what they do own is defense dedicated and may be devalued in the process of the builddown, although because of the curious history of government-owned plants (GOCOs), there is less of this than in other industries. Since defense is a research and management intensive industry, much of their research and management space is relatively marketable. Furthermore, defense firms, especially those in aerospace and naval lines of work, own considerable parcels of very attractive urban land, often adjacent to major airports or harbors. Defense firms also have political capital, which they have assiduously built up over the decades by working with Congress, government agencies and the military, often amplified by hiring former military officers in marketing and management positions. Finally, these firms have considerable money capital which has accrued from past and ongoing success in sales to the government. Of these assets, then, the cash, real estate and a good portion of buildings and equipment are fungible, while political capital and specialized equipment are not.

Defense firms also encompass a large labor force which consists of many long term employees with considerable generic as well as firm-specific human capital. At least four types of employees should be distinguished: managers, with their relatively unique orientation toward defense missions; marketing personnel, with their unique knowledge of Pentagon needs, procurement practices and regulatory procedures; engineers, with their considerable high tech capabilities and specializations; and blue collar workers, many of whom also possess considerable firm-specific and system-specific skills. On the human capital front—and in general, the defense industry is more human capital intensive than most—defense firms enjoy a potential advantage in both the range of skills they possess and, especially for the large system-integration firms, in the internal labor markets they have developed which continually train and integrate diverse skilled employees into teams working on complex and state-of-the-art projects (Yudken and Markusen, 1993).

In addition to these forms of capital and labor, defense firms also have considerable proprietary rights to technologies which are not necessarily embedded in either equipment or people. Broadly referred to as "technology" or know-how, these assets are potentially transformable into other products for non-defense markets.

We can therefore think of a defense company as possessing four forms of capital, employing four forms of labor, and owning rights to proprietary technology, all of which it must consider in fashioning a post-Cold War strategy. The company's behavior will, of course, be further shaped by general business conditions and, in particular, by Pentagon actions that seek to modify its behavior as a member of the defense industrial base or help it bridge the transition from defense to non-defense activities.

What were the dominant strategies among which defense firms could choose? We can summarize these in stylized fashion as consisting of four options, understanding that in practice companies might choose to pursue more than one: (1) redeploy assets, technologies and personnel, facilitated by investment of internally generated cash, into new non-defense product lines (diversification); (2) dominate existing defense segments where possible and liquidate the rest (retrenchment), releasing cash reserves into the external financial markets through stock buyback; (3) sell-off defense (or non-defense) operations as a stand-alone company and become thoroughly civilian or military in orientation, an option open to previously diversified firms; and (4) remain defense-oriented, but eliminate competition and broaden product offerings by spinning off non-defense segments and purchasing competitors and other firms with complementary expertise (consolidation). In this latter case, cash would be indirectly returned to the financial markets for reallocation by reimbursing stockholders of the acquired firms for shares sold. As Oden shows in Chapter 11, the first of these was pursued by TRW, Rockwell, Hughes and Raytheon; the second by General Dynamics and McDonnell Douglas; the third by Honeywell (creating and divesting its defense operations as Alliant Techsystems and by Litton (selling off its non-defense units); and the last by Lockheed Martin, Northrop Grumman and Loral.

Each of these reflects a different judgment of the optimal use of existing assets and personnel. A management team opting for the diversification strategy would be banking on higher returns from internal synergies across existing capital, technology and personnel, perhaps enhanced by bringing in outside expertise in management, marketing and engineering where desirable and using retained earnings to finance this strategy. Managers opting for retrenchment would be returning earnings to stockholders and financial markets for external reallocation, while continuing to rely on monopoly rents to generate profits. In this strategy, political capital and defense-dedicated technology and human capital weigh in heaviest. Firms choosing the third route are betting that reduced complexity will increase profits, permitting savings by retiring substantial portions of fixed capital and cutting the workforce. They may also be counting on enhanced political capital to affect not only the allocation but the level of defense spending in their favor. To comprehend the remarkable diversity in strategy in the early 1990s and the predominance of the fourth option by the mid 1990s, however, one must move outside of defense firms per se and analyze the motivations and behavior of two sets of agents in the financial/business services sector—investment bankers and business consultants.

THE ROLE OF WALL STREET IN RESHAPING THE DEFENSE INDUSTRIAL BASE

In the early 1990s, a group of the nation's most prestigious business consultants and investment bankers, hereafter referred to loosely as "Wall Street," trained their sights on the defense industry. During the 1980s, defense contractors had been spared such scrutiny—their profits were high, and their managers were preoccupied with bidding for, signing and following through contracts. They were thus untouched by the "stockholder revolt" which demanded higher short term returns and had restructured firms through hostile and other takeovers, buying top management acquiescence with stock options, golden parachutes and outright firings. Higher returns were realized chiefly through selling off undervalued units and real estate, subcontracting out more routine portions of production to lower labor costs and avoid unions, and cutting back on longer term, more speculative corporate research (Dymski et al., 1993). By the early 1990s, the defense industry became an attractive target for this type of restructuring, both because of its formidable cash reserves and because of the efficiencies that could be expected from applying similar disciplines to an industry which was widely believed to have hoarded labor and been lax in cost consciousness during the preceding boom.

In reality, the Wall Street offensive was motivated by two relatively short term sets of expectations. First, the architects of the defense mergers were motivated by returns to their services—fees for consulting and successful deals. As I will show, the financial/business service firms involved did an excellent job of conceptualizing and marketing the rationale for the mergers, denigrating conversion and pressing the case for "pure play" defense firms. Second, the mergers were motivated by the promise of higher returns due to cost savings from the sorts of liquidation and relocation mentioned above, including the elimination of most conversion activities, a phenomenon well documented by Oden (forthcoming, 1997a, 1997b). The Wall Street players were cognizant as well of possible future returns to scale economies, enhanced monopoly power and greater political clout. But for the most part, these would accrue in the future, long after the restructurers had moved on to other pastures.

The insights of game theory, with its emphasis on rivalry, imperfect knowledge and small numbers of players, are quite useful in understanding the four year merger spree, although it remains difficult to elegantly model the process which unfolded. In reality, there were a small number of major players in this "game"—the CEOs of the largest defense corporations (many of whom have retired in this period), top managers of the Pentagon (whose ranks turned over three times in this period), and a number of investment bankers, less visible in the public debate. The CEOs in question were heavily divided at the outset of the period, with some strongly embracing conversion and others choosing "pure play." So was the Pentagon. Following the Bush Pentagon, which was hostile to mergers and actively supported vigorous FTC scrutiny, the new Clinton leadership reversed course and welcomed mergers, assuming a hands-off stance. This diversity in strategy and response on the part of the industry and the Pentagon amplified the room for Wall Street initiative.[2]

Understanding the distinctive motivations and behavior of Wall Street actors in this process is important and rarely addressed in either journalistic or academic treatments of defense industry restructuring. Consultants sell advice, and investment bankers sell deal-making and financial packaging. Both have a stake in external restructuring and turnover, per se, and are well-served by new visions, or "fads," in business organization. In the 1970s, Wall Street made money on conglomerate deals, and in the 1980s, they made money on divestitures and slimming down of these same conglomerates. In the 1990s, this group fashioned and then worked assidu-ously to promulgate a view that the defense industry had too many players and that the remedy was the elimination of capacity via marriages among defense specialized firms, financed by cash reserves, new borrowing and sales of non-defense units. In what follows, I document the emergence of this view and evaluate its theoretical and empirical underpinnings, with reliance on both the finance literature and the results of PRIE's four years of research on conversion.

The Evolution of a Strategy

At the outset of the 1990s, the business press was cautiously optimistic about defense firm diversification. In a 1991 editorial, *Aviation Week & Space Technology*, the most prominent trade magazine, heralded the "substantial amount of commercialization activity going on in the defense industry. . . . It is surprising to learn that half the defense firms that decided to offer a commercial product in the past five years succeeded" (*AW&ST*, December 16/23, 1991). The editorial lauded crossovers from the defense side at Raytheon which created its robust appliance and energy businesses. It concluded by suggesting that "the commercial sector deserves a close look, especially when military technology can be applied to products with benefits for all." Rockwell, whose B-1 contract was the first of the 1980s systems to come to a close, was praised for allocating its B-1 generated cash to acquisitions of commercial enterprises with complementary capabilities. Across the spectrum, defense analysts expected that defense firms would suffer some shrinkage, but that use of accrued cash to finance internal diversification and purchase civilian firms and commercial expertise would enable them to survive, lower their defense dependency ratios and integrate across military and commercial markets.[3]

However, members of large consulting firms soon began refining a line of argumentation, writing articles in the business press and talking to journalists about the futility of defense conversion and uncertain returns to diversification.[4] They stressed the permanent downsizing of the industry, failing to note that procurement outlays in real terms remain above the post-Viet Nam trough (Flamm, 1998a). They pooh-poohed conversion, citing one or two anecdotes of failures in the past. They stressed the advantages of narrowing firm focus to profitable defense activities and counseled spinning off units not closely related to these. Less publicly visible but already active in negotiating takeover deals were Wall Street investment banking houses such as Bear Stearns, CS First Boston, Saloman Brothers and Merrill Lynch.

Beginning in 1993, both analysts and the press began to use language quite reminiscent of game theory to characterize the merger dynamic. Two Booz/Allen vice presidents warned against "fence sitters" in a 1993 trade press account on defense mergers.[5] "The notion that management can wait out the coming consolidation is one of the popular myths among many industry executives," they argued, forecasting a rapid surge in activity lasting only a few years, during which "the most attractive partners are taken out of action early in the process" (*Aviation Week & Space Technology*, 1993). Citing consolidation in other industries like printing and publishing, they went on, "We expect a similar pattern to evolve in aerospace, with preemptive moves by aggressive companies foreclosing opportunities for others." Such mergers "could establish world-class leaders and lock out other players from the first-tier. We believe the acquisition lull in recent months represents the eye of the hurricane and will precipitate a sharp increase in consolidation activity."

By early 1996, articles were appearing under headlines such as "Mergers Becoming a Business Imperative" (*Jane's Defence Weekly*, 1996).

As the mergers unfolded, analysts described them in dramatic and gaming terms. Picker (1995) quotes Salomon Brothers' Michael Carr: "Grumman made the classic mistake of wrong assumptions about motivation and level of interest. Northrop was never told they were selling the company. Nor was Northrop asked for its highest and best offer." Picker goes on to summarize the consequences, "Shut out after Grumman agreed to accept MM's offer, an irate Northrop resolved to insert itself into the process by making a higher bid—despite the fact that, should the bid succeed, it would add considerable debt to the company's conservative balance sheet." Or another round: "In the heated competition, Raytheon, after scooping up TI's defense operation for a "costly" $2.95 billion in cash, came out on top for Hughes . . ." (Haber, 1997). As late as 1997, analysts saw the Lockheed Northrop deal as a response to the Boeing/McDonnell marriage and described it as an "endgame" (Egan, 1997; Velocci, 1997b).

Three industry CEOs were influential players in evolution of the "pure play" defense mergers phenomenon, each in quite distinctive ways. Early on at General Dynamics, CEO William Anders charted a course of "sticking to the knitting," selling off some units and concentrating on highly profitable lines such as tanks and nuclear submarines in which his firm had little or no competition. Anders "dismisses the rewards for diversification as largely illusory" (*The Economist*, August 8, 1992). The strategy involved using the firm's considerable cash reserves, still accruing from defense contracts, to buy back its own stock rather than invest in new business generation. Some observers speculate that the concentration of ownership in the Crown family hands played a strong role in this decision. By 1995, General Dynamics had shrunk in size by 58 percent (Pages, 1995: 138).

If Anders' contribution was to repudiate diversification, downsize and bank on dominating market niches, Loral's Bernard Schwartz' pioneering contribution was the aggressive bidding for the assets of medium-sized defense contractors, absorbing them via leveraged buyouts, and treating them subsequently as stand-alone businesses under intense pressure to perform or risk liquidation. In a speech to Johns Hopkins Foreign Policy Institute in 1991, Schwartz "sharply questioned the current Pentagon position that the defense base can be kept if contractors diversify into other areas. . . . 'My experience is that most attempts by the defense industry to diversify into the commercial market have been dismal failures in the past. Defense firms have little experience in the business culture of the commercial world . . .'" (Schwartz, 1997). Schwartz' trademarks, concluded *Business Week*, "are speed and daring coupled with financial discipline. . . . In the past, this approach has meant dispensing with the acquired company's executives and instilling its managers with Loral's gospel of speed and stringent cost controls" (Peterson and Borrus, 1992). Elsewhere, we have described in detail the dismantling of one such acquisition, Los Angeles-based Librascope, subsequent to the Loral takeover (Oden *et al.*, 1996). The merger option, then, was first pursued by Schwartz, who unlike other defense CEOs was a Wall Street financier rather than an internally groomed manager.

The third key defense industry leader to shape the merger movement was Martin Marietta's Norman Augustine, perhaps the best connected of the industry's CEOs inside the beltway. Reputedly reluctant to consider mergers in the early 1990s, Augustine changed course by 1994 and pioneered the market extension merger, going beyond the LBO and individual profit center approach of Schwartz to seek marriages among large firms with complementary defense capabilities where longer term returns could be expected from greater market power and access to the Pentagon as well as savings on management and R&D. Augustine's quip, "Our record at defense conversion is unblemished by success," was heavily cited in press accounts which increasingly disparaged conversion, a characterization patently incorrect given Lockheed Martin's considerable success in commercial satellites. Augustine began publicly stating that his firm was opting for the "grow/consolidate" option, as opposed to what he saw as alternative defense company choices: "shrink/evaporate" or "exit/liquidate" (Picker, 1995: 71).

The repudiation of conversion in these accounts had an intellectual champion in someone far from Wall Street but close to General Dynamics and McDonnell Douglas headquarters—Murray Weidenbaum, a former Boeing Chief Economist and Chair of President Reagan's Council of Economic Advisors. In his book, *Small Wars, Big Defense*, Weidenbaum (1992) relied more on rhetoric than fact in stating that "defense scientists and engineers don't need 'make work,' companies should not be 'forced to convert,' and the government should 'avoid wasteful and fruitless' conversion attempts." In a disingenuous and highly selective review of past periods of American defense downsizing, Weidenbaum argued that conversion efforts have been a disaster. Relying heavily on a number of sobering anecdotes of failure at places like General Dynamics and Grumman, he concluded that military contractors are best at making weapons and if they can't continue to do so, they should simply downsize. Despite powerful, empirically grounded accounts of the success of dual-use and conversion approaches (Alic *et al.*, 1992; Gansler, 1995; Kelley and Watkins, 1995; Markusen and Yudken, 1992), Weidenbaum's view of conversion was readily adopted by the shapers of the emerging defense industrial base.[6]

Takeovers as Cost-cutting Discipline

Defense firms, awash in orders in the 1980s, had added large numbers of employees to their ranks and had been preoccupied with new R&D efforts and the expansion of existing production lines. In such an environment, cost consciousness and efficiency in production received short shrift. Managerial ranks were swollen with marketeers and overseers. By 1990, the industry was indeed ripe for the kind of cost discipline which had been imposed for a decade in civilian industry.

Several historically evolved features of the defense industrial sector made it particularly attractive to financial critics and investment bankers intent on wringing value out of firms. Many of the large defense contractors operated on attractive urban real estate parcels—several on the less polluted west side of Los Angeles and near the Los Angeles airport, for instance. Many were unionized and had evolved pay structures, especially for engineers and managers, considerably more generous than those in the civilian sector.[7] In short order, as the merger phenomenon quickened, top defense contractors disproportionately eliminated jobs—managerial as well as engineering and blue collar jobs—and moved out of higher cost urban settings, often to "right to work" western and southern states with powerful and military-friendly members of Congress.[8]

Large firms also outsourced work to smaller firms with lower wages, enabling them to cut employment much more rapidly than sales. From 1991 to 1995, average revenues in defense electronics dropped by 13 percent while employment fell by 30 percent (Dowdy, 1997: 95). A large portion of the apparent labor productivity gain was actually accounted for by subbing out work from large contractors to smaller ones. McDonnell Douglas increased its sales by 8 percent from 1989 to 1994, but cut its employment by 49 percent. Northrop Grumman sales fell by 4 percent in the same period, and its workforce was cut by 27 percent (Oden *et al.*, 1996). National survey results on small and medium-sized contractors show quite a different pattern, with a small employment increase on average over the period (Feldman, 1997). The prevalence of outsourcing suggests that vertical integration is not apt to be particularly robust.

The cost-cutting and value-liberating motivations for takeovers, however, do not require mergers between competitors or the creation of defense specialized firms. They are a form of stockholder, "external" exercise of discipline which could take place without any merger at all or in the context of a marriage between a defense firm and a commercial firm with similar technologies or complementary business acumen. Indeed, the latter would be more apt to bring the firm immediate access to experience with competitive market mechanics and cost conscious techniques of production. The joining of the expectations of higher returns from cost cutting with the push to create defense specialized or "pure play" firms is linked by some analysts to an academic and Wall Street consensus in favor of "focused" as opposed to conglomerate forms of business organization. Just what does this broader rationale consist of, how good is the evidence that "pure play" firms perform better than more diversified firms and do these arguments apply to the technology-intensive defense firms that are partners in the mergers?

The "Pure Play" Rationale

"Pure play" defense firms may be the product of three of the four strategies identified by Gerald Susman and Sean O'Keefe in Chapter 1. They may result from retrenchment (e.g. General Dynamics) or from mergers among major defense firms (Lockheed Martin). Units spun off in the third strategy generally remain "pure play," either as stand-alone units (Alliant Techsystems) or through acquisition by one of the larger consolidators (e.g. Texas Instruments' military division acquired by Raytheon). In defending the creation of large "pure play" defense firms, Wall Street investment bankers and analysts invoked the superior performance of "focused" rather than conglomerate firms. However, while retrenching firms may indeed be "focused," large firms amalgamating highly disparate activities under one corporate roof—aircraft, missiles, tanks, ordnance—are not. The result is instead a defense conglomerate. In contrast, a diversifying firm which enters civilian markets through internal financing or even acquisition is likely to remain more "focused" on its core competencies. The defense of the dominant mergers as creating "pure play" firms constitutes a misuse of the notion of focus, as we shall see.

Just how should the analyses of focus versus conglomeration be used to evaluate the defense merger phenomenon? This debate, and the evidence on both sides, has a counterpart in a related issue—whether firms should be permitted to retain earnings and reallocate them inside the firm or whether these earnings should be externalized to shareholders, where they can be reallocated via financial markets. While focus is not necessarily linked to externalization (or diversification necessarily linked to internalization), the two often are related in both theoretical and empirical studies.

The Case for Focus and Externalization

A number of scholars have made the theoretical case for focus rather than diversification.[9] Their argument is largely based on the view that owners of capital—shareholders—are the best and most unbiased allocators of capital, responding purely to market signals, whereas corporate managers may have other "satisficing" motivations which introduce distortions into the process—the so-called "agency" problem.[10] Managers may prefer diversification and internal investment to downsizing and returning value to shareholders because of the power and prestige of managing a larger firm and/or because compensation may be related to firm size. They may also prefer diversification because it reduces the risk of their own undiversified portfolio or because it makes them particularly indispensable to the firm (Denis et al., 1997: 136). Since the early 1980s, academics and financial analysts have developed the notion of "shareholder value"—the sum of dividends and expected stock appreciation—as the appropriate maxim for corporate decision-making (Rappaport, 1986).

It is undoubtedly the case that some managers behave in satisficing ways and that some bad decisions have been made by cash rich firms determined to diversify. The record of the large oil companies and a number of food companies such as General Mills illustrates the pitfalls of allowing managers to determine diversification routes. Both moved into unrelated energy fields in years when they were flush with cash and were unable to match the returns to their core businesses with similar results. In the oil industry, Royal Dutch and Exxon used their cash instead to buy back stock and reduce capitalization, with the result that their stocks substantially outperformed oil companies that had invested in coal, solar power and other energy sources. This experience undermined the appeal of internal capital allocation and conglomerate strategies.

A number of studies have been done comparing conglomerate to focused firm performance, and most conclude that the latter have generated greater stockholder returns (Comment and Jarrell, 1995; Denis et al., 1997; Liebeskind and Opler, 1994; Berger and Ofek, 1996). However, these studies rely on a controversial indicator of performance—stock values. Dow and Gorton (1997) show that the link between stock market prices and economic efficiency is tenuous and that relying on stock market signals may lead to sub-optimal investments. The value of a stock is not a measure of return to investment, since it is an evaluation of the entire firm rather than of a marginal investment. Managers may have private information that is not available to stockholders or outside analysts, and thus stock price efficiency is compatible with underinvestment. Nor do all scholars agree on the inferiority of conglomerates. In a careful historical study, Servaes (1996) shows that in the past, conglomerate mergers did not suffer a diversification discount, suggesting that a changing Wall Street consensus holds some sway over stock market responses (see also Matsusaka, 1993).

Institutional changes associated with "shareholder revolt" have accelerated the trend toward the externalization of capital over internal redeployment, particularly the transformation of managers into owners by tying executive compensation to stock performance (Dymski et al., 1993). This shift in managerial affinity has had a corrosive effect on longer term investment strategies, because current managers are tempted to side with stockholders in cashing out, taking their compensation today and reinvesting it elsewhere. William Anders did just this in downsizing General Dynamics and was rewarded with compensation in excess of $200 million for a single year, catapulting him onto the cover of *Business Week* as one of the nation's top ten executives. CEOs of merging companies have received windfall returns associated with large, one-time returns to merger announcements.[11] Additional incentives compensate top managers for their displacement from top positions. Because of Pentagon reimbursement of defense firm merger costs, discussed below, taxpayers picked up the tab for more than $100 million in extra compensation for

Norman Augustine as he moved from CEO of Martin Marietta to Vice President of Lockheed Martin.

The Case for Diversification and Internal Capital Allocation

The case for diversification, made by prominent economists and business analysts such as Alchian (1969), Chandler (1977), Llewellyn (1971) and Williamson (1986) and elaborated upon more recently by Gertner, Scharstein and Stein (1994) remains strong. In general, the case for diversification through acquisition of civilian capabilities and internal financing of growth rests on several key notions: (1) complementarity in research and technical expertise for the development of new products, a variant of the "core competency" principle, (2) information asymmetries which make it difficult for external parties to adequately judge the risk and potential for commercialization, and (3) speed of capital redeployment. Internal allocation of capital may be preferable because it will generally be accompanied by increased monitoring by corporate managers, providing a better flow of information between users and providers of capital. Furthermore, it may permit more efficient asset redeployment: if one unit performs badly, corporate headquarters can redeploy capital rapidly to other, more promising segments (Gertner *et al.* 1994: 1212–13). Internal capital allocation can also enable companies to surmount the short term pressures of Wall Street to foster viable projects with a longer term payoff (Dymski *et al.*, 1993).

Internal capital allocation, Jeremy Stein (1997) argues, makes a corporation a unique form of financial intermediary, one that is capable of creating value by shifting funds from one project to another, even when the firms' overall capital access is constrained. The corporate office does pick winners, but it also "sticks losers," Stein notes, which may be a more expeditious way of sinking bad projects than waiting for an arms length market response. "Stand alone projects can experience credit rationing and therefore underinvestment when attempting to raise financing . . . in the arm's length external capital market" (p. 114). Such credit rationing happens when managers' private information is not exploited. Stein concludes that internal capital markets work best with a small and focused set of projects and where external K markets are relatively undeveloped (p. 129). "Self-interested, empire building type behavior on the part of corporate managers is not always a completely bad thing . . . it can be harnessed" (p. 131).

In contesting the "pure play" convention, some Wall Street analysts argue that it is good management above all that matters. In cases like GE, the stock market is rewarding the past record of its managers, who have shown their ability to make solid longer term investment decisions. Remaining in the defense market are a number of other well-managed corporations with considerable civilian orientation: Allied Signal, TRW, Raytheon and of course, Boeing. Most successful diversifiers, however, as

Susman and O'Keefe suggest in Chapter 1, are firms whose civilian operations build on closely related technologies and expertise.

Diversifiers versus Defense Conglomerates

As Oden (Chapter 11) and Susman and Blodgett (Chapter 15) suggest, diversified firms will be more efficient than "pure play" defense conglomerates when they build on their core competencies in redeploying assets. Pro-merger advocates have conflated defense specialization with the notion of core competency, but this is appropriate only in a very narrow sense. Large "pure play" defense firms may realize economies of scope (Teece, 1980: 240; Radner, 1970: 457), saving on political capital and managerial and marketing efforts to the Pentagon. But they are not taking advantage of technological synergies, since assembling tanks is entirely different from shipbuilding, and building bombers is quite distinct from designing fighter planes. In contrast, firms like TRW that fended off merger pressures (and Rockwell and Hughes before they buckled to such pressures) have built bridges between their aerospace and automotive divisions to build on core competencies. TRW's efforts to apply aerospace guidance systems to motor vehicles or systems integration skills to traffic management systems reach deep into the firm's human capital and technological assets in moving into related civilian fields.

From a societal point of view, diversification is preferable to continued defense specialization, especially when the latter takes the form of mega-merged firms. This is true for several reasons: diversified firms have a better record on innovation and employment retention; mergers are risky and absorb considerable financial capital and managerial attention; and merged firms may rely on political and/or monopolistic position rather than efficiency in operations.

In the longer term, innovation is important to firm performance and economic growth. In a study of leading American, European and Asian firms in 15 industries for the period 1960–86, firm investment in commercially oriented R&D was found to be the principle indicator of subsequent sales growth performance (Franco, 1989). Acquisitions have been shown generally to have a direct negative effect on firms' R&D intensity (Hitt et al., 1990; Hitt et al., 1996: 1085), a finding borne out in Oden's analysis of large defense contractors elsewhere in this volume. Managers may underinvest in R&D in the wake of a merger because their retained earnings have been spent to acquire new units. Or, they may simply be preoccupied with the activities of making the deal and then knitting together the disparate units now under their control.

Studies suggest that transactions costs and deliberate misinformation may hamper the efficiency of merged operations. As Hitt et al. (1996: 1088) put it:

> Acquisition negotiations are often highly complex, with multiple parties involved, including investment bankers, lawyers, and top executives from both firms. To extract the highest price possible for its shareholders, target firm managers attempt to exchange only information that can positively affect the acquisition price. Information asymmetries between the two parties often result. . . . Thus acquiring firms may not easily nor accurately predict potential synergy between the target and acquiring firm assets. This lack of accuracy may lead to problems in integrating the acquired assets into the acquiring firm and to economies of scale and scope that are lower than predicted.

A number of industry watchers have expressed skepticism that the new defense giants will be able to integrate multiple businesses successfully (Velocci, 1997a: 88). In acquired firms, in particular, activities once vigorously pursued often enter a state of "suspended animation" (op. cit., 1089). In work on Grumman, Oden *et al.* (1994) found that conversion and diversification efforts were stopped dead in their tracks by the hostile Northrop takeover.

For defense firms in particular, at least those in higher tech sectors, the "pure play" argument with its anti-diversification subtext, is particularly unsuitable. It is in these sorts of activities that information asymmetries are apt to be largest and thus capital market failures most prominent.[12] We know this from looking at high tech start-ups, where there is good evidence of capital market failure. This view is also consistent with the economics literature on innovation, which stresses the under-funding of research and development with longer term payoffs (Mowery and Rosenberg, 1989). Furthermore, a positive and statistically significant relationship has been demonstrated between diversification and research intensiveness (Gort, 1962: 138).

Why Large Defense Conglomerates and not Smaller, Leaner Firms?

If large "pure play" defense mergers cannot be unambiguously defended on the basis of efficiency gains, is there some other reason for their emergence? The 1990s process of consolidation into a few, very large defense firms has complex drivers. One possibility is that investment bankers, consultants and even CEOs can, through public relations and media work, affect market fashions in ways that are unrelated to underlying efficiencies and are motivated principally by the expectation of large immediate fees and gains from stock appreciation. We have reviewed the active role of these groups above.

Even where specialized firms may be more efficient than diversified ones, there is no particular efficiency reason why such focused firms should grow to be large market dominating firms. In their study of 933 American firms, Denis *et al.* (1997: 137) found no relationship between degree of firm focus and concentration in the industry. Indeed, General Dynamics followed a path of spinning off defense divisions in which it was not dominant and concentrating only on those where it is, becoming a much smaller firm in the process. McDonnell Douglas CEO Harry Stonecipher acknowledged in 1995, before the Boeing merger, that size has its disadvantages and predicted that today's defense mega-firms may someday choose to split themselves into smaller, more manageable and entrepreneurial units (Mintz, 1995). Disparate corporate cultures (at Hughes and Raytheon, for instance) pose formidable problems to successful integration (Velocci, 1997a). The impetus to form very large, specialized defense firms by conglomerating disparate segments of defense capabilities appears to be driven more by expectations of market power and of special access to Pentagon planning than by the efficiencies promised by their architects. Similar arguments were made regarding oil industry intentions in entering related energy fields in the 1970s (Mitchell, 1978).

The defense industrial base consequences of the mergers are the subject of two companion papers (Markusen, 1997c, 1997e). There, I argue that where they fuse competitors, mergers may result in economies of scale and elimination of redundant capacity. Such savings can occur in management and overheads, in R&D and/or in production. However, most mergers to date have not involved firms with competing production lines but are market extension mergers, designed to amplify each firm's portfolio of offerings. These mergers may achieve economies of scope, but to the extent that they eliminate future competition in either R&D or production, they portend loss of competitive discipline, which could mean higher prices for weapons, poorer quality and a diminution in innovation.[13] As one investment analyst put it, "procurement nightmares" could begin emerging as the Pentagon discovers "it cannot go anywhere without running into Lockheed Martin" (Jon Kutler, Quarterdeck Investment Partners, cited in *Jane's Defence Weekly*, 1996). Furthermore, the increasing separation of military from civilian capacity may curtail spinoff and conversion of defense capabilities into civilian uses. High levels of defense dependency will perpetuate political pressures for high levels of defense expenditure and arms exports, resulting in further resource misallocation and perhaps less national security (Markusen, 1997a, 1997b).

THE PENTAGON ROLE

The role of the Pentagon in the merger blitz was not one of leadership but accommodation. At first, it appeared that such mergers would undergo considerable scrutiny. Indeed, under President Bush and Secretary Cheney, mergers were explicitly discouraged, signaled by an emphatic "no" from the FTC in the Alliant Technology/Olin case in 1992. Building on an earlier denial of a Grumman/LTV merger, the FTC found that the elimination of one of the two largest tank ammunition competitors in the Alliant/Olin case would cost the Army millions of dollars. In this industry, "even if a post merger monopolist abused its resulting monopoly power to demand an exorbitant price, entry of a new firm would be unlikely," the court found (Triggs and Heydenreich, 1994: 445). The authors of this study also found the "the merging parties are unlikely to convince a court that an injunction poses a serious threat to national security unless DoD takes an official position that such a risk is likely" (op. cit., p. 462).

Executive branch attitudes in favor of defense mergers shifted with the accession to office of President Clinton. Among Clinton's strong supporters was Bernard Schwartz, an attendee at the Economic Summit held just after the election and an outspoken advocate of mergers among defense contractors. Clinton's first Secretary of Defense, Les Aspin, set up a new Office of Economic Security in the Pentagon, among whose tasks were the evaluation of defense industrial base issues, including how many competitors future weapons procurement could be expected to support, whether arms exports could help take up the slack and whether certain functions should be privatized. The Office was placed under the leadership of a Wall Street investment banker, Josh Gotbaum, who brought with him a mergers and acquisition or "M&A" perspective. William Perry, later Secretary of Defense himself, oversaw procurement issues from the outset of the Aspin period. Perry had most recently been running his own Silicon Valley investment banking/venture capital firm and was apparently merger-friendly on joining the Pentagon.

The Clinton Pentagon, however, never produced a defense industrial base analysis to evaluate whether its future demand for individual weapons systems could reasonably sustain one, two, three or more competitors, despite the advice of experts like the Brookings Institution's Ken Flamm, at the time a staffer in the Office of Economic Security, and Rand economists William Kovacic and Dennis Smallwood (Flamm, 1998b; Kovacic and Smallwood, 1994; Kovacic, 1991). The Rand team laid out a robust procedure for preserving rivalry among contractors by monitoring proposed mergers and preventing them in cases where demand could sustain greater competition. Instead, Perry openly encouraged consolidation, in a famous "last supper" speech in 1993, basing his advocacy on the promise of cost savings to the Pentagon.[14]

Perry embarked on an aggressive program which included special antitrust rules to permit greater consolidation of the defense industrial base (DIB) and succeeded in overcoming the antitrust reservations of the Department of Justice and the Federal Trade Commission, a move welcomed by the Wall Street and business consultants. The Defense Science Board, comprised predominantly of defense contractors and consultants, recommended against formal Pentagon scrutiny of the mergers, which facilitated this backdoor approach—the Pentagon was presumably the most knowledgeable about the competitive impacts of the mergers, but this way it did not have to issue formal reports (Defense Science Board Task Force, 1994), By publicly advocating the need for consolidation of the defense industry, the Defense Department accomplished a number of goals. Perhaps most important is that the government was able to isolate itself from the politically charged task of picking winners and losers by letting the market make those decisions (Dowdy, 1997: 91–92).[15]

At about the same time, the Department of Defense played an active role in creating new financial incentives favoring "pure play" mergers by subsidizing and aggressively promoting liberalized arms exports and by permitting defense contractors for the first time to include the costs of consummating mergers as part of current contracts, on the basis that they should generate future savings. Both of these moves, which I analyze elsewhere at length (Markusen, 1997a, 1997c, and 1997e), provided billions in new subsidies which helped to tilt the balance for contractors in favor of remaining defense specialized and engaging in the merger spree. These initiatives overwhelmed the significance of the Pentagon's efforts to encourage dual-use and civil-military integration. Neither procurement reform nor the TRP project offered the magnitude of subsidies and opportunities available by concentrating on arms exports and mergers.

CONCLUSION

At the end of the Cold War, seven American companies could develop and produce military aircraft: Boeing, General Dynamics, Grumman, Lockheed, McDonnell Douglas, Northrop and Rockwell. Now there are only two: Boeing, which absorbed McDonnell Douglas and Rockwell, and Lockheed, which acquired the aircraft capabilities of General Dynamics, Grumman, and Northrop. Divestitures have broken up more diversified firms, severing the constituent firms' civilian from their military operations at Loral, Litton, Rockwell, Hughes and Texas Instruments. One critic, Representative Bernie Sanders of Vermont, estimated in late 1996 that 58 percent of Pentagon weapons procurement contracts for 1995 went to what are now two just companies: Boeing/Rockwell/McDonnell Douglas and Lockheed/Martin/Loral (Zitner, 1996). This share would be even higher if the Northrop Grumman merger with the latter were taken into account. Altogether, the defense mergers and acquisitions amount to approximately $100 billion (Schwartz, 1997). Further merger activity across international borders is anticipated (Bitzinger, forthcoming), as are pressures for increased vertical integration as large firms absorb smaller subcontractors (Office of the Secretary of Defense, 1997).

This outcome is discouraging for proponents of dual-use and civil-military integration. It would be especially troubling if a round of vertical integration were now to develop, engulfing yet more firms under the defense specialized umbrella. Many smaller firms have successfully redeployed internal capital into new projects while they continue to cut costs and fulfill defense contracts.[16] At this point, the mergers having been concluded, the Pentagon has little choice but to live with its giant suppliers. One team characterizes this problem as "America's private arsenal" and suggests that we think about renationalizing firms which are a virtual monopoly (Sapolsky and Gholz, forthcoming). Others suggest that we consider importing weapons from competitors abroad, perhaps working towards an international division of labor, at least among the major NATO allies (Gansler, 1995). Yet neither of these approaches would in any way restore an emphasis on dual-use and civil-military integration.

My suggestions are the following. First, the Pentagon should build capability for analyzing the defense industrial base and actively scrutinizing the impact of proposed mergers on that base (see also Alic, 1998; Flamm, forthcoming; Office of Secretary of Defense, 1997). Second, the merger reimbursement policy should be overturned, as a number of members of Congress on both sides of the aisle have urged. In its place, the Pentagon might consider just the opposite—a merger reimbursement policy for firms that mate with commercial firms offering them skills and capability for diluting their defense dependency and moving into civilian markets. Third, mergers between the newly formed giants and smaller defense contractors and subcontractors should be discouraged and disapproved if they do indeed threaten to eliminate competition (Pages, 1998, forthcoming). Fourth, the Pentagon could fashion a policy to induce companies to spin off certain of their capabilities where the market can support additional competitors and/or to encourage new startups and entry in various market segments. New entrants could be particularly sought among the ranks of civilian firms that have similar expertise in systems integration, component systems, and so on.

Dual-use advocates might also want to participate more pointedly in the debate about science and technology policy in this country. President Clinton's efforts to fashion new policies like the TRP and ATP were savaged by the 1994 Republican Congress (Stowsky, forthcoming; Pages, 1998, forthcoming). It is particularly important to engage in research on and debate about the role of short-term financial pressures in curtailing corporate R&D and in siphoning off retained earnings that managers might otherwise invest in conversion and diversification. Successful large corporations that are diversified and have exploited strong internal synergies across divisions—Boeing is an outstanding example—should be studied and showcased (Bertelli, 1997; Feldman, 1998; Oden, 1997, forthcoming). The evidence suggests that neither short-term nor longer-term real efficiency gains are unambiguously associated with these mergers and that the social costs of monopoly can be quite high (Cowling and Mueller, 1978). Large defense firms are a very big part of the picture—the fate of dual-use inside each should not be left unscrutinized. Finally, there appears to be no choice at this juncture but for the Pentagon to assume a greater management and oversight role with respect to industrial base management and procurement practices.

NOTES

1. For instance, Lockheed Martin decided to shutter its highly productive East Windsor, NJ facility communications satellite facility which was 50 percent civilian, 50 percent government (DOD, NASA, NOAA) to build a new "commercial only" facility in Silicon Valley while consolidating its military satellite production in a separate plant. See Rutgers Defense Conversion Project, 1996. See Jacques Gansler's discussion in his book *Defense Conversion* for the argument favoring within-plant integration.

2. I am not attempting to construct a conspiracy theory here, nor do I have evidence to support one. I am demonstrating that multiple paths were possible and that, as a result of a campaign by a group of Wall Street entrepreneurs, an outcome which is not supported by strong evidence of gains in either economic efficiency or national security resulted. The outcome has been highly profitable for the Wall Street players and the CEOs who participated most enthusiastically (Finnegan, 1997a; 1997b).

3. Consider, for instance, the following prognosis by Anthony Velocci, *AWST* industry correspondent, in 1991: "A growing number of suppliers are leaving the defense business for more stable and profitable commercial work. . . . It is unlikely there will be any failures among major U.S. defense companies. All top 10 contractors have investment-grade ratings. . . . Contractors will have expanded their shares of nonmilitary government business by a substantial amount. As a result, more of the Pentagon's major suppliers will be less vulnerable to the vagaries of defense contracting and do a better job of leveraging their technology strengths. Lockheed Corp., Raytheon Corp and TRW Corp. will be at the forefront of this diversification effort. . . . Wholesale mergers among major defense contractors are unlikely . . ." pp. 140–142). See also Miller, 1991 and Schine, 1991.

4. See for instance, Lundquist, 1992. The article was published, "Hang on to what you've got to have and throw the rest away." Subheadings in Lundquist's article include, "Commercial applications of defense products are very hard to come by," and "For any company in any industry, diversification outside of core businesses is prone to failure." Lundquist, a former Air Force officer, White House Fellow and staff assistant to former Senator Sam Nunn, was a principal at McKinsey & Company's New York Office at the time he wrote this article.

5. In another press account, one of the same consultants is quoted as citing "countless reasons why defense businesses should steer clear of conversion. . . . Only the smallest businesses and the most basic technologies can readily move between one and the other. Big defense companies can no more adapt to the commercial world than can the products they churn out" (*The Economist*, pp. 63–64, January 16, 1993.

6. Weidenbaum's account ignores the dramatic creation of entirely new high tech industries out of defense-oriented firms and technologies. His own former employer, Boeing, is an outstanding example of a company that transformed a military tanker into the highly profitable 707. The computer, semiconductor and communications industries consist chiefly of firms that first bred themselves on defense contracts and went on to create successful commercial products

that now dwarf their defense work. For a fuller review, Weidenbaum, 1992; Alic *et al.*, 1992; Gansler, 1995; Markusen, 1997f.

7. See Markusen and Yudken, 1992, Chapter 6.
8. See the evidence in Oden *et al.*, 1996 and Oden, 1997.
9. See the review in Denis *et al.*, 1997.
10. One might offer an alternative hypothesis: external investors prefer focus because it makes firms more transparent, decreasing information asymmetries but devaluing important information and technology transfer potential in the process.
11. Often, company stock soars dramatically in value at the announcement of the merger. In January of 1996, for instance, the announcement of the Lockheed purchase of Loral sent Loral's share price up 23 percent, while Lockheed's rose 4 percent (Keefe, 1996).
12. One researcher has recently argued that in general, information asymmetries have become less of an issue in corporate financing and so the disadvantages of diversification have started to outweigh the benefits (Bhide, 1990). This may be true for some types of firms, but is not likely to be the case with high tech defense firms.
13. See also the work of Pages, 1996, 1998 and forthcoming.
14. Perry was explicit, saying he hoped that "several aircraft firms would disappear through mergers, as well as three of the five satellite firms in business then, and one of three missile companies" (Mintz, 1995).
15. Dowdy's article is the only one which has appeared in a major foreign policy journal of defense mergers—Dowdy is a partner in the Los Angeles office of McKinsey & Company.
16. See for instance our account of Leach Corporation in Oden *et al.*, 1996.

REFERENCES

Alchian, A. "Corporate Management and Property Rights." In Mann, H., ed., *Economic Policy and the Regulation of Corporate Securities*. Washington, DC: American Enterprise Institute, pp. 337–360, 1969.

Alic, J. "The Perennial Problems of Defense Acquisition." In Susman, G. and O'Keefe, S., eds., *The Defense Industry in the Post-Cold War Era: Corporate Strategy and Public Policy*. Oxford: Elsevier Science, 1998.

Alic, J., Branscomb, L., Brooks, H., Carter, A. and Epstein, G. *Beyond Spinoff: Military and Commercial Technologies in a Changing World*. Cambridge: Harvard Business School Press, 1992.

Aviation Week & Space Technology, "Tread Carefully with Commercialization", pp. 16–23, December 1991.

Berger, P. and Ofek, E. "Bustup Takeovers of Value—Destroying Diversified Firms." *Journal of Finance*, Vol. 51, pp. 1175–1200, 1996.

Bertelli, D. "Restructuring Defense Industries from the Ground Up." Paper presented at Second Klein Symposium on the Management of Technology, September 15–17, 1997.

Bhide, A. "Reversing Corporate Diversification." *Journal of Applied Corporate Finance*, Vol. 3, pp. 70–81, 1990.

Bitzinger, R "Globalization in the Post-Cold War Defense Industry." In Markusen,

A. and Costigan, S., eds., *Arming the Future: A Defense Industry for the 21st Century*. New York: Council on Foreign Relations, forthcoming.

Chandler, A. *The Visible Hand*. Cambridge, MA: Belknap Press, 1977.

Comment, R. and Jarrell, G. "Corporate Focus and Stock Returns." *Journal of Financial Economics*, Vol. 37, pp. 7–87, 1995.

"Consolidation Myths Pose Risks to 'Fence Sitters.'" *Aviation Week & Space Technology*, p. 46, August 30, 1993.

Cowling, K. and Mueller, D. "The Social Costs of Monopoly Power." *Economic Journal*, Vol. 88, pp. 727–748, 1978.

Defense Science Board Task Force. *Antitrust Aspects of Defense Industry Consolidation*. Washington, DC: Office of the Undersecretary of Defense for Acquisition and Technology, 1994.

Denis, D. Denis, D. and Sarin, A. "Agency Problems, Equity Ownership and Corporate Diversification." *The Journal of Finance*, Vol. 52(2), pp. 135–160, 1997.

Dow, J. and Gorton, G. "Stock Market Efficiency and Economic Efficiency: Is There a Connection?" *The Journal of Finance*, Vol. 52(3), pp. 1087–1130, 1997.

Dowdy, J. "Winners and Losers in the Arms Industry Downturn." *Foreign Policy*, pp. 88–101, Summer 1997.

Dymski, G., Epstein, G. and Pollin, R. *Transforming the U.S. Financial System: Equity and Efficiency for the Twenty First Century*. New York: M. E. Sharpe, 1993.

Economist "The Defence Industry Jettisons its Excess Baggage." pp. 57–58, August 8, 1992.

Economist "Slimming the General." pp. 61–62, January 16, 1993.

Egan, M. "Lockheed, Northrop to Merge in $11.6 Billion Deal." *Reuters News Service*, July 3, 1997.

Electronic News "Loral Chairman Warns Firms May Exit DOD Market". *Electronic News*, Vol. 37, p. 25, December 2, 1991.

Feldman, J. "Conversion and Diversification after the Cold War: Results from the National Defense Economy Survey." Working Paper, Project on Regional and Industrial Economics, Rutgers University, 1997.

Feldman, J. "The Conversion of Defense Engineers' Skills: Explaining Success and Failure through Reorganization, Teaming and Managerial Integration at Boeing-Vertol." In Susman, G. and O'Keefe, S., eds., *The Defense Industry in the Post-Cold War Era: Corporate Strategy and Public Policy Perspectives*. Oxford: Elsevier Science, 1998.

Finnegan, P. "Lockheed Martin Sets Pace with Massive Pay Packages." *Defense News*, pp. 12–15, September 1–7, 1997a.

Finnegan, P. "Censure Fuels U.S. Pay Debate." *Defense News*, pp. 12–15, September 1–7, 1997b.

Flamm, K. "U.S. Defense Industry Consolidation in the 1990s." In Susman, G. and O'Keefe, S., eds., *The Defense Industry in the Post-Cold War Era: Corporate Strategy and Public Policy Perspectives*. Oxford: Elsevier Science, 1998.

Flamm, K. "Redesigning the Defense Industrial Base." In Markusen, A. and Costigan, S., eds., *Arming the Future: A Defense Industry for the 21st Century*. New York: Council on Foreign Relations, forthcoming.

Franko, L. "Global Corporate Competition: Who's Winning Who's Losing and the

R&D Factor as One Reason Why." *Strategic Management Journal*, Vol. 10, pp. 449–474, 1989.

Gansler, J. *Defense Conversion: Transforming the Arsenal of Democracy*. Cambridge: The MIT Press, 1995.

Gertner, R., Scharstein, D. and Stein, J. "Internal versus External Capital Markets." *Quarterly Journal of Economics*, Vol. 109, pp. 1211–1230, 1994.

Gort, M. *Diversification and Integration in American Industry*. Princeton: Princeton University Press, 1962.

Haber, C. "New Defense Giant Born." *Electronic News*, January 20, 1997.

Hitt, M., Hoskisson, R. and Ireland, R.D. "Mergers and Acquisitions and Managerial Commitment to Innovation in M-form Firms." *Strategic Management Journal*, Vol. 11, pp. 29–46, 1990.

Hitt, M., Hoskisson, R., Johnson, R. and Moesel, D. "The Market for Corporate Control and Firm Innovation." *Academy of Management Journal*, Vol. 39(5), pp. 1084–1119, 1996.

Jane's Defence Weekly "Mergers Becoming a Business Imperative," pp. 23, January 12, 1996.

Keefe, R. "Lockheed to Acquire Loral for $9.1 Billion." *St Petersburg Times*, p. IE, January 9, 1996.

Kelley, M. E. and Watkins, T. "The Myth of the Specialized Military Contractor." *Technology Review*, Vol. 98(3), pp. 52–58, April 1995.

Kovacic, W. "Merger Policy in a Declining Defense Industry." *Antitrust Bulletin*, Vol. 36, pp. 544–553, 1991.

Kovacic, W. and Smallwood, D. "Competition Policy, Rivalries, and Defense Industry Consolidation." *Journal of Economic Perspectives*, Vol. 8(4), pp. 91–110, Fall, 1994.

Llewellen, W. "Pure Financial Rationale for the Conglomerate Merger." *Journal of Finance*, Vol. 26, pp. 521–537, 1971.

Liebeskind, J. and Opler, T. "Corporate Diversification and Agency Costs: Evidence from Privately Held Firms." Working Paper, Ohio State University, 1994.

Lundquist, J. "Shrinking Fast and Smart in the Defense Industry." *Harvard Business Review*, November–December, 1992.

Markusen, A. "America's Military Industrial Makeover." In Lo, C. and Schwartz, M., eds., *Social Policy and the Conservative Agenda*. Oxford: Basil Blackwell, 1997a.

Markusen, A. "The Downside of Boeing–McDonnell Merger." *St Louis Post–Dispatch*, January 5, 1997b.

Markusen, A. "The Economics of Defence Industry Mergers and Divestiture." *Economic Affairs*, Vol. 17(4), pp. 28–32, 1997c.

Markusen, A. "The Foolish, and Costly, Defense Merger Mania." *The International Herald Tribune*, January 11, 1997d.

Markusen, A. "Understanding American Defense Industry Mergers." Working Paper, Project on Regional and Industrial Economics, Rutgers University, June, 1997e.

Markusen, A. "Why We Lost the Peace Dividend." *The American Prospect*, July–August, 1997f.

Markusen, A. and Yudken, J. *Dismantling the Cold War Economy*. New York: Basic Books, 1992.

Matsusaka, J. "Takeover Motives During the Conglomerate Merger Wave." *Rand Journal of Economics*, Vol. 24, pp. 357–379, 1993.

Miller, W. "After Desert Storm: What Next for Defense?" *Industry Week*, Vol. 240, pp. 48–53, July 1, 1991.

Mintz, J. "Going Great Guns." *The Washington Post*, October 2, 1995.

Mitchell, E. J., ed., *Horizontal Divestiture in the Oil Industry*. Washington, DC: American Enterprise Institute for Public Policy Research, 1978.

Morrocco, J. "Uncertain U.S. Military Needs Hamper Industry Restructuring." *Aviation Week & Space Technology*, Vol. 134, pp. 62–63, June 17, 1991.

Mowery, D. and Rosenberg, N. *Technology and the Pursuit of Economic Growth*. Cambridge: Cambridge University Press, 1989.

Oden, M. "Defense Mega-Mergers and Alternative Strategies: The Hidden Costs of Post-Cold War Defense Restructuring." In Susman, G. and O'Keefe, S., eds., *The Defense Industry in the Post-Cold War Era: Corporate Strategy and Public Policy Perspectives*. Oxford: Elsevier Science, 1998.

Oden, M. "Cashing-in, Cashing-out and Converting: Restructuring of the Defense Industrial Base in the 1990s." In Markusen, A. and Costigan, S., eds., *Arming the Future: A Defense Industry for the 21st Century*, forthcoming.

Oden, M. "Regional Adjustment to Post-Cold War Defense Reductions: The United States' Experience." Paper presented at the International Conference on Regional Aspects of Defense Conversion, Bonn, Germany: Bonn International Center for Conversion, 1997.

Oden, M., Markusen, A., Flaming, D., Feldman, J., Raffel, J. and Hill, C. *Post Cold War Frontiers: Defense Downsizing and Conversion in Los Angeles*. New Brunswick, NJ: Rutgers University, Project on Regional and Industrial Economics, 1996.

Oden, M., Mueller, E. J., and Goldberg J. *Life After Defense: Conversion and Adjustment on Long Island*, New Brunswick, NJ: Rutgers University, Project on Regional and Industrial Economics, 1994.

Office of the Secretary of Defense. *Defense Science Boards Task Force on Vertical Integration and Supplier Decisions*. Washington, DC: OSD, May 1997.

Pages, E. *Responding to Defense Dependence*. Westport, CN: Praeger, 1996.

Pages, E. "The Future U.S. Defense Industry: Smaller Markets, Bigger Companies, and Closed Doors." *SAIS Review*, pp. 135–151, 1995.

Pages, E. "The American Defense Industry: Consolidating Back to the Future." In Markusen, A. and Costigan, S., eds., *Arming the Future: A Defense Industry for the 21st Century*. New York: Council on Foreign Relations, forthcoming.

Pages, E. "Defense Diversification: Can We Return to the Road Not Taken." In Susman, G. and O'Keefe, S., eds., *The Defense Industry in the Post-Cold War Era: Corporate Strategy and Public Policy Perspectives*. Oxford: Elsevier Science, 1998.

Peterson, T. and Borrus, A. "Loral Could Wind Up Being the Biggest Gun of All." *Business Week*, pp. 23–24, August 10, 1992.

Picker, I. "Defense Merger War Games." *Institutional Investor*, pp. 70–71, January 1995.

Radner, R. "Problems in the Theory of Markets under Uncertainty." *American Economic Review*, May 1970.

Rappaport, A. *Creating Shareholder Value*. New York: The Free Press, 1986.

Rutgers Defense Conversion Project. *Planning for Defense Conversion: Evaluating the Closure of Lockheed/Martin Astrospace*. New Brunswick: Rutgers University, Project on Regional and Industrial Economics, December 1996.

Sapolsky, H. and Gholz, G. "Private Arsenals: America's Post-Cold War Burden." In Markusen, A. and Costigan, S., eds., *Arming the Future: A Defense Industry for the 21st Century*, forthcoming.

Schine, E. "Defenseless Against Cutbacks". *Business Week*, p. 69, January 14, 1991.

Schwartz, B. "Defense Industry Consolidation: Where Do We Go From Here?" Talk given at Johns Hopkins School of Advanced International Studies, Washington, DC, October 6, 1997.

Servaes, H. "The Value of Diversification during the Conglomerate Merger Wave." *The Journal of Finance*, Vol. 51(4), pp. 1201–1225, 1996.

Stein, J. "Internal Capital Markets and the Competition for Corporate Resources." *The Journal of Finance*, Vol. 52(1), pp. 110–134, 1997.

Stowsky, J. "America's Technical Fix: The Pentagon's Dual Use Strategy, TRP and the Political Economy of US Technology Policy." In Markusen, A. and Costigan, S., eds., *Arming the Future: A Defense Industry for the 21st Century*, forthcoming.

Teece, D. "Economics of Scope and the Scope of the Enterprise." *Journal of Economic Behavior and Organization*, Vol. 1, pp. 223–247, 1980.

Triggs, C. and Heydenreich, M. "The Judicial Evaluation of Mergers Where the Department of Defense is the Primary Customer." *Antitrust Law Journal*, Vol. 62, pp. 435–463, 1994.

Velocci, A. "Ill-Defined U.S. Defense Priorities Making Industry a 'Gambler's Paradise.'" *Aviation Week & Space Technology*, pp. 141–142, June 17, 1991.

Velocci, A. "Competitive Advantages of Scale Could Elude Aerospace Giants." *Aviation Week & Space Technology*, pp. 99–89, February 10, 1997a.

Velocci, A. "U.S. Plays Out Merger Endgame." *Aviation Week & Space Technology*, pp. 63–65, July 14, 1997b.

Weidenbaum, M. *Small Wars, Big Defense: Paying for the Military after the Cold War*. New York: Oxford University Press, 1992.

Williamson, O. *Economic Organization: Firms, Markets and Policy Control*. New York: New York University Press, 1986.

Yudken, J. and Markusen, A. "The Labor Economics of Conversion: Prospects for Military-Dependent Engineers and Scientists." In MacCorquodale, P., Gilliland, M., Kash, J. and Jamneson, A., eds., *Engineers and Economic Conversion*. New York: Springer-Verlag, pp. 135–191, 1993.

Zitner, A. "Mergers Could Endanger Defense." *The Boston Globe*, p. Al, December 23, 1996.

11

DEFENSE MEGA-MERGERS AND ALTERNATIVE STRATEGIES: THE HIDDEN COSTS OF POST COLD-WAR DEFENSE RESTRUCTURING

MICHAEL ODEN

INTRODUCTION

The breathtaking merger movement among large defense companies has led to a highly concentrated defense industrial base. If the mega-mergers proposed in 1996 and 1997 are all approved, the U.S. defense industrial base will be dominated by three consolidated companies receiving about one-third of total defense contracts. Although some anxiety is beginning to build in the DoD and Congress, the overwhelming opinion is that the merger wave is a natural and virtuous adjustment to the realities of the post-Cold War defense marketplace. Fewer firms serving a significantly smaller sales base is a sign of efficient market adjustment. Combined companies will be able to operate at higher levels of capacity utilization, reduce overhead costs and bring weapons development and production costs down. However, a major paradox in the "efficient adjustment" story immediately emerges from the recent record of large defense company performance: why are defense companies making so much money and enjoying such high share price valuations in a severely declining market? If supply was adjusting efficiently to falling demand, firms, at best, would be expected to register average profits, and share price changes would be following the Dow Jones Industrial Index.

There could be several reasons for such robust returns and high share prices. Defense firms may be successfully diversifying out of defense into high growth-high margin civilian markets. It will be shown below that defense firms have had modest but significant success at diversifying into non-DoD markets. Diversification success, however, does not explain the stock premiums on defense specialized firms or divisions.

Another possible explanation is that companies may be getting high margins on ongoing defense production because they have reduced unit costs and overhead expenses more than unit prices. With budgets constrained and few large-scale weapons development projects being funded, the industry has become somewhat more sensitive to cost and price issues. But newly merged contractors are maintaining high margins from sales in diverse portfolios of mature programs (both to the DoD and to export markets). Many of these mature programs involve sole-source production contracts which have traditionally yielded high profits. But in the Cold War acquisition system, the economic profits on sole-source production franchises were a reward to stimulate company innovation and investment. In the post-cold war era, the real needs for weapons innovation are less immediate. Since fewer major development projects are on the horizon, firms are under less pressure to invest internal funds for research and development and hence have more free cash for distribution to shareholders or for further acquisitions. This explanation suggests that savings that may have resulted from recent mergers and consolidations are not being passed on to prices for defense products. Perhaps some other goals are being met on the customer side of the market, but such a process would not represent competitive efficiency in the classic sense.

This chapter first seeks to explain why the DoD is not trying more diligently to capture savings by pushing for price reductions in ongoing programs or putting more pressure on firms to invest internal funds in research and development. It will be argued that the merger wave and high short-term profits for major contractors are in part explained by the underlying objectives of the DoD. The types and scale of mergers have been influenced by DoD industrial base management policies devised in a rather *ad hoc* fashion to deal with a fundamental mismatch between an ambitious post-Cold War security strategy and a constrained defense budget. In this environment, the DoD implemented policies directed more at keeping capacity in weapons production and design than in reducing unit costs on existing systems. High margins have been allowed on mature programs, while mergers are encouraged to retain production and development capacities across a wide spectrum of weapons systems and technologies. This helps explain why weapons exports and defense firm combinations were promoted and subsidized while few incentives were offered for diversification into non-defense markets.

In the second section, the restructuring and re-orientation of major prime contractors will be examined in the context of the range of corporate strategies available to adjust to declining defense sales. Evidence will be presented that most mergers were not necessary for corporate "survival" or a need to acquire "critical mass" as some executives have claimed. It will be shown that a number of large firms which pursued riskier defense diversification strategies were actually quite successful.

In the third part, the motivations of merging defense companies will be examined in more detail. It will be shown that mergers between defense units that had attractive portfolios of mature contracts represented a relatively low-risk means to secure short-term profits. The dominant strategy became diversification within defense markets achieved primarily through classic market extension type mergers. It is unclear whether such defense combinations represent a superior adjustment strategy over the long term. However, defense industrial base policies and financial market preferences tipped the balance in favor of market extension type mergers between defense firms or divisions. High valuations by financial markets made the sale of defense divisions irresistible for all but a few diversifying firms.

In the final part of this chapter, the risks of the merger and consolidation process will be detailed. The new, more concentrated defense industry base may turn out to be more isolated, technologically sluggish and expensive to maintain. Mergers among major primes in the broad aerospace and electronics segment may formally preserve weapons modernization capability by allowing firms high short-term profits. But to date, the high profits of consolidating firms have not led to healthy levels of company R&D to keep development capacities sharp. The DoD may actually be responsible for funding an even greater share of future R&D if new development programs are urgently needed due to a major change in the geopolitical environment. The more fundamental problem is that the logic and political support for the status-quo security policy is suspect. There are currently no compelling arguments for either significant increases in modernization or the maintenance of the large, heavy force structure presumably necessary to fight two regional conflicts. Because the Bottom-up Review (BUR) and subsequent Quadrennial Defense Review (QDR) seem based more on inertia and a lack of alternatives than on a strong consensus about real national defense needs, we may be restructuring the defense industrial base in the wrong way. In the absence of clear threats or a popular consensus there is danger that the three giants (Lockheed Martin, Boeing, and Raytheon) will have inordinate power over choices about security strategy, priorities within the defense budget, technology trajectories and the weapons systems chosen for future development.

THE ROLE OF DOD POLICY IN DEFENSE INDUSTRY RESTRUCTURING

The new framing strategy for U.S. security and defense priorities detailed in the Pentagon's 1993 BUR forced the DoD into a difficult balancing act

(DoD, 1993). The BUR's call for the retention of large, heavy forces for two major interventions had far-ranging implications for defense acquisition options. The BUR ensured that manpower and O&M costs would squeeze overall procurement spending. The plan further signaled that the need for increased mobility would be met by already developed platforms such as carriers and heavy air transports. If the two war scenario in the BUR was to be taken seriously, warm production capacity had to be maintained in current generation tactical systems to replace possible losses in major conflicts. By subtraction, large development programs for next generation weapons would be extremely limited in the 1990s (GAO, 1994; 1995).

The composition of the budget cuts over the 1989–1998 period demonstrates how the BUR priorities led to a squeeze on procurement and R&D spending. Total budget authority has fallen by nearly 34 percent in real terms over the 1989–98 period. Spending for Personnel, O&M, and Other (mostly infrastructure) taken together has only declined by about 24 percent. But, Procurement and R&D budget authority together fell by roughly 48 percent (DoD, 1996a; Finnegan, 1997).

The requirements of the BUR have obviously influenced (although not strictly determined) the composition of contract awards (see Table 11.1). Contract spending has obviously moved from major systems to force

Table 11.1. Domestic defense contacts over $25,000 by major category

	1989		1996		
	Total contracts (millions of 1996$)*	Percent of total	Total contracts (millions of 1996$)*	Percent of total	Percent change 1989–96
Aircraft	33,563	23.0%	25,941	23.7%	−22.7%
Missile & space systems	25,149	17.2%	11,554	10.6%	−54.1%
Ships	12,307	8.4%	7,520	6.9%	−38.9%
Tanks & vehicles	4,794	3.3%	2,979	2.7%	−37.9%
Weapons	2,020	1.4%	1,509	1.4%	−25.3%
Ammunition	4,820	3.3%	1,980	1.8%	−58.9%
Comm. & elec.	23,583	16.2%	13,499	12.3%	−42.8%
Total hard goods	106,236	72.8%	64,982	59.4%	−38.8%
Other supplies and services	39,772	27.2%	44,425	40.6%	11.7%
Total contacts	146,009	100.0%	109,407	100.0%	−25.1%

*Includes only domestic contracts over $25,000, excludes classified programs equalling 5% of total in FY 1989 and 2.8% of total in FY 1996
Source: DoD, *Prime Contract Awards by Region and State, FY 1996, 1989*, (Washington D.C: DIOR), Tables 11-1, 11-2, III-1 (1996).

readiness and O&M activities as the real dollar value of Other Supplies and Services actually increased by 11.7 percent over the 1989–96 period. It must be noted that the O&M and Other Supplies and Services categories include much procurement and even R&D on computer software development. Hence cuts in weapons and battlefield logistics and communications may be somewhat overstated because software and systems integration have become an increasingly important component of systems and weapons upgrading and modernization. Nevertheless the major hard goods categories where the major primes dominate, Missiles & Space Systems, Ships, Ammunition, and surprisingly, Communications and Electronics declined faster than average. A significant share of overall Communications and Electronics contracts (which exclude most purchases of software) were linked to strategic forces which have experienced the most drastic funding cuts. The Aircraft category has been the least affected by post-Cold War downsizing with developed systems, the F-22, F-18 upgrade, and C-17 moving into production over this period.

This pattern of demand change helps explain certain merger and consolidation moves among the major prime contractors. The impetus behind the numerous combinations that resulted in the current Lockheed Martin group came from the desire of the original firms to extend their portfolios centered on strategic systems and electronics into more promising tactical aircraft, operations and maintenance, and systems upgrading. The relative health of the aircraft sector also made McDonnell Douglas' mature, and cash-rich fighter and transport programs an attractive prize for the Boeing Company.

The Modernization Issue

Major modernization programs planned through the turn of the century consist of only one major new fighter aircraft program (JSF), one new aircraft carrier, and a set of potentially large theater-missile defense programs (Bischak, 1996). There are a number of smaller projects and significant funding for Command, Control, Communications, Computers and Intelligence or C^4I and computer and software innovation, but the pipeline for major new weapons system development has been severely crimped by budget constraints and the production and readiness requirements flowing from the BUR and QDR. The dearth in major weapons development programs over the 1990s created an especially difficult problem for defense planners. Although there are few real threats on the horizon, the DoD does not want to lose the capacity to rapidly develop new systems in every major category, or at least all categories of tactical systems.

In the Cold War acquisition system, sustaining constant technological advance in weaponry was based on a particular incentive system. The procurement cycle was managed through a system of "prizes for innovation" whereby firms or groups would compete vigorously in the design and early development stages of new projects, investing at least some internal funds in the competition. The DoD then awarded a sole source contract which gave the winner high returns (economic profits) on the production franchise. Through this process, major prime contractors were given adequate revenues to maintain research and production teams over the procurement cycle (Rogerson, 1989, 1993). Since there are so few new weapons development projects to serve as prizes to keep company research and development teams intact, how can the capacity to develop next generation technologies be secure?

In the short-term, procurement of unneeded weapons has been used to maintain, variously, the submarine industrial base, the armored vehicle industrial base, the aircraft carrier industrial base, the bomber industrial base, etc. But this is an expensive stop-gap approach. Moreover, justifying expensive weapons buys to sustain an industrial base opened the door for Congressional preferences. The expensive continuation of the B-2 program was based on a presumed need to maintain capacity. This and other Congressional top-offs simply drew off additional scarce resources from the already shallow modernization pot.

The recently released QDR, which was supposed to offer a more long-term solution to the modernization problem, provided little hope for more development funding. With the balanced budget agreement in force and no convincing threats looming, the QDR's stubborn emphasis on the need to fight two major conflicts doomed any significant increase in weapons spending. Development and procurement spending will be stuck at current real dollar levels over the next two years, while increased funding for modernization is put off for out-year budgets in 2001 and 2002. As John McCain, a powerful Republican on the Senate Arms Services Committee observes, "[the] Pentagon is deluding the contractors with promises of sales and the contractors are deluding themselves that the money will be available" (Fulghum, 1997).

DoD Industrial Base Policies

With the BUR and now QDR strategies in place, those in the DoD responsible for managing the industrial base have been groping for a way to simultaneously maintain hot production capacity and the ability to develop next generation systems in all major weapons categories. DoD leadership publicly argued that it could meet BUR requirements and residual modernization needs through major efficiency gains. Mergers were touted as a way to reduce unit acquisitions costs so that, supposedly, more weapons could be produced and more modernization funded. In the 1993–94 years there was also a strong emphasis on procurement reform and dual-use. The latter elements promised to yield savings and aid modernization by incorporating better and cheaper commercial components into defense systems and by generating economies of scale through combining commercial and military production. However, the strongest incentives embodied in industrial base management policies in the immediate post-Cold War period encouraged merger and consolidation among large defense firms.

The first element of policy that encouraged diversification within defense markets was the continued allowance of high profits on sole-source production contracts. This was not a policy decision per se, but simply a continuation of a normal contract management process designed for the Cold War years. In the early 1990s there was a considerable production backlog from the intense weapons development cycle of the Reagan era. Most of the larger companies had a number of systems under sole-source contract that were midstream in their production cycles—prizes for earlier innovations and prior contract wins. Volumes on many of these production contracts were shaved back in the early 1990s, but few were canceled. Moreover, in many cases, cuts in domestic orders were offset by foreign arms sales, sustaining highly profitable production and service franchises.

As modernization needs diminished, the motivation for the DoD to allow these high returns should have been reduced. However, the DoD, in its contract management systems had little experience in trying to push down, or substantially renegotiate, unit prices on ongoing programs. In the contract management process itself, indirect or overhead costs (which constitute 50 percent or more of total costs in many weapons contracts) are managed on a contract or program basis. There is no way under current procedures to estimate overhead and other indirect costs on a plant-wide or company-wide basis.[1] It is therefore difficult to drive down unit costs if there is no way of determining how overheads are being allocated across the various contracts or programs in a specific plant or across the company. This is a continuing challenge, since most of the presumed savings from defense mergers are associated with company-wide reductions in operating and overhead expenses.

A second, related policy that encouraged merger and consolidation was the permissive attitude toward arms exports. After a short period of contraction after the Persian Gulf War, the DoD and State Department de-emphasized fashioning new arms export control agreements or discouraging sales of certain systems. Indeed, the marketing effects of the Persian Gulf War combined with the supportive stance of DoD and the State Department allowed U.S. firms to increase their post-Cold War share of the international market substantially. It is also important to remember that arms exports are supported by a range of direct and indirect government subsidies estimated to be in the neighborhood of $5–7 billion per annum (Hartung, 1996). Arms exports helped address the DoD's industrial base problem in two related ways. Ongoing arms export contracts yielded high margins allowing firms to retain production and development capacity. And in terms of the U.S. defense budget, it was a cheap way to keep certain mature production lines open for the tactical systems called for in the BUR. Although not a primary motive, attractive profits on many export deals made companies with big export contracts attractive acquisition targets.

The third thrust in DoD industrial base management policy was the direct subsidy of defense mergers. Beginning in July of 1993, the DoD began to allow restructuring costs including severance pay, plant closure costs, and unemployment payments to be added to allowable costs on a firm's defense contracts (GAO, 1995). In return for these subsidies, defense companies promise to pass on future savings in the form of lower prices. Although restructuring costs incurred in all the various mergers are likely to be substantial, a comprehensive estimate of the subsidies that might be associated with the policy has not emerged (Korb, 1996). In a study of five earlier large mergers, the GAO estimated that restructuring costs could equal about $1.3 billion, with costs reimbursed by the government amounting to about $180 million to date. The GAO report claims that the DoD has enjoyed savings of $347 million or about $1.93 for every dollar in subsidy so far (GAO, 1997). However, this is a partial report which doesn't estimate future costs for the much larger number of mergers that are only now being certified for reimbursement. Moreover, the DoD in its own report to Congress on *Payment of Restructuring Costs Under Defense Contracts* acknowledges, "it is not feasible to completely isolate the effect of restructuring from other complex determinants of the difference between projected and actual costs over a long period of time." In sum, no comprehensive estimate exists of overall subsidies or associated cost savings associated with large defense mergers.

It is noteworthy that these arms export and merger subsidies together dwarf the annual $300 million or so in assistance offered to companies over the 1993–95 period to create dual-use development and production capacity and to diversify into non-DoD markets. Merger and consolidation subsidies have, in effect, swamped incentives to diversify.

The results of this policy mix in terms of acquisition costs on existing programs and the reduction in competition resulting from defense mergers suggests that the DoD may be more concerned with keeping capacity in weapons production and design, not in reducing unit costs on existing systems. In newly combined and diversified companies, high profits are allowed on production contracts in one weapons area in order to maintain the capability to carry-out new development programs, or to keep less profitable production going in other defense segments. It is hoped that the restructured industry base can continue to provide at least latent development and modernization capability across the spectrum while overall R&D and procurement spending remain constrained. This approach stems from the way the acquisition system was managed in the broadest terms during the Cold War. Hence, the emphasis on maintaining capacity may reflect institutional inertia as much as a coherent strategy for the new era.

Prizes for innovation have been supplanted by prizes for merger. Instead of winning profitable production contracts from design competitions, it is now possible to simply buy profitable production backlog through acquisition. When McDonnell Douglas lost out on the Joint Strike Fighter program, it simply married one of the winners, Boeing. Boeing, in turn, obtained a number of mature production programs that it could milk for short-term revenues.

ALTERNATIVE ADJUSTMENT STRATEGIES OF MAJOR PRIME CONTRACTORS

A look at the 25 top prime contractors confirms that merger and consolidation has been a dominant strategic trend over the 1989–96 period. The data presented in Table 11.2 offer a profile of the top 25 contractors in 1989 and 1996. The estimate of defense sales and defense dependence for each company is based on an annual estimate in *Defense News* which includes sales (versus contracts) to the DoD *and* military sales to foreign governments (Finnegan and Marburger, 1990 and 1997). Hence, these defense dependence estimates are different than other published sources and provide a more accurate picture of actual dependence on government defense markets.

Overall real sales and defense dependence fell significantly among the top 25. Real defense sales fell by $41.3 billion, non-DoD sales increased by a modest $6.2 billion, yielding a fall in overall sales of about $35.1 billion. The considerable reshuffling and reorganization among the top primes as a result of merger and consolidation is clearly in evidence. Major combinations include Lockheed and Martin, Northrop and Grumman, and the absorption of defense units of Rockwell, Westinghouse, Loral, Unisys (defense) and Honeywell (defense) into other entities. Moreover, this list does not account for the most recent mergers of Hughes and Texas Instruments with Raytheon, McDonnell Douglas with Boeing or the pending merger of Northrop Grumman with Lockheed Martin.

There were a number of strategies besides merger available to large defense companies to deal with falling contract revenue. These include: internal consolidation into remaining defense markets; diversifying into non-DoD markets through internal product development or acquisition; or selling-off defense divisions and exiting defense altogether. According to defense company executives, large defense mergers were necessary for "survival" (Norman Augustine, Lockheed Martin) or to obtain "critical mass" (Michael Armstrong, Hughes). These statements suggest that large defense companies or divisions would have faced losses or would have seen their market shares slip away if they did not merge with other large defense firms.

However, as the cases of McDonnell Douglas and General Dynamics show, prime contractors choosing to stay focused in defense markets can

Table 11.2. Sales and defense dependancy Top 25 U.S. prime contractors

	Company (1989)	Total sales 1989 (Millions 92 $)	Defense sales 1989 (millions 92 $)	Defense/total sales 1989 (a)		Company (1996)	Total sales 1996 (millions 92 $)	Defense sales 1996 (millions 92 $)	Defense/total sales 1996 (a)
1	McDonnell Douglas	14,928	10,799	72%	1	Lockheed Martin	23,843	12,875	54%
2	General Dynamics	11,308	10,130	90%	2	McDonnell Douglas	12,308	8,985	73%
3	General Electric	60,841	9,851	16%	3	Northrop Grumman	7,206	5,960	83%
4	Lockheed	11,027	8,796	80%	4	Hughes	14,146	5,602	40%
5	Hughes	12,663	6,585	52%	5	Boeing	20,463	5,136	25%
6	Boeing	22,604	6,349	28%	6	Raytheon	10,943	3,589	33%
7	Martin-Marietta	6,462	5,688	88%	7	United Technologies	20,907	3,032	15%
8	Raytheon	9,806	5,333	54%	8	General Dynamics	3,186	2,937	92%
9	Northrop	5,851	5,278	90%	9	Litton	3,213	2,638	82%
10	United Technologies	21,775	5,207	24%	10	TRW	8,770	2,017	23%
11	Rockwell	13,955	3,924	28%	11	Texas Instruments	8,843	1,574	18%
12	TRW	8,183	3,507	43%	12	Newport News Shipbuilding	1,664	1,566	94%
13	Westinghouse	14,319	3,404	24%	13	General Electric	70,445	1,479	2%
14	Litton	5,600	3,344	60%	14	ITT Industries	7,756	1,396	18%
15	Grumman	3,929	2,648	67%	15	Allied Signal	12,429	1,119	9%
16	Allied Signal	13,313	2,609	20%	16	Computer Sciences Corp.	4,996	964	19%
17	Texas Instruments	7,271	2,407	33%	17	SAI Corporation	2,137	932	44%
18	Unisys	11,256	2,326	21%	18	United Defense	899	899	100%
19	Tenneco	14,328	2,161	15%	19	Textron	8,274	894	11%
20	Textron	8,294	1,892	23%	20	Alliant TechSystems	969	853	88%
21	ITT	22,302	1,761	8%	21	Tracor	963	825	86%
22	Honeywell	6,755	1,405	21%	22	Lucent Technologies	21,352	662	3%
23	FMC Corporation	3,832	1,226	32%	23	Harris Corporation	3,203	609	19%
24	GTE Corporation	20,347	1,221	6%	24	Rockwell	9,253	537	6%
25	Loral	1,323	1,048	79%	25	GTE	18,950	531	3%
	Total/Weighted Percentage	332,270	108,899	33%			297,118	67,609	23%

Sources: Company Annual Reports, Investment Analysts Reports, Company Interviews, GAO, *Defense Contractors*, October 1995, GAO/NSAID-96-19BR, *Defense News* "Top 100 Reports" 1990–94.
(a) estimates from *Defense News*, "Top 100 Reports", include sales to DOD and estimate of arms export sales. Does not include other government sales such as NASA and DOE.

deal with falling sales and excess capacity without merging. Prior to 1996, these firms simply shed labor and capital assets to become profitable at a lower overall volume of sales. McDonnell Douglas successfully executed this strategy and saw the profitability of its defense operations soar before they sold out to Boeing (Oden et al., 1993). While their firm-wide profit performance has been unspectacular, this was due in large measure to their weak commercial aircraft operations. It is somewhat hard to fathom why $10 billion companies such as McDonnell Douglas or Hughes don't have the "critical mass" to compete for projects, or at least be major team members with other multi-billion dollar firms.

The other option was diversification into non-DoD markets. Research on major prime contractors confirms that diversification has been a real, but limited phenomenon among multi-billion dollar prime contractors (Oden, forthcoming). In one study of 25 large prime contractors over the 1989–94 period, it was found that 12 companies had better than average sales performance and also reduced significantly their defense dependence (Hughes, Martin Marietta, Raytheon, TRW, Allied Signal, Textron, Texas Instruments, United Technologies, FMC Corporation, GTE, ITT and Computer Sciences) (ibid). While several companies such as Martin Marietta gained commercial sales through acquisition, many of these companies had moved into alternative markets primarily through internal product development.

Technology and product specialization strongly conditioned how firms were positioned to enter non-DoD markets. The type of firms that pursued serious diversification suggests that systems integrators in the communications and electronics segments tend to find more opportunities to diversify than platform or weapons makers. But large firm strategies did not seem entirely predictable on the basis of market or industry segment alone. Platform specialists such as FMC and United Technologies (not on the list) pursued vigorous diversification, while Loral received little commercial market entré from its electronics and information technology capabilities prior to its absorption by Lockheed Martin. Commercialization by major prime contractors generally occurred in closely related markets. Platform makers more commonly targeted civilian aerospace and space launch, while electronics and surveillance companies focused on such markets as telecommunications, environmental sensing or satellite communications.

The Oden study found that over the 1989–94 period, six of 25 major contractors successfully carried out an adjustment strategy centered on commercialization (Hughes, TRW, Textron, Texas Instruments, FMC Corp, and Computer Sciences) (ibid). It was found that successful commercialization required intense management devotion to the strategy, creative reorganization of corporate operations, and a major commitment of resources to internal product development. For example, Hughes has focused for years on integrating their commercial and military satellite operations, creating a real dual-use capacity from design to production. In the bigger automotive electronics segment, Hughes-Delco drew on technology and expertise from across the company to respond to GM and other auto clients. The division has carried out more than 150 product development projects for GM. To encourage technology transfer from its defense and space segments to automotive applications, TRW set up an Automotive Technology Center at its defense division. Automotive segment personnel, funded by their own business units, came to the center to extract technologies from the defense and space group. Similar examples of inter-segment integration and product development were also occurring at Textron and Texas Instruments.

Studies suggest that market diversification through internal product development tends to be more successful than diversification through acquisition (Ravinscroft and Scherer, 1988; Dunne *et al.*, 1989). The six successful firms were found to be relying primarily on internal development in their efforts. However, because of the traditional wall of separation between defense and commercial operations, companies must often develop new sales, distribution and servicing capabilities through teaming with companies already active in target markets. For example, Hughes has teamed with the major consumer electronics companies to distribute Direct TV.

The question is: were subsequent decisions on the part of many of these firms to jump on the merger bandwagon a result of poor performance due to the failure of commercialization strategies? To get a tangible measure of company performance, I first look at productivity, R&D, investment commitment, and profitability for six companies that have basically followed a merger or consolidation strategy and six companies that have pursued a diversification strategy.

These data strongly suggest that diversifying companies were more committed to the internal development of their business lines. They committed a significantly larger share of company resources to R&D and capital expenditures than the consolidating firms. Surprisingly, diversifying companies appeared to be more, rather than less successful at increasing productivity and profitability.

As shown in Table 11.4, the diversifying companies also had relatively decent sales growth over 1989–94, the period when defense cuts were most severe. They successfully offset defense sales losses with commercial sales gains. In the process, they retained significantly greater employment than the merging and consolidating companies as a group. These diversification efforts did not lead to disaster, but instead yielded reasonably good results, especially when compared to other military prime contractors.

Among the consolidators, Loral and Lockheed Martin are the only firms that have expanded sales and employment. This has occurred primarily through acquisition. However, in the case of Lockheed Martin, there have been significant internal pushes into new non-DoD government markets such as air traffic control and information systems as well as success in commercial space launch.

Table 11.3. Performance measures of consolidating and diversifying firms (%)

	Growth in sales per employee 1889–94	Company R&D/sales 1994	Capital Investment/ sales 1994	Profit/sales(a) 1994	ROI(b) average 1989–94
Consolidating companies:					
Northrop Grumman	7.70	2.60	2.00	0.50	4.10
Litton	2.50	1.70	2.30	1.90	7.20
General Dynamics	23.60	1.00	1.20	7.30	5.80
McDonnell Douglas	63.50	2.30	0.90	4.50	7.40
Loral	13.10	4.30	2.50	5.70	10.20
Lockheed Martin	12.60	2.90	2.30	4.40	7.80
Average (consolidators)	20.50	2.50	1.90	4.10	7.10
Diversifying companies:					
Hughes	26.20	5.00	5.30	6.60	7.30
TRW	22.40	6.10	5.60	3.70	8.50
Textron	21.70	1.90	3.10	4.50	7.40
Texas Instruments	77.10	6.70	10.40	6.70	6.10
FMC Corporation	13.70	4.10	8.80	4.30	10.90
Computer Sciences Corp	24.40	na	4.60	3.50	11.40
Average (diversifiers)	30.90	4.80	6.30	4.90	8.60

Sources: Company Annual Reports, Investment Analysts Reports, Company Interviews, GAO, *Defense Contractors*, October 1995, GAO/NSAID-96-19BR.

(a) Calculated as real 1992$ net earnings over sales
(b) Calculated as net operating income (before interest payments and taxes) over total assets.

Table 11.4. Sales changes and employment in consolidating and diversifying firms

	Change in defense sales 1989–94 (millions 92 $)	Change in non-defense sales 1989–94 (millions 92 $)	Change in total sales 1989–94 (millions 92 $)	Change in employment 1989–94	Percentage change in employment
Consolidating companies:					
Northrop Grumman	(1,658)	-616	(2,274)	(27,500)	-39%
Litton	-335	(1,983)	(2,318)	(21,700)	-41%
General Dynamics	(7,416)	-979	(8,395)	(80,900)	-79%
McDonnell Douglas	(2,010)	-369	(2,379)	(62,166)	-49%
Loral	1,928	567	2,495	19,700	155%
Lockheed Martin	1,400	3,106	4,506	17,300	12%
Total/weighted percentage	(8,090)	-275	(8,365)	(155,266)	-30.40%
Diversifying companies:					
Hughes	-585	1,349	764	(15,000)	-16%
TRW	(1,864)	2,336	471	(10,100)	-14%
Textron	-939	1,867	928	(5,000)	-9%
Texas Instruments	-762	3,315	2,553	(17,521)	-24%
FMC Corporation	-198	224	26	(2,766)	-11%
Computer Sciences Corp	69	897	966	7,000	32%
Total/weighted percentage	(4,280)	9,989	5,709	(43,387)	-12.50%

Sources: Company Annual Reports, Investment Analysts Reports, Company Interviews, GAO, *Defense Contractors*, October 1995, GAO/NSAID-96-19BR.

This evidence suggests that diversifying companies such as Hughes and Texas Instruments were not compelled by poor performance to merge. Why, then, did the sale and acquisition of defense units become so prevalent as a means to deal with defense reductions? Wall Street pressures, government incentives, and the combination of healthy short-term profits and relatively low risk available to merging firms are the reasons why diversifying firms chose in the end to sell-off defense operations. Quick payoffs from selling off defense divisions or acquiring defense units discouraged companies from making the wrenching changes and investments necessary to move to alternative markets.

THE MERGER WAVE CLEARS THE DECKS

The defense priorities and DoD industry policies noted above have had a powerful impact on the strategies large companies implemented to deal with defense reductions. But big defense companies react to market opportunities and shareholder desires as well as new defense policy signals.

A defense firm can improve its position through a horizontal merger with a competing company producing similar defense products. In theory, the acquiring firm can reduce unit costs and limit destructive competition. If the efficient scale for producing tactical missiles of a similar type is 1000, and there are two contractors each producing 500, combining operations reduces unit costs. Real savings and efficiency gains can accrue in horizontal mergers, such as Hughes' purchase of General Dynamics missile division where several production lines in California were shuttered and combined in a large production facility in Tucson, Arizona. If other competitors remain, and the DoD pushes for price reductions, the affordability of weapons can be also enhanced by this form of merger.

The remaining reasons for merger are to reduce the cost or improve the quality of supply by acquiring a producer of a major input (vertical merger), or to diversify into other markets to stabilize and expand company revenues (market extension merger). There is strong evidence that in most of the biggest defense mergers the primarily motivation has been market extension, not horizontal combination (Oden, forthcoming). The importance of market extension, or diversification within defense markets, as the primary motive for many defense mergers was confirmed in a recent Defense Science Board report. They noted, ". . . defense firms are generally acquiring (or merging) with other firms that have been producing a variety of defense products. . . . Defense firms are seeking to increase revenue by buying other firms' existing or backlog orders, to improve profit margins and stock market performance, and to reduce excess capacity" (Office of the Secretary of Defense, 1997). Combinations of large primes with diverse portfolios of production and development contracts reduces the total number of major contractors, but the surviving firms are active in more segments of the defense market.

The financial community began to recognize that short-term profit opportunities were very attractive when one defense firm specializing in one weapons segment bought another firm or division specializing in another segment with an attractive portfolio of mature production contracts. The newly combined firm could cut certain overheads, reduce R&D and capital investment (since few new programs would be up for competition) and reap high margins on mature programs where R&D and investment had already been paid off. When profitable backlog began to decline in an important segment, the firm could simply acquire more backlog by executing another merger.

Wall Street's recognition that "pure play" defense combinations could yield high short-term profits led to relatively high valuations if defense firms or divisions were put into play. The defense industrial index has significantly out-paced the Dow Jones Industrial Index in the 1990s, a surprising fact given big defense cuts (Summers, 1997). Boeing, for example, recently paid 22 percent over market valuation for McDonnell Douglas (Velocci, 1996). However, if companies wanted to maintain a diverse portfolio of commercial and military production, or were focusing on internal product development, they were penalized with low share prices for "not sticking to their knitting" or not focusing on their "core competencies." Management of companies trying to buck the merger trend encountered extreme pressure from institutional shareholders and Wall Street deal-makers. If they sold off their defense divisions, they could reap immediate windfalls. If they patiently invested in new product and market development, their share prices suffered.

Add in the sweeteners provided by the government policy of subsidizing arms exports, merger costs for defense combinations, and the sell-off of profitable defense units became irresistible for all but a few major primes. Industry was already engaged in major merger deals in 1992, before the BUR and explicit policies supporting and subsidizing merger were formulated. However the merger movement, involving predominantly defense acquisitions did accelerate through 1993 and culminated in a series of mega-mergers in 1994, 1995 and 1996.

Table 11.5 details the most significant mergers in the defense industry since 1989. This time-line shows that in the early post-Cold War years some mergers were motivated by the desire to diversify out of defense. Northrop's purchase of LTV's commercial aircraft components business and Textron's purchase of Cessna from General Dynamics were primarily diversification moves. However, all subsequent mergers involved building more diverse portfolios in defense by market extension mergers or, much less commonly, classic horizontal mergers to improve positions in specific segments and rationalize excess capacity.

These combinations of large primes with diverse portfolios of production and development reduced significantly the total number of major contractors.

Table 11.5. Merger and acquisition activity in the defense industry 1990–97

Year	Buyer	Unit acquired	Seller	Buyer motivation	Seller motivation
1990	Loral	Ford Aerospace	Ford Motor Corp.	Market extension/defense	Cash-in/exit
	Northrop	LTV Aircraft	LTV	Commercial diversification	Cash-in/exit aerospace
1991	Texton	GD-Cessna	General Dynamics	Commercial diversification	Cash in/exit commercial
1992	Hughes	GD-Missiles Systems	General Dynamics	Horizontal merger/consolidation	Cash-in
	Loral	LTV Missiles Division	LTV	Market extension-defense	Cash-in/exit aerospace
	Martin Marietta	GE-Aerospace	General Electric	Market extension-defense horizontal merger/consolidation	Exit defense electronics
1994	Lockheed	GD-Military Aircraft	General Dynamics	Market extension-defense	Cash-in
	Northrop	Grumman	Grumman	Market extension-defense	Hostile takeover
	Loral	IBM-Federal Systems	IBM	Market extension-defense/commercial	Cash-in/exit defense
	Martin Marietta	GD-Space Systems	General Dynamics	Horizontal merger/consolidation	Friendly takeover
	Martin Marietta	Lockheed	Lockheed	Market extension-defense horizontal merger/consolidation	Friendly takeover
1995	Loral	Unisys-Defense	Unisys	Market extension-defense	Cash-in/exit defense
	Litton	Teledyne-Electronics	Teledyne	Horizontal merger/consolidation	Exit defense electronics
	Raytheon	E-Systems	E-Systems	Market extension-defense	Friendly takeover
	Hughes	Magnavox Electronic Systems	Carlyle Group	Vertical merger-space GPS	Cash-by investment group
	General Dynamics	Bath Iron Works (shipyard)	Fulcrum II Ltd.	Market extension-defense	Cash-by investment group
1996	Northrop Grumman	Westinghouse-Defense	Westinghouse	Market extension-defense	Exit to buy TV network
	Lockheed Martin	Loral-Defense	Loral	Market extension-defense horizontal merger/consolidation	Exit to commercial space
	Boeing	Rockwell-Defense & Space	Rockwell	Market extension-defense	Exit defense to commercial
	Boeing	McDonnell Douglas	McDonnell Douglas	Market extension-defense horizontal merger/consolidation	Friendly takeover
1997	Raytheon	Texas Instruments-Defense	Texas Instruments	Market extension-defense horizontal merger/consolidation	Exit to commercial electronics
	Raytheon	Hughes Aircraft & Electronics	GM-Hughes	Market extension-defense horizontal merger/consolidation	Cash-in/exit defense
	Lockheed Martin	Northrop-Grumman	Northrop-Grumman	Market extension-defense	Friendly takeover

Sources: Company Annual Reports, *Defense News*, January 8–14. p. 7, *Aviation Week*, September 15, 1995, *Aviation Week, December 1996* .

But the surviving firms have strong positions in more segments of the defense market. In defense, each major program tends to resemble a small independent company with its own research, production, marketing, contractor and supplier teams (Lundquist, 1992). Having fewer major companies, more diversified across programs might reduce company-wide management overheads, marketing costs, and costs associated with keeping reserve R&D and development capacity. But these types of savings would only lead to reduced acquisition costs if the DoD forced companies to integrate and consolidate management and research functions across business segments and appropriated a share of these savings by reducing allowable overhead costs on existing contracts. As noted above, procurement agencies have little experience operating in an activity-based costing mode. Absent a new effort to manage costs, the primary gain for the DoD from the merger wave will be in retaining capacity rather than in obtaining significant savings in ongoing contracts.

THE DEFENSE INDUSTRIAL BASE CIRCA 1997—THE POTENTIAL COSTS OF CONCENTRATION

Another way to gauge the restructuring of the defense industry base is to follow the top 25 contractors during the most intense years of the merger movement, 1993–97. As Table 11.6, shows, 13 of the top 25 contractors in 1993 were merged or absorbed by 1997 (assuming the mergers announced in late 1996 and 1997 all go through). If the most recent 1997 structure (including proposed and approved mergers) is superimposed over 1996 contact awards (the last available year as of this writing) the top four contractors would control nearly one-third of total prime contracts over $25,000. If the Lockheed Martin/Northrop Grumman merger is consummated, three newly formed giants, Lockheed Martin, Boeing, and Raytheon will control 30 percent of prime contracts, a major concentration at the top.[2]

The Issue of Market Competition

The impact of recent mega-mergers on competition in defense markets is a cause for concern, yet the federal government has so far approved all mergers with only minor adjustments. Aside from having a large share of the overall market, the three giants (Lockheed Martin, Boeing and Raytheon) will have dominant positions in broad aerospace and electronics markets. The only major segments where the big three do not have a strong position are shipbuilding and armored vehicles, relatively stagnant markets.

The trend toward market extension mergers has meant that competition, while narrowed, has been formally preserved in all major weapon segments (Table 11.7). No firm has acquired a monopoly position in these broad weapons system categories. In aircraft and helicopters, there will be only

Table 11.6. Consolidation of Major Prime Contractors Since 1993

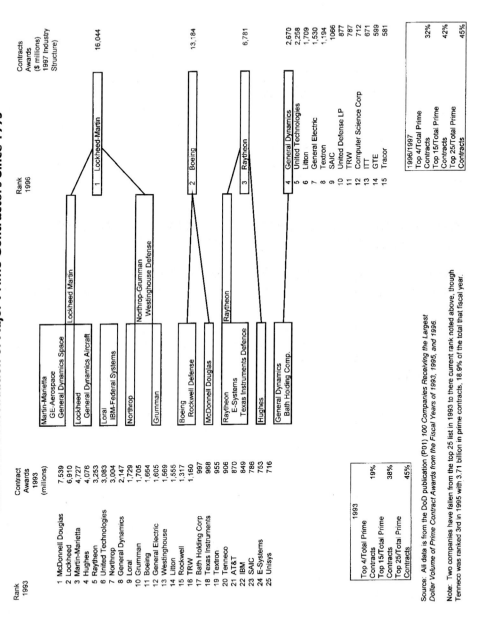

Rank 1993	Contract Awards 1993 (millions)
1 McDonnell Douglas	7,539
2 Lockheed	6,910
3 Martin-Marietta	4,727
4 Hughes	4,076
5 Raytheon	3,253
6 United Technologies	3,004
7 Northrop	2,147
8 General Dynamics	1,729
9 Loral	1,705
10 Grumman	1,664
11 Boeing	1,605
12 General Electric	1,569
13 Westinghouse	1,555
14 Litton	1,317
15 Rockwell	1,160
16 TRW	997
17 Bath Holding Corp	968
18 Texas Instruments	955
19 Textron	906
20 Tenneco	870
21 AT&T	849
22 IBM	786
23 SAIC	753
24 E-Systems	716
25 Unisys	

Rank 1996	Contracts Awards ($ millions) 1997 Industry Structure)
1 Lockheed-Martin	16,044
2 Boeing	13,184
3 Raytheon	6,781
4 General Dynamics	2,670
5 United Technologies	2,258
6 Litton	1,709
7 General Electric	1,530
8 Textron	1,194
9 SAIC	1066
10 United Defense LP	877
11 TRW	787
12 Computer Science Corp	712
13 ITT	671
14 GTE	599
15 Tracor	581

	1993
Top 4/Total Prime Contracts	19%
Top 15/Total Prime Contracts	38%
Top 25/Total Prime Contracts	45%

	1996/1997
Top 4/Total Prime Contracts	32%
Top 15/Total Prime Contracts	42%
Top 25/Total Prime Contracts	45%

Source: All data is from the DoD publication (P01) *100 Companies Receiving the Largest Dollar Volume of Prime Contract Awards from the Fiscal Years of 1993, 1995, and 1996.*

Note: Two companies have fallen from the top 25 list in 1993 to there current rank noted above, though Tenneco was ranked 3rd in 1995 with 3.71 billion in prime contracts, 18.9% of the total that fiscal year.

Table 11.7. Key competitors in major defense segments (1997)

Missiles	Aircraft	Naval vessels	Helicopters	Satellites	Armored vehicles	Defense electronics	Defense services/software
Raytheon	Lockheed Martin	Newport News	United Technologies	Lockheed Martin	General Dynamics	Lockheed-Martin	Lockheed Martin
Boeing	Boeing	Litton	Boeing	Raytheon	United Defense	Raytheon	Tracor
Lockheed Martin		General Dynamics		TRW	Textron	Litton	Logicon
		Avondale				ITT	BDM
						GTE	TRW
						Alliant Techsystems	ITT
						Harris	Litton
							Raytheon

Source: Anthony Velocci, "U.S. Industry Poised for Further Realignment," *Defense News*, December 23–30, 1996.

two major competitors if the Northrop Grumman/Lockheed Martin merger goes through.

However, a more fine-grained analysis suggests that rivalry in certain technologies and sub-segments has been significantly reduced. Only Lockheed Martin (if it successfully combines with Northrop Grumman) will have production experience with stealth technology, only Boeing (after its acquisition of McDonnell Douglas) makes heavy air transports, and General Dynamics is the sole submarine producer. Raytheon (after its acquisition of Hughes) will have a truly dominant position in tactical missiles. Since competition has been only nominally preserved in a number of areas, the new, highly concentrated structure brings with it the danger of increasing market power and continual cost escalation. It could be that divisions or units within large firms will compete against each other for contract awards. But this intra-firm competition will be constrained by broader profit maximizing objectives of the parent firm. In addition, the existence of several units within a large defense firm capable of competing for large scale contracts may actually be a sign of underlying inefficiency. Such intra-firm competition could indicate that there is considerable excess capacity remaining in the giant firms which the DoD is paying for in the end through overhead allowances on existing contracts.

The Giants Must be Fed

More troubling than the issue of market power is the influence the big four (or if the Lockheed Martin/Northrop Grumman merger goes through, the big three) will have over defense policy in general and acquisition policy in particular. The big three are unusually intertwined, with an array of contracting and subcontracting relationships and partnerships on major weapons production and development contracts. The concerted pressure of the big firms for modernization and increased procurement will intensify as backlogs of mature production contracts that are a major driver of firm profits continue to fall. Furthermore, these firms will push, individually and collectively, for a certain type of modernization centered on large systems linked to force projection models. The often overlooked insider influence of defense contractors on strategy, doctrine and technology will likely become more transparent. A recent report on a marketing meeting between Rockwell sales representatives and those of their new parent, Boeing, offers a sobering glimpse of how new defense strategy might be shaped (Opall, 1996: 26):

> Company marketing representatives will be instructed to focus not only on the sale of a specific system, say the ACU-130 gunship, but also on the battlefield notion of force projection. According to this philosophy, marketing representatives will be planting the seeds for future products within the corporate force projection portfolio, such as the F-22 stealth fighter, RAH-66 Comanche and Joint Strike Fighter.

There would be less to fear if the current security strategy and budget priorities were firmly grounded in more tangible security threats or a popular consensus about the U.S. role in the world. But the absence of a truly serious debate over fundamental security and defense issues leaves decision-making about strategy and weapons acquisition needs vulnerable to the concerted interests of the large contractors and their allies in the services and in Congress. The remaining big contractors will have considerable power over export policies, the domestic development projects chosen and the associated costs. Also, the range of technical approaches and choices available in defense technology will be reduced within the few, more isolated defense giants dominating new development work.

The Road Not Taken

The somewhat surprising success of defense diversification suggests that an alternative path existed which would have allowed more transfer of technology and know-how into rapidly growing non-DoD markets. A strong technology and defense industry base could have been sustained by large companies retaining profitable defense and commercial divisions and achieving some integration between the two markets. Smaller firms, left out of the merger wave have been relatively successful at expanding commercial sales (Feldman, 1997; Oden, forthcoming; Vernez et al., 1995). Many also continue to produce for defense. This also suggests that a strong industrial capacity could have been sustained through diversification.

The emphasis on merger and consolidation has certainly slowed the transfer of defense technologies and other assets to the commercial sector. The debt service pressures and internal disruptions associated with merger and consolidation slow down technology transfer and entry into alternative markets. Research has shown that by destroying teams and networks within companies, merger and downsizing slow down the product development process in general (Dougherty and Bowman, 1995; Baily et al., 1994; Challenge, 1995).

The idea of basing residual modernization needs on more commercial–military integration has also been set back by the merger wave. While improved defense production practices and increased use of lower cost commercial components is occurring in select cases, the overall result of most market extension mergers is to increase the segregation between defense and non-defense activities. There will be fewer opportunities to transfer technologies from commercial divisions within the firms because many commercial divisions have been sold off. The exposure of managers in the newly concentrated defense companies to commercial practices will also be reduced in many cases. Hence, accessing and designing around leading commercial technologies (spin-on) will likely become more difficult in the new, more segregated structure.

The process of defense restructuring seen over the last eight years only makes sense in the context of the overly ambitious objectives of the BUR, and now QDR. Each plan implies an immediate capability to produce all major weapons types and to develop new generation systems. While a more robust conversion policy or a more determined effort to integrate commercial and military suppliers through dual-use production could have conceivably supported a long-term latent capacity to develop next generation systems, maintaining a "hot" capability in each major weapons segment could not be readily ensured through these means. However, experts have failed to convincingly isolate a threat that would not allow significant response time for the relevant segments of a more integrated industrial base.

The cynical conclusion is that modernization proponents within the DoD and the contractor community are betting that defense budget plans for the early 2000 years can be pushed up to support a higher volume of new weapons development. The QDR simply put the big debate off for another year or two. The fight over money to sustain the extensive force structure versus more money for new weapons development will intensify in the next few years. The big contractors and their supporters within government will face-off against field commanders and base-dependent Congressional representatives. Hopefully, the real debate can also finally be joined—why exactly do we need either such a large force structure or major new weapons development?

NOTES

1. I want to thank Gene Porter of the Center for Naval Analyses for bringing this issue to my attention.
2. The analysis of industry structure in the text and tables was done prior to the final rejection of the proposed Lockheed Martin/Northrop Grumman merger by the DoD and the Justice Department in mid-1998. This intervention halted this final consolidation between giant contractors, however the top four (versus the top three) contractors still control close to one-third of all defense contracts with the private sector.

REFERENCES

"Federation of the American Scientists." *Arms Sales Control*, February 1995.
Baily, M., Bartelsman, E. and Haltiwange, J. "Downsizing and Productivity Growth: Myth or Reality." Working Paper No. 4741, Cambridge, MA: National Bureau of Economic Research, 1994.
Bischak, G. "The Implications of Alternative Security Doctrines and Policies for the Defense Science, Technology and Industrial Base." New York: Council on Foreign Relations, April 1996.
Challenge "Corporate Surveys Can't Find Productivity Revolution Either." *Challenge*, pp. 31–34, December 1995.
Department of Defense "100 Companies Receiving the Largest Dollar Volume of

Prime Contract Awards." Washington, DC: Department of Defense DIOR, FY 1990, Exhibit A, 1990.

Department of Defense, *Report on the Bottom-Up Review*. Washington, DC: Office of the Secretary of Defense and Joint Chiefs of Staff, September 1, 1993.

Department of Defense, *Defense Budget Estimates for FY 1997*. Washington, DC: Office of the Undersecretary of Defense (Comptroller), March 1996a.

Department of Defense, "100 Companies Receiving the Largest Dollar Volume of Prime Contract Awards." Washington, DC: Department of Defense DIOR, FY 1996), Exhibit A, 1996b.

Dougherty, D. and Bowman, E. "The Effects of Downsizing on Product Innovation." *California Management Review*, Vol. 37(4), pp. 28–44, Summer 1995.

Dunne, T. R., Roberts, M. and Samuelson, L. "Patterns of Firm Entry and Exit in U.S. Manufacturing Industries." *The Rand Journal of Economics*, Vol. 19, pp. 495–515, 1989.

Feldman, J. "Diversification after the Cold War: Results of the National Defense Economy Survey." Working Paper 112, Center for Urban Policy Research, Rutgers University, 1997.

Finnegan, P. *Defense News*, pp. 3, 42, February 10–16, 1997.

Finnegan, P. and Marburger, L. "Top 100 Worldwide Defense Firms." *Defense News*, July 1990, 1997 (various years).

Fulghum, D. "Critics Slap Review for Fiscal Timidity." *Aviation Week & Space Technology*, p. 22, May 19, 1997.

Gansler, J. *Defense Conversion: Transforming the Arsenal of Democracy*. New York: Twentieth Century Fund, 1995.

Hartung, W. *Welfare for Weapons Dealers: The Hidden Costs of the Arms Trade*. New York: World Policy Institute, 1996.

Korb, L. "Merger Mania." *The Brookings Review*, pp. 22–25, Summer 1996.

Lundquist, J. "Shrinking Fast and Smart." *Harvard Business Review*, pp. 74–85, November–December 1992.

Opall, B. "Boeing To Indoctrinate Rockwell on New Marketing Path." *Defense News*, p. 26, December 2–8, 1996.

Oden, M. "Cashing-in, Cashing-out and Converting: Restructuring of the Defense Industrial Base in the 1990s." In Markusen, A. and Costigan, S., eds., *Arming the Future: A Defense Industry for the 21st Century*, forthcoming.

Oden, M., Hill, C., Mueller, E., Feldman, J. and Markusen, A. "Changing the Future: Converting the St. Louis Economy." Working Paper 59, Center for Urban Policy Research, Rutgers University, 1993.

Oden, M., Markusen, A., Flaming, D. and Drayse, M. "Post Cold War Frontiers: Defense Downsizing and Conversion in Los Angeles." Working Paper 105, New Brunswick, New Jersey: Center for Urban Policy Research, Rutgers University, 1996b.

Office of the Secretary of Defense. *Vertical Integration and Supplier Decisions*. Washington, DC: OSD, May 1997.

Ravinscroft, D. and Sherer, F. *Mergers, Sell-offs, and Economic Efficiency*. Washington, DC: Brookings Institution, 1988.

Rogerson, W. "Profit Regulation of Defense Contracts and Prizes for Innovation." *Journal of Political Economy*, No. 97, pp. 1284–1389, December 1989.

Rogerson, W. "Economic Incentives and the Defense Procurement Process." Working

Paper 93-33, Center for Urban Affairs and Policy Research, Northwestern University, 1993.

Shenon, P. "Pentagon Urges Trims in Military and More Base Closings." *New York Times*, p. A-18, May 20, 1997.

Summers, H. "Are U.S. Forces Overstretched? Operations, Procurement and Industry Base." *Orbis*, Vol. 41(2), pp. 199–207, Spring 1997.

U.S. General Accounting Office (GAO) "Future Years Defense Program, Optimistic Estimates Lead to Billions in Overprogramming." GAO/NSAID-94-210, July 1994.

U.S. General Accounting Office (GAO) "Defense Restructuring Costs: Payment Regulations Are Inconsistent With Legislation." GAO/NSAID-95-106, August 10, 1995.

U.S. General Accounting Office (GAO) "Defense Restructuring Costs: Information Pertaining to Five Business Combinations." GAO/NSAID-97-97, 1997.

Velocci, A. "Defense Firms Show Financial Prowess." *Aviation Week and Space Technology*, pp. 40–60, May 30, 1994.

Velocci, A. "U.S Industry Poised for Further Realignment." *Aviation Week and Space Technology*, pp. 10–11, December 23–30, 1996.

Vernez, G. *et al*. "California's Shrinking Defense Contractors: Effects on Small Suppliers." Santa Monica: Rand Corporation, MR-687-OSD, 1995.

12

DEFENSE DIVERSIFICATION: CAN WE RETURN TO THE ROAD NOT TAKEN?

ERIK R. PAGES

INTRODUCTION

Touting his new "Operation Restore Jobs," President Clinton announced his administration's defense conversion initiatives to great acclaim in early 1993. At the time, many observers expected that Clinton's $20 billion package of programs would help encourage diversification among defense-dependent firms. Indeed, the flagship of "Operation Restore Jobs," the Technology Reinvestment Project (TRP), was widely touted as a model for government–industry partnerships to enhance American competitiveness.

Nearly five years later, the great enthusiasm engendered by the TRP seems ill-conceived. While many defense firms have diversified, the primary response of U.S. business has been consolidation and/or exit from the defense sector. While Washington dickered over the politics of technology policy, industry acted, initiating a painful downsizing and restructuring process that has completely transformed the American defense sector.

This chapter examines the factors that produced a consolidated (as opposed to diversified) American defense sector. Was diversification a feasible outcome? Could government and industry have pursued an alternative approach that emphasized a stronger corporate commitment to diversification and dual-use? Would such a strategy make sense?

I conclude that government encouragement of defense diversification was a case of "too little, too soon." Financial constraints (in both Washington and Wall Street) created unavoidable pressures in favor of a consolidation strategy. Even though programs like the TRP were popular and

173

well-managed, their funding levels were simply too minimal and their objectives too proscribed to replace a collapsing defense market.

In hindsight, consolidation seemed inevitable. Yet, this business strategy has helped foster an industry structure where vertical integration could pose significant problems in the future. In contrast to the early 1990s, a diversification/dual-use strategy may be more successful in today's consolidated and vertically integrated defense sector. Building on the lessons of the TRP, a new dual-use effort can effectively promote diversification while strongly supporting military needs.

TRENDS IN DEFENSE CONSOLIDATION

The process of industry consolidation has received extensive press coverage (Dowdy, 1997: 88–96; Velocci, 1996c: 10–12; Grant, 1997). Such heavy coverage makes sense as both the pace and extent of change are truly breathtaking. Within a short period of five years, the American defense industry has been completely transformed.

The most prominent changes have occurred in the aerospace sector where players have consolidated down to three major firms. Three "gorillas" have emerged: Boeing, Lockheed Martin, and Raytheon. They replace an industry that numbered twenty-six major contractors in 1946, and, more recently, numbered eight major players in 1985.

Other sectors have also been touched. In defense electronics, 25 major industry participants in 1985 have been replaced by 13 major players in 1997. In 1996 and 1997 alone, 14 major defense electronics firms merged down to nine (Dowdy, 1997: 90). Many of the old "household names" of defense, such as Westinghouse, and Rockwell, have exited the business.

The value of acquisitions has been similarly impressive. When the merger and acquisitions wave first began in 1992, the total value of all mergers reached $2.5 billion. With the recently announced merger of Northrop Grumman by Lockheed Martin, the value of M&As in 1997 will exceed $53 billion (Donnelly and Clark, 1997: 1, 13).

WHY DID CONSOLIDATION OCCUR?

The reason for consolidation is fairly simple: market shrinkage in the range of 50 percent. While overall spending on defense procurement remains at levels similar to average Cold War spending levels, the period between 1985 and 1995 represents the longest consistent decline in defense procurement since World War II (see U.S. Congress, GAO, January 1997). With such large cuts in defense procurement, companies could not persist with business as usual.

Huge declines in U.S. defense budgets were a necessary cause of industry consolidation. These budget reductions were probably a sufficient cause as well. However, basic market trends were reinforced by strong U.S. government encouragement. This support took several forms:

Exhortation: Active encouragement of consolidation from top Pentagon officials.

Flexible antitrust rules: Approval of all proposed mergers and acquisitions.

Merger subsidies: Financial support to encourage rationalization.

The Clinton Pentagon has been remarkably aggressive in its encouragement of market consolidation. This effort started at the top with former Secretaries of Defense Les Aspin and William Perry. In one of his first interviews as Deputy Secretary of Defense, Perry claimed that: "the difference between the previous Administration and ourselves is [that] . . . we're willing to commit to a defense industrial policy" (Perry, 1993: 40–43).

The famed "Last Supper" of 1993 is the best known of these efforts to "jawbone" the industry toward consolidation. At a small Pentagon dinner with leading defense company CEOs, Aspin and Perry prophesied that, within five years, at least half of the companies represented at the dinner would no longer be needed to support defense needs. They further noted that they expected firms to go out of business, and that DoD would not intervene to prevent such actions (described in Augustine, 1997: 85).

The Pentagon's strong support for mergers and consolidation has remained remarkably consistent, and top DoD officials remain pleased with the results of this process. Paul Kaminski, former Undersecretary of Defense for Acquisition and Technology, recently provided an apt summary: "I am happy with what's happened in the restructuring of the industry. We will be better serviced by two or three robust primes (contractors) competing than by five or six unhealthy primes" (Pearlstein and Mintz, 1997).

The use of the "bully pulpit" was supplemented by two other policies: use of the carrot through subsidies for mergers, and removal of a potentially powerful stick via more flexible enforcement of antitrust rules affecting mergers.

The use of subsidies for mergers proved most controversial. This initiative began in July 1993 when DoD issued a new policy that provided reimbursement to defense firms for restructuring costs. Through a complex procedure, the policy authorizes the DoD to share cost savings that a contractor generates through organizational restructuring and downsizing (for a detailed description, see U.S. Congress, GAO, April 1996).

This subsidy policy was derided by opponents as "payoffs for layoffs." As expected, the defense industry has applauded these efforts as a rational means to encourage needed consolidation. In fact, in industry's view, rationalization might not occur without such subsidies. As James Skaggs, Chairman of Tracor, a major defense contractor noted, efforts to eliminate these subsidies "could dampen the enthusiasm of some (companies) to participate in the consolidation process" (Velocci, 1996b: 30). The reimbursement of restructuring costs appears to have been especially important in finalizing Raytheon's purchase of Hughes Electronics (Bulkeley, 1997: C2).

While opposition to this policy remains quite strong, the program has survived several Congressional challenges. Most recently, in July 1997, the Senate soundly defeated (by a vote of 83 to 15) an amendment that would have ended the reimbursement policy; a similar proposal was defeated in the House National Security Committee.

As of March 31, 1997, DoD had estimated its share of projected restructuring costs at approximately $755 million. These costs included the disposal and relocation of facilities and equipment, relocation of employees, benefits and services for laid-off workers, and other services. Estimated savings from consolidations were projected to exceed $3.3 billion (see figures cited in U.S. Congress, GAO, April 15, 1997).

In addition to the carrot of these subsidies, the Clinton team also encouraged consolidation through more flexible antitrust rules. Antitrust regulators have recognized that defense is different. With only one buyer—the Department of Defense—traditional concerns about competition and consumer prices prove less compelling in the defense sector. Thus, a less aggressive stance has been used in reviewing defense M&As.

The Clinton team's revised approach to antitrust rules for defense grew out of an extensive debate begun in the last years of the Bush administration. This controversy was triggered by the Federal Trade Commission's (FTC) November 1992 decision to block Alliant Techsystems, Inc., from acquiring Olin Corporation's Ordnance Division. The FTC justified this action by contending that the new firm would become a monopoly ammunition supplier for the Army.

The FTC's decision sparked a huge outcry by the affected industries, prompting the creation of a special Defense Science Board (DSB) Task Force, headed by current FTC Chairman Robert Pitofsky, to create new guidelines for regulating defense M&As (U.S. Department of Defense, April 1994). Prior to the DSB report, the DoD had little role in reviewing or commenting on industry mergers on military and defense industrial base grounds. As a result of this initiative and pressures from industry, the Clinton administration expanded DoD's voice in reviewing M&As and antitrust authorities generally assumed a more relaxed posture on reviewing defense M&As.

These procedural changes had significant effects. Between 1994 and 1997, DoD and the antitrust agencies reviewed 34 different defense M&As. While seven cases required some alterations and remedies to address antitrust concerns, all of these transactions were approved (U.S. Department of Defense, May 1997).

DIVERSIFICATION: THE ROAD NOT TAKEN

In its first two years, the Clinton administration pursued a Janus-like policy toward the defense industry. The U.S. government did more than simply promote consolidation. It also made a concerted effort to promote diversification among defense firms. In fact, its defense conversion programs received much greater public attention than the efforts to support consolidation.

First announced in March 1993, Clinton's defense transition program contained carrots for a host of constituencies: business, labor, peace activists, and impacted communities. Moreover, the program appeared to be backed with real money. Nearly $22 billion was earmarked for defense reinvestment and conversion between FY1993 and FY1997.

This $22 billion included funding for a host of programs, including retraining assistance, military base closure support, and civilian R&D programs. For defense businesses, however, expansive rhetoric was not followed with significant real benefits. In fact, new funds to support business diversification were relatively limited. The cornerstone of the Clinton program, the Technology Reinvestment Project, received only $605 million at its highest level.

TRP's basic strategy was to "leverage commercial market size, technological know-how, and investments for military benefit . . . by supporting R&D projects to develop dual-use technology" (U.S. Department of Defense, 1995). The program supported these objectives through R&D funding for projects that promoted both spin-off and spin-on. Spin-off referred to projects that helped shift defense-funded technologies into new commercial applications. For example, one TRP project supported the use of advanced composites for bridge repair. Spin-on projects supported commercial technologies that might provide superior benefits to the military. Examples of spin-on technologies included flat panel displays, software, and high-density data storage.

TRP managers, based at DoD's Defense Advanced Research Projects Agency (DARPA), were originally overwhelmed with demand for its limited funds. More than 12,000 companies entered into partnerships to compete for the first round of TRP funding. Of this total, 212 projects supporting 1600 companies ultimately received funding. Because of this strong competition and political pressures to widely disperse funds, companies saw very little immediate benefit to their corporate bottom lines. For example, GM/Hughes was a partner in 16 successful TRP projects valued at $208 million. After sharing funds with various partners and reimbursing costs for proposal preparation, the firm received $11 million (Stowsky, 1996).

In addition to these financial limits, TRP was further hampered by its focus on R&D funding. Limiting TRP grants to R&D support made sense on a variety of grounds, however, such restrictions further reduced the utility of these grants in promoting diversification. Because successful projects would take years to bring products to markets, R&D grants did little to assist in promoting diversification. More basic forms of business assistance, such as access to financing, marketing assistance, and other programs, have proved more helpful in aiding diversification.[1]

TRP survived for a period of three years (1993 to 1995). It was suspended in February 1995 in the face of strong opposition from the new Republican Congress.[2] Congressional critics emphasized two critiques of the TRP. First, they contended that TRP was part of a Clinton administration "industrial policy" that falsely relied on government intervention to replace the power of market forces. Second, they claimed that TRP was an example of "non-defense" spending inside the defense budget. With tight defense budgets, the Pentagon could ill afford R&D efforts with limited direct benefits to the military.

These various forces presented defense firms with a relatively simple decision calculus. On the one hand, the defense market was imploding and ultimately lost more than 50 percent of its value since the late 1980s. Meanwhile, top DoD officials were actively encouraging consolidation and promoting policies that solidly backed this rhetoric. On the other hand, firms saw a strong rhetorical commitment to diversification but little hard backing. While the TRP was a well-managed and designed program, it was like the proverbial finger in the dike when stacked up against pressures for consolidation and downsizing.

Given the difficulties of diversification in any industry and even under the best economic circumstances, the decision for defense prime contractors was relatively simple. Consolidate, downsize and merge to achieve a new structure that is more optimal for a smaller but still robust defense market.

WHERE ARE WE TODAY?

As noted above, the pace of defense industry consolidation has been breathtaking. After nearly seven years of aggressive downsizing and consolidation, a new industry structure has emerged. Three huge defense conglomerates—Boeing, Lockheed Martin, and Raytheon—dominate prime contracting. A host of other large firms, such as Tracor, United Technologies, General Dynamics, fall in a second tier below these behemoths. Finally, as in the past, tens of thousands of subcontractors continue to do business with the Pentagon, supplying everything from computer chips to potato chips.

Although defense has always been characterized by heavy concentration of suppliers, today's concentration levels are at historic highs. In 1995, prior to many of the defense mega-mergers, only eight firms accounted for two-thirds of all defense product sales. Concentration levels have subsequently grown as four of these eight firms—Boeing, McDonnell Douglas, Raytheon, and Hughes—were involved in major post-1995 mergers. By contrast, 17 firms accounted for two-thirds of sales in 1989.[3] Citing another measure of concentration, Paine Webber analyst Jack Modzelewski notes that Lockheed Martin now receives one out of every four DoD procurement dollars (Donnelly and Clark, 1997: 1, 13).

U.S. industry's rapid consolidation has been applauded in both government and business circles. It is also envied overseas, where Western European contractors complain that home governments block their ability to rationalize and consolidate across borders (Tigner, 1996: 36; Gregory, 1997: 20–24).

The American defense industry has now achieved a somewhat stable end-state. The July 1997 announcement of Lockheed Martin's purchase of Northrop Grumman is likely to be the last of the mega-mergers. With the bulk of consolidation completed, industry analysts and DoD officials are now beginning to examine the ramifications of this new industry structure.

While consolidation was unavoidable, several dangers could emerge in the newly merged and concentrated defense industry: These include:

- *an increase in defense costs*
- *a loss of technological competition*
- *further erosion of the subcontractor base.*

Industry consolidation will reduce the positive cost effects of contractor competition. With only one or two major suppliers in each market segment, market pressures to reduce costs will prove less powerful. DoD has already acknowledged this problem in its 1996 Report to the Congress which stated: "Consolidation carries the risk that DoD will no longer benefit from the competition that encourages defense suppliers to reduce costs, improve quality, and stimulate innovation" (U.S. Department of Defense, 1996: 74).

At present, many analysts believe that sufficient competition can be maintained over the short-term (Kovacic and Smallwood, 1994: 91–10). Because of time lags between the initiation of new programs and the beginning of actual production, potential problems may be delayed for several years. However, the financial effects of this declining competition are likely to pose a growing problem in certain industrial sectors over the next decade. Given the realities of reduced competition and less stringent regulation, one should not expect major long-term cost savings to emerge from the process of industry consolidation.

While recognizing that cost and price effects of reduced competition are important, defense planners express more serious concerns about consolidation's effects on technological innovation. As Kovacic and Smallwood have noted: "The main potential hazard of mergers is the danger that technological competition will diminish, and that specific technologies may become entrenched as the one or two remaining suppliers freeze out innovative design approaches that threaten their vested interests or defy conventional wisdom" (ibid: 102–103.)

The real danger of reduced competition results from the potential for "technological lethargy, myopia, or error" (ibid). Since American military prowess has traditionally relied upon technological superiority, a continuous flow of new ideas and technologies is required to maintain military effectiveness. Scherer (1964) has shown that the presence of competition is crucial to the development of new ideas and concepts in the defense arena. The absence of competition may "freeze" military technologies and inhibit innovation.

Problems related to cost and technological competition are largely related to horizontal integration, i.e., the merger of previously competing firms. DoD also faces potential dangers related to vertical integration, which refers to the process whereby firms are able to produce many of the subsystems or components used in their products.[4]

As large prime contractors become more vertically integrated, many DoD officials fear that contractors' preference for internal suppliers will have a negative ripple effect on smaller subcontractors. Such a process could lead to an erosion of the defense industrial base, worsening a loss of the subcontractor base already triggered due to budget cutbacks. This trend poses a serious potential problem. DoD has accepted the prospect of minimal competition at the prime contractor level. However, acquisition officials do hope, as Paul Kaminski has noted, that, "if we have limited competition among our prime contractors, then what we will want is vigorous competition at the sub-tiers" (Meadows, 1997: 18).

At present, these problems appear manageable. Indeed, the recent Defense Science Board study on vertical integration concluded that there is limited evidence that integration now poses a "systemic problem." However, the report also noted that the situation warrants caution.

Today, many prime contractors are gaining capabilities that represent an "inter-related system of systems."[5] In other words, firms maintain expertise in all of the various components and technologies that comprise a weapons system. For example, Lockheed Martin is now capable of supporting space launch, launch ground services, satellites, and related communications links. In the past, different contractors specialized in various system areas.

The real dangers of vertical integration may emerge if firms use their market power to disadvantage competitors. The Defense Science Board Integration Study lays out two potential scenarios:[6] (1) Firms refuse to purchase or sell important technologies or products to rivals, or (2) Firms exclude outside suppliers and contract in-house for the supply of components and sub-tier work in major systems contracts.

While some analysts have pointed to examples of such behavior, no clear trends of exclusionary behavior have yet emerged. However, the potential for problems does exist. Such concerns are particularly appropriate when one considers the time-line for consolidation efforts. It has generally taken 18–24 months for large-scale mergers to impact lower-tier suppliers. Since consolidation is a relatively new phenomenon, we can only speculate on some of its longer-term effects. For example, the DSB report was published prior to the completion of four of the largest defense mergers—Boeing-McDonnell Douglas, Raytheon-Texas Instruments, Raytheon-GM Hughes, and Lockheed Martin-Northrop Grumman. Thus, the DSB's present sanguine views may require reassessment after these firms complete their internal reorganization and restructuring efforts.

THE CASE FOR DIVERSIFICATION

All of these potential problems have a similar cause: too few suppliers chasing too few contract opportunities. Given declining defense budgets, this problem was unavoidable. However, accepting its consequences is not. DoD and the industry must begin to take steps to ensure a competitive defense marketplace in the future.

Industry analysts have proposed a number of means for encouraging greater competition in the defense marketplace. These suggestions have run the gamut of possibilities. For example, Sapolsky suggests that expanded inter-service competition could serve as a proxy for competition between contractors (Sapolsky, 1997: 50–53). Creation of independent design bureaus or laboratories offers another means for introducing new ideas into the weapons development process. Aggressive use of prototyping might also help support these objectives.

Competition can also be enhanced through the contracting process itself. Program managers must consider acquisition strategies that help foster the creation of new competitors. In some cases, DoD might be able to separate major platform contracts into discrete units where separate competitions could be held for major subsystems, training, and support services. Under this proposal, DoD would shift away from the current practice of bundling procurements into one solicitation. Unbundling procurements would create new opportunities for small and medium-sized firms.

Finally, DoD should consider returning to the concept of dual-use and beginning an active effort to promote diversification among defense firms.

WHY DIVERSIFICATION?

When industry began its consolidation drive, diversification was a dirty word. A 1991 statement of William Anders, then CEO of General Dynamics, was typical: "Defense industry management teams have little commercial experience and market savvy. . . . most don't bring a competitive advantage to non-defense business. Frankly, sword makers don't make good and affordable plowshares" (Anders, 1991).

Anders' points were echoed by his colleagues in the boardrooms of major contractors, and aggressive consolidation was the adjustment strategy of choice. However, diversification's bad reputation was not fully deserved. Many defense-dependent firms successfully pursued diversification efforts.

Small and medium-sized firms proved especially nimble in moving into non-defense markets (for background on some of these success stories, see Oden, 1996; Feldman, 1997; Pages, 1993). Since these firms form an extremely heterogeneous group, the causes of their successes vary. However, survey data point to several important causal factors. Firms with the greatest diversification success were more likely to retain in-house R&D capacity, collaborate with outside firms, and seek outside (public and private) assistance with marketing and R&D (Feldman, 1997).

Diversification successes are not limited to small and medium-sized firms. Many of the largest prime contractors also made progress in entering non-defense markets. For example, Hughes enjoyed great success with its Direct-TV home satellite dish systems. Similarly, Rockwell International completed its transition from a defense firm to a commercial company with its August 1996 sale of its aerospace and defense divisions to Boeing (Forenski, 1997). Rockwell's semiconductor division is now thriving, moving from an emphasis on centralized mainframe computers to the production of communications chips for modems and other devices.

Now that consolidation efforts have reached their end-state, contractors appear more willing to consider diversification strategies. Under CEO Phil Condit, Boeing has been a leader on this front, marking a major change from its company strategy of only a few years ago (Proctor, 1997: 20–21; Bryant, 1997: D1, 19). Under its current 20-year plan, called Vision 2016, Boeing expects to be an integrated aerospace company with three core competencies when it reaches its 100th anniversary in 2016. These three core competencies—detailed customer knowledge and focus, complex system integration, and lean and efficient design and production—are based on an aggressive program to move Boeing into new markets beyond its traditional focus on commercial and military aircraft production. Boeing is particularly interested in service businesses, such as airline maintenance, airport development, and data-handling.

Boeing's new strategy has not been uniformly applauded on Wall Street, where analysts cite past diversification failures. Nonetheless, Condit's strategy has received a somewhat warmer reception because of its emphasis on related diversification. Instead of moving into new markets, such as furniture (1920s) or wind turbines (1970s), Boeing is now focused on new markets that build on existing aerospace-related competencies.

Boeing is not alone in actively considering diversification. Even Lockheed Martin, the industry's leader in consolidation, expects its future growth to occur outside of its defense businesses. Norman Augustine, Lockheed Martin's former Chairman and CEO, admitted in July that "the ability to grow in a major way domestically (in defense) will be difficult (Finnegan, 1997: 6). Like Boeing, Lockheed Martin expects to focus on related diversification, with an emphasis on government contracts in areas like information technology, training and technical services.

Just as consolidation made sense in the early 1990s, diversification makes sense today. With little prospect for major defense budget increases, contractors cannot expect major growth in defense. Contractors now face a major dilemma as they consider significant structural changes. As one analyst put it: "Management (must) show that they can exploit their company's talent, technology, and financial strength to add a factor of growth that investors can identify and underwrite . . . (Defense) can(not) generate enough incremental business to stimulate real growth" (Velocci, 1996a: 21).

This pressure to grow has two outlets: defense export markets and related diversification. American defense contractors are moving aggressively to enter foreign markets, and have succeeded in capturing a large chunk of the world's defense business. Nonetheless, this export outlet appears limited as a long term proposition. Indeed, the prospects for a transatlantic defense trade war appear to be growing (Grant, 1997: 111–137; Pages, 1995: 135–152). Should this gloomy scenario come to pass, interest in diversification will increase.

DIVERSIFICATION—NOW MORE THAN EVER

These two trends—consolidation, and a growing interest in diversification— create a new environment that may be more conducive to dual-use programs like the TRP. These programs were first introduced in an environment where the benefits to industry were uncertain and the incentives for Pentagon support were limited. For industry, aggressive diversification made little sense when consolidation appeared both inevitable and very profitable. For DoD and the military services, support for dual-use also made little sense. With R&D budgets declining, their support for using R&D funds to foster diversification was limited.

These two conditions no longer exist. Industry consolidation has reached its endpoint. As a result, interest in diversification is growing. At the same time, the DoD is recognizing potential problems due to a lack of competition in its supplier base. While R&D budgets are not growing rapidly, a renewed commitment to dual-use might help create new sources of supply and technological ideas.

In effect, this new dual-use initiative would constitute a partial return to the original pre-Clinton vision for the Technology Reinvestment Project. When first conceiving the program, TRP designers focused on the program's ability to foster commercial–military integration, the creation of a single unified industrial base for commercial and military technology development. Ultimately, TRP sought to create new sources of (affordable) supply for defense needs. Because these suppliers would be commercially based (and thus less dependent on defense), DoD could benefit from the lower costs achieved through more efficient commercial production and economies of scale.

A new dual-use initiative would explicitly focus on creating new sources of technology and ideas for DoD. This approach might involve spin-on or spin-off strategies, but concerns regarding the transition of technology from military to civilian markets (or vice versa) would remain a secondary concern for the program. In other words, diversification would be a byproduct of the program, not its explicit purpose. Instead, dual-use would serve as the linchpin of a strategy to counteract the potential downsides of industry consolidation.

HOW CAN WE BRING BACK DUAL-USE?

This new dual-use initiative would build on the many useful lessons derived from our experiences with the TRP. While the TRP was short-lived, it did foster a number of achievements. Most importantly, it proved that the concept of dual-use makes sense. Funding for dual-use R&D has been shown to create new technologies, processes, and products that would not otherwise happen. TRP served as a "proof-of principle" for the concept of dual-use.

TRP also introduced several important innovations into the management of technology policy. Its emphasis on cost-sharing with industry, partnerships between private firms, and on focused competition have all proved effective and should be replicated in other programs. Finally, DARPA has pioneered various tools for acquiring research by non-traditional means. Using cooperative agreements and "other transactions" authority helps reduce barriers between defense and civilian work. This flexibility makes defense research programs more attractive to private firms which have traditionally feared the heavy paperwork burdens and tight intellectual property restrictions found in government contracts.[7]

The program's structure will require some alterations. One of the more persistent criticisms of many TRP projects concerned their apparent lack of direct ties to military needs. Moreover, since TRP funds were managed by DARPA, the military services gave TRP a lukewarm reception. When budget cuts were proposed, any dual-use initiatives were the first on the cutting block. The absence of the military's "buy-in" was a critical TRP shortcoming.

Any future dual-use initiative must be fully integrated into the R&D initiatives of various military services. The recent creation of a Joint Dual-Use Program Office (JDUPO)[8] within the Office of the Secretary of Defense makes sense. At present, JDUPO's primary responsibility is to oversee the Commercial Operations and Support Savings Initiative (COSSI) which seeks to insert commercial products and processes into fielded systems to help reduce costs.

While COSSI is a laudable program, JDUPO should assume a more aggressive role in support of dual-use and diversification. Funds should be expended to help support R&D and production projects proposed by individual military services that help promote dual-use and diversification. Instead of simply improving the operations of existing systems, JDUPO must also consider means to help support new sources of technology and ideas.

For much of this effort, program categories could remain purposely broad to encourage a wide search for solutions to somewhat generic problems. Existing DARPA programs operate according to this principle. DARPA's new biodefense project offers a case in point (Marshall, 1997: 744–746). This program seeks projects that help detect biological agents on the battlefield, promote body shielding techniques, and manage responses to attacks. DARPA does not specify technologies to be employed; it simply requests technological ideas for tackling these generic problems.

As noted earlier, a new dual-use program would emphasize creation of new supply sources by funding projects to tackle emerging generic problems. It would not seek to promote diversification through support for financing, marketing, and other needed services. These programs make sense, but should be operated out of non-defense agencies such as the Department of Commerce, the Small Business Administration, and, most importantly, their state and local counterparts.

The JDUPO, would not actually initiate programs, but would instead sponsor projects started with the military services. Actual project funding would be a mix of private, military service, and separate DoD dual-use funds. The dual-use office's primary roles would be to serve as the Pentagon's voice for dual-use by providing education and training to the military services, to identify future problem areas worthy of investigation, and to use separate funding as a carrot to encourage more aggressive support for dual-use throughout the military.

CONCLUSIONS

One of the primary lessons of TRP is that dual-use works. However, a variety of external pressures doomed TRP from the start. Financial pressures left defense contractors with few options but to downsize and consolidate. Even if firms had aggressively pursued diversification efforts, it is unlikely that such consolidation could have been avoided. In this sense, TRP's critics had a valid point. The program's limited pool of R&D resources was simply overwhelmed by the myriad pressures leading down the path to defense consolidation. Strong political opposition to the Clinton administration's technology policies reinforced these market pressures and ultimately doomed any aggressive dual-use programs.

Today, both the political market and economic market have changed. Politically, we can still expect opposition to dual-use (and any activist technology policies) within some parts of the Congress. However, as concerns about defense industry consolidation develop, support for dual-use inside the military and among its supporters in Congress may grow. Military support for dual-use may help counteract any ideological opposition to more aggressive defense industrial base policies. As dual-use becomes more institutionalized as a defense program—as opposed to a high-profile Clinton–Gore initiative—the political heat may cool down.

The economic marketplace for dual-use is also more favorable. With the age of defense mega-mergers complete, contractors are searching for new ways to create shareholder value. Defense markets, both at home and abroad, may not offer sufficient growth opportunities. As a result, interest in commercial markets will continue to grow. A new dual-use initiative cannot force this shift; however, it can reinforce existing market trends.

If this new dual-use initiative is carefully structured and designed, it will bring benefits to all interested parties. For the military, it can help create competition for new ideas and technologies. Similarly, it may contribute to slowing the erosion of the defense subcontractor base. For industry, dual-use will not directly create new markets. However, it can stimulate new ideas and help support risky, but potentially lucrative, new research projects. In today's changed defense environment, it is a win-win proposition for all.

NOTES

1. These types of business development initiatives fall far outside of DARPA's traditional focus on supporting cutting edge R&D. Other federal agencies, such as the Economic Development Administration, Small Business Administration, and state and local governments are better suited to this task. Indeed, some TRP critics advocated expanded support for these economic development initiatives and a reduction of funding for new R&D (see Oden *et al.*, 1995).
2. TRP has been replaced by the Dual-Use Assistance Program (DUAP). DUAP, which was funded at $185 million dollars in FY1997, is explicitly focused on supporting military R&D needs.
3. DoD analysis presented in Report of the Defense Science Board Task Force on Vertical Integration and Supplier Decisions, Washington, DC: U.S. DoD, p, 10, May 1997; hereafter referred to as DSB Integration Study. These data are drawn from a Defense Contract Audit Agency (DCAA) historical database and are based on company-reported sales accounting data. The database includes all firms which, in a single fiscal year, have claimed more than $10 million in defense IR&D costs or who have auditable DoD costs exceeding $70 billion. These data include firms' sales of defense products but do not represent all sales (products without direct military connections, like medical services, petroleum, and transportation are excluded) to DoD, or all firms selling to DoD.
4. DSB Integration Study, p. 2–3.
5. DSB Integration Study, p. 9.
6. Ibid.
7. These various contracting tools enable more flexible terms and conditions than standard financial management and intellectual property provisions found in most DoD contracts. For background, see U.S. General Accounting Office, *Acquiring Research by Non-Traditional Means*, NSIAD-96-11, March 1996.
8. JDUPO runs the Dual-use Applications Program, the successor to TRP. The JDUPO includes representatives from the Undersecretary of Defense (Acquisition and Technology), the Service Acquisition Executives, and the Director, Defense Research and Engineering.

REFERENCES

Anders, W. A. "Rationalizing America's Defense Industry." Keynote Address to Defense Week 12th Annual Conference, October 30, 1991.

Augustine, N. R. "Reshaping an Industry: Lockheed Martin's Survival Story." *Harvard Business Review*, pp. 83–94, May–June 1997.

Bryant, A. "Boeing Has Its Feet on the Ground." *New York Times*, pp. D1, 19, July 22, 1997.

Bulkeley, W. M. "Loss of Reimbursement Program May Threaten Raytheon Deals." *The Wall Street Journal*, p. C2, April 14, 1997.

Donnelly, J. and Clark, C. "Merger Mania Hits $53 Billion This Year—So Far". *Defense Week*, pp. 1, 13, July 7, 1997.

Dowdy, John J. "Winners and Losers in the Arms Industry Downturn." *Foreign Policy*, pp. 88–96, Summer 1997.

Feldman, J. M. "Diversification after the Cold War: Results of the National Defense Economy Survey." New Brunswick, NJ: Center for Urban Policy Research, February 1997.

Finnegan, P. "Lockheed Martin's Growth Lies Outside of Defense." *Defense News*, p. 6, July 21–27, 1997.

Forenski, T. "Bold Move by Rockwell." *Financial Times*, April 2, 1997.

Grant, C. "Linking Arms" (Survey of the Global Defence Industry). *The Economist*, June 14, 1997.

Grant, R.P. "Transatlantic Armament Relations Under Strain." *Survival,* Vol. 39(1), pp. 111–137, Spring 1997.

Gregory, Bill. "Time is Running Out." *Armed Forces Journal International*, pp. 20–14, 1997.

Kovacic, W. E. and Smallwood, D. "Competition Policy, Rivalries, and Defense Industry Consolidation." *Journal of Economic Perspectives*, Vol. 8(4), pp. 91–110, Fall 1994.

Marshall, E. "Too Radical for NIH? Try DARPA." *Science*, Vol. 275, pp. 744–746, February 7, 1997.

Meadows, S. "Pentagon Begins to Wrestle with Industry Consolidation Aftermath." *National Defense*, p. 18, July–August 1997.

Oden, M. "Cashing-in, Cashing-out, and Converting: Restructuring the Defense Industrial Base in the 1990s." Paper presented to the Council on Foreign Relations Study Group on Consolidation, Downsizing and Conversion of the U.S. Military Industrial Base, February 1996.

Oden, M., Bischak, G. and Evans-Klock, C. The Technology Reinvestment Project: The Limits of Dual-Use Technology Policy. Washington, DC: National Commission For Economic Conversion and Disarmament, July 1995.

Pages, E. R. "How Defense Contractors Can Survive in a Peaceful World." *Business and Society* Review, No. 86, Summer 1993.

Pages, E. R. "The Future U.S. Defense Industry: Smaller Markets, Bigger Companies, and Closed Doors." *SAIS Review*, Vol. 15(1), pp. 135–152, Winter–Spring 1995.

Pearlstein, S. and Mintz, J. "Too Big to Fly?" *The Washington Post*, pp. H1, H6, May 4, 1997.

Perry, W. "Guarding the Base" (Interview). *Government Executive*, pp. 40–43, August 1993.

Proctor, P. "New Strategic Focus Drives Boeing Transformation." *Aviation Week & Space Technology*, pp. 20–21, April 28, 1997.

Sapolsky, Harvey M. "Inter-service Competition: The Solution, Not the Problem." *Joint Forces Quarterly*, pp. 50–53, Spring 1997.

Scherer, F. M. *The Weapons Acquisition Process: Economic Incentive*. Cambridge, MA: Harvard University Press, 1964.

Stowsky, J. "America's Technical Fix: The Pentagon's Dual-Use Strategy, TRP, and the Political Economy of U.S. Technology Policy." Paper prepared for the Council on Foreign Relations Study group on Consolidation, Downsizing and Conversion of the U.S. Military Industrial Base, January 6, 1996.

Tigner, B. "EU Officials Fear U.S. Threat to Aerospace, Defense Base." *Defense News*, p. 36, November 11–17, 1996.

U.S. Congress, General Accounting Office. Defense Contractor Restructuring: First Application of Cost and Savings Regulations. Washington, DC: GAO, April 1996.

U.S. General Accounting Office. Acquiring Research by Non-Traditional Means, NSIAD-96-11. Washington, DC: GAO, March 1996.

U.S. General Accounting Office. Defense Industry: Trends in DOD Spending, Industrial Productivity and Competition. Washington, DC: GAO, January 1997.

U.S. General Accounting Office, Defense Industry Restructuring: Cost and Savings Issues, GAO/T-NSIAD-97-141, April 15, 1997.

U.S. Department of Defense, Report of the Defense Science Board Task Force on Antitrust Aspects of Defense Industry Consolidation. Washington, DC: DoD, April 1994.

U.S. Department of Defense, Advanced Research Projects Agency. The Technology Reinvestment Project: Dual-Use Innovation for a Stronger Defense. Washington, DC: U.S. DoD, 1995, Available at http://www.jdupo.darpa.mil/jdupo/annual_95/

U.S. Department of Defense. Annual Report to the President and the Congress. Washington, DC: DoD, March 1996.

U.S. Department of Defense, Report of the Defense Science Board Task Force on Vertical Integration and Supplier Decisions. Washington, DC: U.S. DoD, May 1997.

Velocci, A. L. Wolfgang Demisch of Bankers Trust. "Profit Surge Poses Prickly Dilemma." *Aviation Week & Space Technology*, p. 21, February 12, 1996a.

Velocci, A. L. "U.S. Industry in Vigorous Fight against Proposed Policy Reversal." *Aviation Week & Space Technology*, p. 30, July 22, 1996b.

Velocci, A. L. "U.S. Industry Poised for Further Realignment." *Aviation Week & Space Technology*, December 23–30, 1996c.

13

REVITALIZATION STRATEGIES FOR SECOND-TIER DEFENSE COMPANIES

WILLIAM L. SHANKLIN

INTRODUCTION

With the easing of world tensions, the American defense industry is being buffeted by reduced governmental spending levels that it has not seen the likes of since the lull between the two World Wars. World politics have always wrought changes that set the stage for making or losing fortunes, running the gamut from recurring fighting in the Middle East for control of oil to more subtle forms of economic warfare in international trade. The disintegration of the former Soviet Union and the Iron Curtain countries of Eastern Europe that culminated in 1989 is the root cause of the downsizing of the American defense establishment.

The largest defense companies have been consolidating to take advantage of economies of scale. However, closer examination reveals that some of this consolidation is actually diversification within the defense industry. An example is when a firm whose chief business is electronics combines with a company whose main business is airplanes or missiles. These giants of the defense industry are also seeking business in the civil sector of government.

Subcontractors that have relied on prime contractors for business are also seeking new kinds of customers in order to survive. These second-tier companies have little experience in diversification because they have not had to venture outside of defense. Fortunately, they do not have to fly blind, as other companies have been confronted with a decaying of their primary markets and have transformed themselves.

Much can be learned from examining some of these case histories. An observation by Henry Kissinger is right on target in what management can and cannot expect to derive from so doing: "History is not, of course, a cookbook offering pretested recipes. It teaches by analogy, not by maxims. It can illuminate the consequences of actions in comparable situations, yet each generation must discover for itself what situations are in fact comparable."

Just as studying bygone battles can help military students to grasp strategy and tactics, so can managers understand from evaluating business strategies that were employed by other companies. Since no two battles ever have the same elements and no competitive situations in business are identical, Kissinger's point about analogies and maxims needs to be kept in mind. Top managers in second-tier defense-related companies can draw on what worked for firms that diversified in the past, but there are no guarantees.

Diversification from defense to commercial ventures is risky by any measure or means, yet is achievable. For instance, the Department of Defense examined the effects on employment of 97 military bases closed in the 1960s and 1970s. The study revealed that while the closings eliminated 87,600 civilian jobs over a 20-year period, by 1993, non-defense industries located on the former military facilities were employing 171,100 people. The new businesses ranged from a small brew pub to a giant biotechnology firm (Spiegel, 1997).

If a company's management is properly to evaluate potential business strategies for entering non-defense markets, the place to start is with an assessment of its technology portfolio. How transferable is it to commercial products and services?

TECHNOLOGICAL INCONGRUITY

A company may have state-of-the-art technology yet still face revenue problems, because the cutting-edge technology is incompatible with market demands. Richard N. Foster (director of McKinsey & Company) chronicled in his book *Innovation: The Attacker's Advantage* how this was a commonly made error in the years beginning immediately after World War II and lasting until the late 1950s. Prevailing business thinking in this period did not recognize the importance of closely aligning corporate R&D projects with customer needs. Foster (1986) pejoratively called this era in American industrial history the "lab in the woods" phase.

During the war, remote government research sites like Oak Ridge, Tennessee and Los Alamos, New Mexico, developed and perfected weapons, which were then used on the front lines. Subsequent to the war, American industry mimicked this technique. It became customary for corporate management to build a communications wall between the research and development and business functions so that, in Foster's words, "scientists [would] have time to think and be free from the 'distractions' of senior management" (ibid, p. 54).

A remote R&D department would develop a product and turn it over to marketers for commercialization, in a process akin to what had been done with R&D and the front-line troops in World War II. This strategy remained popular until management realized that much of the leading-edge technologies being developed in their laboratories were incompatible with their business missions. It might be fascinating technology, but what did you do with it commercially?

Many defense-related firms today find themselves in the same position. Management has sophisticated technology at its disposal, but the dilemma is that much of it has uncertain commercial value.

The tendency is to deduce that a second-tier defense company's revenue predicament can be mitigated by simply finding new kinds of customers to replace its military business. Yet that response is easier suggested than achieved. The task at hand may involve far more than finding new customers for existing technology. In the first place, if a commercial market is attractive, the likelihood is that companies with established reputations and a reservoir of experience are already serving it. What reason is there to believe that a neophyte like a defense company can dislodge these incumbents, which will fight any overtures to their customers? In fact, defense firms might have a better chance to compete in emerging markets, which are not yet well understood or dominated by any of the entering players.

Furthermore, a defense company's technologies can be exceptional in a scientific or engineering sense while being incongruent with commercial needs. What consumer or industrial market potential is there for contractor United Defense, which has state-of-the-art expertise in high-powered guns for tanks, destroyers, and battleships? A company official commented, "You can't sell Navy guns in Wal-Mart" (Davis, 1997). For sure, some companies will not be able to find non-defense customers as long as they view their output largely in terms of defense products or services like tanks, destroyers, and battleships. A distinction needs to be made between and among knowledge and skill, physical assets, and final products. In this context, the question is: How can existing knowledge and skill from the defense business be redeployed to create different products and services with economic value in the marketplace?

The U.S. National Aeronautics and Space Administration annually publishes a list of successful commercial products that are offshoots of its aerospace programs, such as insulin infusion pumps, improved wheelchairs, reading machines for the blind, and scratch-resistant glasses.[1] Although laudable, the reality is that these spinoffs represent a small number of products when NASA's vast research and development expenditures are weighed.

An objective situation analysis necessitates that a defense company's management ascertain which one of two scenarios more closely approximates its state of affairs. Either the company has technological know-how that can be applied to the product and service needs of commercial customers or the firm has an esoteric technological portfolio that makes finding non-defense applications virtually impossible.

Management's answer sets parameters for what strategies they can reasonably pursue to reverse revenue decline. Assuming that a board of directors elects to stay in business, the first scenario requires only concentric (related) as opposed to conglomerate (unrelated) diversification. In the former, there is a commonality between the old and the new—in technology or manufacturing or some important aspect of marketing. For example, when Philip Morris acquired Miller Brewing they already knew how to sell hedonistic products; cigarettes and beer use very similar distribution channels and advertising appeals.

In concentric diversification in the defense industry, the assignment is one of finding new customers and using the company's existing technological knowledge to develop products and services to fulfill their needs. By contrast, where compatibilities are few and far between, the more radical and precarious conglomerate diversification is necessary. In this case, the dual chores are to locate markets unrelated to defense and sell products and services based on technologies and skills that the company does not already have in its portfolio.

STRATEGIES FOR VENTURING INTO NON-DEFENSE MARKETS

Whenever a firm has a technological portfolio that can be redirected to commercial products and services, it may be able to reach out to non-defense customers by adding a marketing capability, either by bringing aboard sales and marketing associates who are knowledgeable and experienced in the targeted industrial sectors or by entering into strategic alliances with well-established companies. On the other hand, if management concludes that it has a technology too arcane to be applied in commercial products and services, then merger and acquisition is likely to be the only realistic course of action, especially if the company's defense business is in a freefall. In that case, a corporation must eventually liquidate or close up shop as a defense company and use the shell corporation to make non-defense acquisitions. The latter option is particularly attractive whenever the company has a stock already listed on a major exchange and therefore has ready access to capital for financing acquisitions.

Historically, the stock performances of widely diversified companies are inferior when compared to stock returns of corporations that adhere to well-defined business missions. For example, one study found that "In each year from 1978 through 1990, highly diversified firms had below-average market valuations per dollar of assets" (Stock, 1993). Statistical analysis in another investigation demonstrated that the effects of both focus and diversification are realized almost immediately: "The firms that experienced significant increases in focus out-performed the general market by about 2 to 4 percentage points in the year of focus change, while the firms that decreased their focus underperformed the market by about 4 percentage points" (Jarrell, 1991).

For every successful conglomerate, there are multifold prominent firms—Litton Industries, PepsiCo, and ITT—that have been forced by under-performing stock prices to spinoff operations in search of focus. Wall Street refers to this as a "pure play" strategy. Starting with the tenure of Jack Welch at General Electric in 1981, this most successful conglomerate of all began selling businesses that did not fit into one of three categories—services, technology, and manufacturing—and acquired businesses that did. This winnowing process followed Welch's mission of achieving "integrated diversity, the principle that GE's varied businesses can maintain their operating independence while working closely as a team, sharing everything from financial data to people to best practices" (Tichy and Sherman, 1993). Welch also stipulated four main criteria that each business must meet in order to be retained in the GE portfolio or acquired: the business must have the first or second largest market share in its industry; yield well-above-average real returns on investment; have a distinct competitive advantage; and leverage from GE's particular strengths (the company does best in large-scale pursuits necessitating tremendous capital expenditures, patience, and management expertise and is disadvantaged in rapidly changing industries dominated by fast-moving entrepreneurs (ibid, pp. 89–90). By the early 1990s, General Electric was still a conglomerate, but one with more definition and profitability.

The only circumstance that should prompt management to consider a risky diversification strategy is when its firm's core market looks to be in a longtime and possibly permanent decline. Second-tier defense companies fit that description in that many of them will go out of business unless sources of non-defense revenue are found soon.

Once the diversification decision is made, the question becomes how to achieve it—through internal entrepreneurship, strategic alliances, merger and acquisition, or a mixture? The trouble is, all of these options are laden with risk. Internal entrepreneurship takes considerable ramp-up time before revenues and especially profits are realized. A study by McKinsey & Company, which looked at acquisitions by 116 companies over an 11-year period, found that only 23 percent of them earned their cost of capital. The minority of corporations that are successful "know exactly what they want from the companies they're buying—be it technology, market access, or distribution" (Henkoff, 1996). Consultant Jordan Lewis' survey of 2000 managers in North America and Asia asked executives to assess how well their own strategic alliances have met corporate goals. The average grade was B minus for achieving strategic goals, but for financial performance it was C minus. Lack of shared objectives was the main explanation given for such mediocre results (ibid, p. 88).

Other companies have faced the same bleak futures as many second-tier defense companies do today, so there are some remarkable examples to learn from. For instance, in a hundred-year evolution of internal growth coupled with acquisition, Warren Featherbone Company went from making featherbone (split turkey quills bound to form a cord) for stiffening and shaping corsets, collars, bustles, and gowns, in the 19th century, to plastic baby pants for sealing diapers in 1938, and finally to baby clothes in the last half of the 20th century (Morgenthaler, 1989). Throughout its history, the company leveraged its knowledge of the clothing business as changing circumstances dictated. Similarly, Houston-based Stewart & Stevenson's odyssey took it from being in the "power business," circa 1900—shoeing horses and building carriages—to manufacturing and selling worldwide another form of horsepower—gas turbines—now. Company president C. Jim Stewart remarked: "We're still harnessing power" (Cook, 1989).

A mere two decades ago, Bell & Howell was an American icon, holding a virtual monopoly on home movie cameras, microfilm machines, and film projectors. As home video cameras and videocassette recorders displaced Bell & Howell's products, the company floundered with an unfocused acquisition spree that encompassed audio products, banking services, and proprietary technical education. A leveraged buyout group purchased Bell & Howell and devoted its resources to two new businesses—information management (of financial services, education and library materials, and the transportation vehicle market) and mail processing (for companies and the U.S. Postal Service). The turnaround team's initial actions at Bell & Howell were to streamline its bureaucracy and change the culture by outsourcing almost half of its manufacturing, doing away with senior managers' reserved parking places at headquarters, and instituting other significant departures from past practices (Quintanilla, 1997). Revenues and profits responded to the focus strategy and cultural alterations, and the buyout group took the company public again.

A lesson in all of these cases is that results do not come without bold action in changing the existing corporate culture. A current example is Eastman Kodak. In 1993, George M. C. Fisher left Motorola to recreate Eastman Kodak. Fisher saw digital photography (which records images on computer chips instead of film) as Kodak's future core business. However, four years into his tenure, the preponderance of consumers still preferred conventional film. Consequently, Kodak was losing market share to Fuji, and sales and profits suffered (Nelson and White, 1997). Like his predecessor, Kay Whitmore, Fisher's planned-for turnaround was thwarted by failure to act decisively. While Whitmore did not do enough to move Kodak into the digital age, Fisher resisted the massive downsizing that was needed to make the company lean enough to remain attractive to investors during the longtime transition from film to digital photography. Finally, in late 1997, Fisher initiated the necessary restructuring.

Another lesson is that large revenue levels are hard to replace. Consider the case of defense contractor Alliant Techsystems, which is known mostly for its rocket motors and land mines. In the wake of government cuts in spending for weapons and equipment, Alliant Techsystems sought out commercial markets. The company, for instance, applied its military technologies to needs in law enforcement and environmental safety, teamed with IBM to offer RoughWriter, a rugged laptop computer for demanding field applications, and used its pool of knowledge to develop DocMaestro electronic use software. However, the problem for Alliant Techsystems and defense firms in general is that the revenue potential of commercial ventures often pales in comparison to what the companies were formerly garnering from defense contracts. Alliant Techsystems' largest business unit, Conventional Munitions, accounts for 44 percent of all corporate revenues, but commercial end-users constitute only 14 percent of this total. Commercial customers contribute 22 percent and 4 percent, respectively, of revenues in Alliant Techsystems' other major divisions, Space and Strategic Systems and Defense Systems. Alliant Techsystems' Emerging Business Group comprises 4 percent of corporate revenues, yet even in this new-ventures division, commercial end users contribute just 17 percent of sales.[2]

IMPORTANCE OF EXECUTIVE LEADERSHIP

James Champy (1997), co-author of *Reengineering the Corporation* and chairman of consulting for Perot Systems, wrote for *Forbes* magazine an article on coping with change titled "Mark Twain, Business Consultant." Champy quoted Twain, as follows: "We should be careful to get out of an experience only the wisdom that is in it, and stop there, lest we be like the cat that sits down on a hot stove-lid. She will not sit down on a hot-stove-lid again—but also she will never sit down on a cold one anymore." Champy said: "Twain . . . has smart advice for executives in the computer age. More than a century ago he understood the dangers of blindly trusting past experience for dealing with the future."

Indeed, sometimes experience is of little value or even a hindrance in solving a problem, because the immediate circumstances are much different than in previous situations that are serving as the frame of reference. Executives with outstanding records of performance in one industry often do not replicate their successes when they change industries; what works best in one market environment or corporate culture does not necessarily transfer to another. In a three-year study of 40 industry leaders, Michael Treacy and Fred Wiersema (1993) found:

> Companies that pursue the same value discipline [low cost or product leadership or customer service] have remarkable similarities, regardless of their industry. The business systems at Federal Express, American Airlines, and Wal-Mart are strikingly similar because they all pursue operational excellence [low cost]. An employee could transfer from FedEx to Wal-Mart and, after getting oriented, feel right at home. Likewise, the systems, structures, and cultures of product leaders such as Johnson & Johnson in health care and pharmaceuticals and Nike in sports shoes look much like one another. But across two [value] disciplines, the similarities end. Send people from Wal-Mart to Nike and they would think they were on a different planet.

Louis Gerstner, Jr. of IBM and Michael Jordan of Westinghouse Electric both began their careers as McKinsey & Company consultants. Their backgrounds in a plethora of companies doing business in a multiplicity of industries undoubtedly gave them the broad perspectives needed to size up what strategies have the best chances of working in various situations. Everyone can read from the same strategy books, but it is knowing under what conditions to apply a certain plan or scheme that counts.

Gerstner turned in a stellar performance in companies as different as American Express, RJR Nabisco, and IBM. Jordan transformed Westinghouse Electric into CBS and sold off all non-broadcasting businesses. Compared to executives like Gerstner and Jordan, managers who have worked twenty or thirty years in a particular industry have plenty of experience, but it is industry-specific and often company-specific as well. When industry conditions change markedly, or in the event that executives switch industries, it is difficult for them to unlearn what they have learned.[3] Even then, that is only half the task. They must quickly develop an understanding of how an industry or corporate culture unfamiliar to them does business.

IBM's policy of hiring from within and providing lifetime employment was cited as a model until the personal computer revolution came along. IBM's then-chief executive officer, John Akers, and his top aids were so deeply imbued with a mainframe mentality that they were virtually incapable of saving "Big Blue." These hardware-oriented career IBMers could not or did not choose to see that personal computers would capture much of the market for mainframe "big iron." Compounding their error, IBM literally conceded Microsoft co-founder Bill Gates the rights to the personal-computer operating system that secured Gates' historical status as a figure comparable to John D. Rockefeller.

Examples like these argue strongly for the proposition that companies looking to diversify in the face of erosion in their traditional markets are well advised to look to new leadership from outside of the declining industry. Second-tier defense companies closely fit this profile. These firms are confronted by curtailed military spending by government and with the reality that they are too small to consolidate like the big defense companies are doing.

The main obstacle, of course, is that existing boards of directors and upper-level managers are understandably resistant to replacing themselves. Knowing what should be done in the best interests of stockholders and actually choosing to do so are not always one and the same. Only after Apple Computer was in jeopardy of bankruptcy and its common stock was in a tailspin did a new board of directors assume command.

The issue in defense companies comes down to whether a corporate culture created with military needs in mind is capable of bringing a stream of commercial products to market successfully. Moreover, is dual-use best achieved by separating the defense and commercial strategic business units, so that the corporate cultures do not impede one another? One way to answer this question is to consider the circumstances that have enhanced shareholder value in conglomerates which have spun-off subsidiaries into separate companies. *The Wall Street Journal* consulted executives, analysts, and management consultants and asked them for guidelines pertaining to when spinoffs have made shareholders better off because "the sum of the parts will be greater than the whole" (Guyon, 1996). The consensus about when to break up the company is as follows:

- Do it when product lines do not share a technology base and have distinctly different R&D approaches, but do not if spinning-off businesses destroys a common R&D process.
- Do it whenever businesses have different marketing or distribution approaches, but do not if products use the same channels or the same marketing strategies.
- Do it if businesses require different debt-to-equity ratios to compete in their industries.
- Do it if investors will have a clearer view of how to value individual businesses, but do not if they place a premium on the parts (ibid).

In virtually all instances, defense businesses and commercial enterprises will have different marketing strategies and distribution channels, which is a compelling rationale for separating the units. One option is to establish different corporations with their own managements for defense and commercial, but share a common R&D investment through some form of partnering, such as a joint venture or strategic alliance. Another option is to keep defense and commercial under the same corporate umbrella, establish a central R&D facility, and isolate the marketing efforts for defense and commercial.[4]

Faced with the severe cuts in governmental defense spending that followed the end of the Cold War, Northrop Grumman Corporation searched for commercial markets for its military radar technology. Five of its initiatives failed: multi-sensor radar surveillance systems for airports; residential security; commercial security and access control; law enforcement; and police mobile information systems. The company did find a profitable commercial application in parcel handling. According to John Steulpnagel,

Northrop Grumman's Director of R&D Operations, the failed initiatives resulted because management did not understand buyer behavior and distribution channels in the intended new markets. For example, in the effort to develop multi-sensor radar surveillance systems for airports, the company discovered that dealing with the FAA is much different than dealing with the Pentagon. Stuelpnagel said that the necessary conditions for diversification are a coherent corporate strategy, focused resources, and customer understanding.[5] It is difficult or even impossible to achieve even one of these conditions if management's attention is divided between defense operations and commercial ventures.

The president of the Electronics Systems Sector of the Harris Corporation, Albert Smith, explained how his company adopted a much different dual-use strategy. Because 70 percent of Electronics Systems Sector employees are engineers, who are accustomed to dealing with the Pentagon, top management concluded that the requisite core competencies for effective ventures into commercial markets was lacking. Hence the decision was made that defense and commercial businesses should not be housed in the same company. Rather, Electronics Systems Sector staff would identify promising technologies for commercial applications, link-up with venture capitalist firms to underwrite startups, and do initial public offerings for the successes. Harris has followed this process in several cases and maintains a minority ownership (about 20 percent) in the new companies.[6]

STEPS TO THE TURNAROUND

For many—perhaps most—defense companies that have grown and prospered as subcontractors to major firms, the prognosis is bleak unless management takes aggressive action to diversify. The patient is in dire straits and a more conservative wait-and-see alternative is too passive. The fact is, U.S. inflation-adjusted defense spending is likely to remain flat or decline as a percentage of gross domestic product. Defense subcontractors either must find alternative commercial sources of revenue or become progressively smaller.

Any successful turnaround strategy must begin at the upper-most echelon of the company, with the board of directors and top management, because only they possess the authority to shift course dramatically and take decisive actions. A strong leader, usually an outside member of the board of directors, must initiate the process whereby the incumbent board and CEO are evaluated against a single criterion: Do they have the requisite perspectives and skills needed to engineer a major modification in corporate focus? A long and extensive precedent in turnaround situations, covering a variety of companies and industries, points to the need for new leadership at the top. The necessary insights and operating experiences in commercial markets call for an infusion of fresh talent on the board of directors and in the senior-executive suite.

A study of substantial organizational makeovers among 25 U.S. minicomputer firms found that "major changes in environmental conditions and succession of a CEO significantly and positively influence revolutionary transformations" of companies. A new chief executive, particularly one recruited from outside of a company, is more likely to initiate the necessary radical alterations in strategies, organizational structure, and cadre of senior executives (Romanelli and Tushman, 1994). He or she is not committed to the company's long-standing strategies and policies and may have different perspectives than previous CEOs about appropriate organizational actions (Helmich and Brown, 1974).

Another critical step is for management to evaluate objectively the company's technological portfolio. Is it so esoteric that radical (unrelated) diversification is required? If radical diversification is indicated—wherein the company must seek both new kinds of customers and also secure new products and services to sell—the board has to decide whether the shareholders are better off by pursuing this high-risk diversification strategy or basically liquidating assets. Galileo Corporation completely exited from the defense business and diversified into commercial ventures,[7] whereas Rockwell Corporation sold its defense operations (the MX missile and the B-1 bomber) and aerospace ventures (the space shuttle) to Boeing. Whenever diversification into entirely new ventures is selected over liquidation of defense operations, the board is essentially making an investment decision for shareholders that the stockholders might prefer to make for themselves.

Provided that the technological portfolio is not limited predominantly to defense applications, then it may be possible to find commercial outlets. If this is the case, management has a realistic chance to engineer a turnaround by means of a concentric (i.e., related) diversification; that is, by finding new markets for existing technologies.

The place to commence is with a plan to search for and attempt to identify potential commercial uses for the firm's technological know-how. Having engaged in this kind of "We have this expensive technology, now what can we do with it" exploratory process personally, this author knows that a search and identify procedure is difficult. The end result is the company has so many "leads" that nearly all of them must be appraised initially on an intuitive a priori basis; the firm has neither the time nor money to explore all of the possibilities in an in-depth manner.

The most promising ideas should then be thoroughly investigated with market research, so as to eliminate much of the guesswork before a firm proceeds to the product-development stage. Smaller defense firms are literally betting their futures on the commercial markets they choose to diversify into. Moreover, product development is an expensive undertaking.

The required market research is typically not the phone or mail survey variety, although these methods are useful in the exploratory phases. Rather, time-consuming, in-person visits are usually necessary to discuss likely applications with engineers and executives in companies that plausibly may have a use for the defense firm's technologies.

If management is successful in locating prospective commercial markets for its technologies, the problem left to solve is how to proceed to enter. Drawing on the experiences of giants like Bell & Howell and small firms similar to Warren Featherbone, the guideline is that, when a company's traditional core market is eroding fast, aggressive action on multiple fronts is essential. Most CEOs in this situation do not have the luxury of enough time to grow the company from within. Thus a second-tier defense company that is seeking entry into one or more commercial markets needs to consider an aggregation of methods. What the defense company requires above all else is an intricate knowledge of customer needs in its new markets. That intelligence can be learned and procured by coupling internal entrepreneurship with some combination of strategic alliances, joint ventures, and mergers and acquisitions.

MOUNTAINS TO CLIMB

At the outset, Henry Kissinger's observation was noted that "History . . . can illuminate the consequences of actions in comparable situations, yet each generation must discover for itself what situations are in fact comparable." If the industrial past is an accurate guide, the future does not bode well for the vast majority of defense subcontractors. Companies faced with deterioration of their core businesses have usually been ineffective in diversifying into new industries.[8] This fact does not mean that diversification in second-tier, defense-related firms is not worth the effort. However, whether it is or not should be the basic issue addressed by a company's board of directors. As fiduciaries for shareholders, board members must assess the odds of success, which vary across companies and individual circumstances, and decide whether the risks of diversification are acceptable.

Once the board of directors makes a determination that a company will remain in business, its members will need to approve a turnaround plan. Generally, this blueprint should be infused with fresh thinking from newly hired executives whose expertise is from outside of the defense industry. A determination must be made early on concerning the extent to which the company's defense and the commercial businesses should be partitioned.

The outcome depends on how well management succeeds in achieving two formidable assignments: matching defense-oriented technologies to commercial market demands and changing corporate culture. The turnaround team needs to include at least one executive with the instincts of a high-technology entrepreneur and another with the leadership qualities to effect change. Even with these assets, a company's transition from defense to commercial markets is highly problematic and takes time.

A company seeking a new identity certainly needs a clear strategy to guide its efforts, but there is usually no "magic bullet" solution. Whether the firm succeeds or not depends as much or more on the creativity and diligence of its associates in searching for and pursuing promising opportunities.

NOTES

1. *Spinoff* (Washington, DC: U.S. Government Printing Office, annual update of spinoff technology).
2. *Alliant Techsystems 1997 Annual Report.*
3. For more discussion on "unlearning", see Shanklin, W. L. "Offensive Strategies for Defense Companies." *Business Horizons*, pp. 53–60, July–August, 1995.
4. For a discussion of how to interface marketing and R&D in high-technology businesses, see Shanklin, W. L. and Ryans, Jr., J. K. "Organizing for High-Tech Marketing." *Harvard Business Review*, pp. 164–171, November–December, 1984.
5. Presentation by John Stuelpnagel, Director, R&D Operations, Northrop Grumman Corporation, at the Second Klein Symposium on the Management of Technology, Pennsylvania State University, September 17, 1997.
6. Presentation by Albert Smith, President, Electronics Systems Sector, Harris Corporation, at the Second Klein Symposium on the Management of Technology, Pennsylvania State University, September 17, 1997.
7. Presentation by William T. Hanley, Chief Executive Officer, Galileo Corporation, at The Second Klein Symposium on the Management of Technology, Pennsylvania State University, September 17, 1997.
8. For an insightful study of the experiences of 22 firms in seven different industries that were partially or wholly supplanted by superior technology, see Cooper, A. C. and Schendel, D. "Strategic Responses to Technological Threats." *Business Horizons*, pp. 61–69, February 1976.

REFERENCES

Champy, J. "Mark Twain, Business Consultant." *Forbes*, p. 103, August 11, 1997.
Cook, J. "We're Still Harnessing Power." *Forbes*, May 29, 1989.
Davis, R. A. (Knight-Ridder Newspapers). "Defense Companies Try Crimefighting." (Akron) *Beacon Journal*, p. D7, March 13, 1997.
Foster, R. N. "Innovation." *The Attacker's Advantage*. New York: Summit Books, 1986.
Guyon, J. "Hanson Spinoff Plans Haven't Raised Shareholder Value." *The Wall Street Journal*, p. B8, September 26, 1996.
Helmich, D. L. and Brown, W. B. "Successor Type and Organizational Change in the Corporate Enterprise." *Administrative Science Quarterly*, Vol. 17, pp. 371–381, 1974.
Henkoff, R. "Growing Your Company: Five Ways to Do It Right." *Fortune*, p. 84, November 25, 1996.
Jarrell, G. A. "For a Higher Share Price, Focus Your Business." *The Wall Street Journal*, p. A14, May 13, 1991.
Morgenthaler, E. "An American Story: A 19th-Century Firm Shifts, Reinvents Itself and Survives 100 Years." *The Wall Street Journal*, pp. A12–A13, May 9, 1989.

Nelson, E. and White, J. B. "Blurred Image: Kodak Moment Came Early for CEO Fisher, Who Takes a Stumble." *The Wall Street Journal*, pp. A1–A6, July 25, 1997.

Quintanilla, C. "Bell & Howell is Seeing Its New Image More Clearly." *The Wall Street Journal*, p. B4, May 28, 1997.

Romanelli, E. and Tushman, M. L. "Organizational Transformation as Punctuated Equilibrium: An Empirical Test." *Academy of Management Journal*, Vol. 37(5), pp. 1141–1166, 1994.

Shanklin, W. L. "Offensive Strategies for Defense Companies." *Business Horizons*, pp. 53–60, July–August 1995.

Shanklin, W. L. and Ryans, Jr. J. K. "Organizing for High-Tech Marketing." *Harvard Business Review*, pp. 164–171, November–December 1984.

Spiegel, P. "Close a Base, Create a Job." *Forbes*, pp. 110–112, September 22, 1997.

Stock, C. "Study Shows Diversification is Usually Not Good for Business." (Akron) *Beacon Journal*, pp. B8 and B12, August 23, 1993.

Tichy, N. M. and Sherman, S. *Control Your Own Destiny or Someone Else Will*. New York: Currency Doubleday, p. 62, 1993.

Treacy, M. and Wiersema, F. "Customer Intimacy and Other Value Disciplines." *Harvard Business Review*, pp. 84–93, January–February 1993.

14

STRATEGIC DIVERSIFICATION AND THE TECHNOLOGY INTENSIVE DEFENSE FIRM: "ON THE CONVERSION ILLUSION"

MICHAEL RADNOR AND JOHN W. PETERSON

INTRODUCTION

Defense conversion is a "politically correct" misnomer. The former defense company is not being converted. It is undergoing a generic mission shift from defense supplier to that of free market competitor. That requires a time-phased disengagement from the historical mission, reconfiguring of the enterprise value-added chains and regeneration of the enterprise core strategies.

When diversifying, a typical first strategy step in entering civilian markets, defense-companies-in-transition are confronted with opportunities in both established and emerging markets. Unfortunately, the commercial opportunities are typically already being served by leaner, more technologically adept, and more focused civilian players. As a result, the defense-companies-in-transition quickly run into difficulty, in part because the emergence of real diversification opportunities happens slower than the collapse of their core defense businesses and, in part, because their mission capabilities remain dominated by their historical competencies. A typical result is organizational entropy because the enterprise was unable to successfully complete the change of mission.

It is only by creating an appropriate business vision and supporting strategies, realigning the elements of the technology value-chain,[1] effectively leveraging the business value-chain,[2] and by attaining intense commitment at every level of the business can such companies regenerate organizational vitality, establish a commercial markets "beach head," and become the desired engines of economic recovery and growth.

BACKGROUND

Reigning conventional wisdom has been that, due to the perceived technology density of military weapons platforms and large systems, military and defense suppliers had specific competencies, core and otherwise, in the application of sophisticated technologies. Apparently it was thought that successful conversion from supplier in the military-industrial complex to free market competitor in the global manufacturing sector required only that these application competencies be primed with technology that aligned with civilian applications. Once that happened, the world would beat a path to the supplier's door.

The harsh reality is that applied "leading-edge" technology-based systems and major weapons platforms represent only a portion of defense spending. Additionally, such leading-edge systems tend to be sourced from the surviving defense meta-suppliers. The critical mass of American defense and military systems, however, are built on platforms and product device technology that is essentially obsolescent when compared to free market civilian equivalents. Throughout American history, with the possible exception of the War in the Gulf, the sons and daughters of the Republic have been sent in harm's way to buy the time for the American industrial complex to be bought to bear. Leveraged by an effective logistics system, the American military-industrial complex has been able to supply sufficient **objets de guerre** for the National Command Authority to force enemies to engage in wars won or lost based on strategic attrition.

Defense suppliers have evolved within that paradigm. Their survival as individual enterprise indicates that they have been effective within the scope of the designs, components and systems they provided to the military. Both their survival and historical profitability reflect, in part, their efficiency in aligning specific capability packages with their defense-related missions. However, in the last twenty years, the strategic nature of competition in the rest of the global industrial manufacturing sector has evolved much more significantly than in the defense sector. The emergence of the "lean" production model has contributed to more efficient enterprises whose rapid response times make it more difficult to maintain a sustained strategic advantage (NCMS, 1996). Effective defense sector mission capability packages that evolved over the past five decades (Samad, 1996) have been rendered obsolete in just a matter of months. This then is the competitive environment in which the mission shift must take place and not all the firm's key players are necessarily capable or willing to recognize the need to retask.

Many of the larger defense suppliers have purposely diversified to include related civilian businesses in their portfolio to mitigate the effects of cyclical military acquisition budgets. Such companies, given time for transition, have been able to substitute civilian sales for declining military revenues. Rockwell International, for example, reduced sales to the Department of Defense from 50 percent of its revenues to 18 percent (*Economist*, 1997) before selling off the majority of its defense-related assets in 1997. Monitoring the signals of change and openness to alternative strategies allows such anticipatory planning responses. Separating the defense-related assets and sale or spin-off at decision time are appropriate strategies for larger diversified enterprises in the West. Unfortunately, such enterprises represent only a small portion of the global defense industrial base.

The multiplier effect of the number of defense-related contractors and subcontractors (most with similar skills and capabilities) and their geographical density tends to magnify the negative economic impacts of defense cuts. Much has been written and many suggestions offered about how to ameliorate the macro-economic impacts and effects of "defense conversion" on local, regional and national communities. Both the effects and the concerns are real and well intentioned (Bertelli, 1994). However, macro-economic solutions engineered by central governments are at best temporary. Some argue that the historical record suggests that government job creation and simple product substitution at the plant level simply do not work (Samad, 1996).

THE CONCEPTUAL MODEL

Huang T'ai Kung, the ancient Han dynasty strategist once said that "**before the warrior can face the enemy, the warrior must face himself**." This holds true today for the enterprise-in-transition. Successful re-tasking requires exceptional vision, remarkable objectivity, a very thorough understanding of the firm's competencies and capabilities, a viable strategy and much, much more.

The term "vision" as applied to the business enterprise was a much overworked term during the 1980s. Nonetheless, it is potentially one of the more powerful tools a firm can leverage in defining its strategy. "It is perhaps the core element in binding together and leading dispersed components of today's most successful intelligent enterprises" (Quinn, 1992). The dilemma for the chief executive lies in creating a "virtual" embodiment of the future enterprise, and then leading the management team in selecting, aligning and prioritizing, monitoring and achieving the competencies and operating objectives necessary to attain that embodiment. It has to be achieved while at least "satisficing"[3] the current minimum expectations and requirements of powerful internal and external constituencies.

The strategy itself assumes a long planning horizon and provides the basic guiding principles and expectations that are going to guide its behavior during an "N" year quest to attain its "virtual" future identity. Key to its mission are the explicit definitions of its future competencies and the strategic objectives (scope, agenda resources and priorities) to be achieved, either developed internally or acquired, during its mission horizon. Among the critical drivers in the mission tasking process, the most influential identified in the recent past has been the concept of core competence (Thouati, *et al.*, forthcoming).

By shifting the focus from the traditional sustainable competitive advantage of Porter's (1985) five forces model to internal skills and focus of the enterprise, Prahalad and Hamel (1990) apparently appealed to the heart and soul (and perhaps egos) within corporate America. They helped re-legitimize the idea of corporate self-determination. They established that the CEO and the dominant management team could make decisions that provided intrinsic value and brought advantage to the enterprise.

This has since been expanded to include the additional idea of a company's dynamic capability. Unfortunately, "there are almost as many definitions of organizational capabilities as there are authors on the subject" (Collis, 1994). The core competence emphasizes "technological and production expertise at specific points along the value-chain, (whereas) capabilities are more broadly based, encompassing the entire value-chain" (Stalk *et al.*, 1992). The transition from current to future embodiment is illustrated at a very high level in Fig. 14.1 where the future vision of the enterprise is bounded by the realities of the cost of capital, the customer value proposition and the technology and business value-chains. (More detailed discussions of the breakthrough concept of the technology value-chain are provided below.)

In the defense sector, both technology and innovation tend to be managed as separate and independent activities. The typical defense firm's enterprise business value-chain is design and engineering dominated. In addition, it has been optimized to meet the procurement practices of the military and the reporting and accounting practices (and other legislative "impedi-ments") of central government regulators. Its distribution and logistics systems have been designed to support the military depot systems for maintenance and upgrades (Gansler, 1995). Such practices are inconsistent with civilian trends of cutting time to market and simplifying the customer interface. In spite of helping to meet the archaic practices and regulatory needs of the defense bureaucracy, such practices add significant extra cost burdens in multiple steps of the business value-chain.

In a similar manner, the heavy defense sector dependency on mature and well-proven device level technologies adds additional direct costs to maintain less cost-effective (obsolescent) technologies. It also strands the technical skills and competencies of the defense systems integrators a generation or two behind their civilian counterparts. The result is embedded impedance (price, delivery, and technology) to commercial market competitive parity and to successful mission change. What therefore is required for

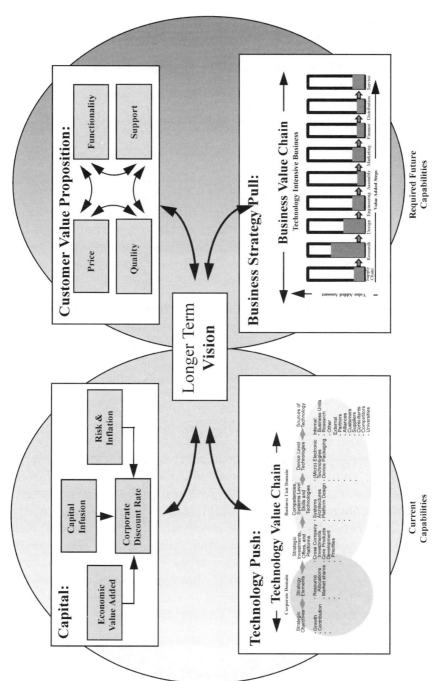

Fig. 14.1. Future vision bounded by "realities"

retasking is a cognitive reinventing of the enterprise. We suggest there are at least six prerequisites to be addressed:

- real and incremental incentives to diversify;
- time-phased disengagement from the historical mission;
- regeneration of the core strategies;
- reconfiguring and possible outsourcing of selected elements of the "historic" value-chains;
- redefinition to a global business model;
- a concerted assault on the traditional economics of targeted industries.

IN CONTEXT

Business strategy for the enterprise-in-transition exists at least at two levels and both must be addressed simultaneously. The first level is macro. It reflects the directions and intentions at the top, perhaps beginning with the board of directors—this generally translates to an enterprise or corporate strategy. At this level, the primary business "battlespace" includes such global, national, regional and local considerations as:

- political leverage and positioning of the firm and its directors
- economic considerations, especially the availability and cost of capital
- social (cultural and societal) conditions such as education and employment levels, and
- technological sophistication and local value-added.

At the macro level, the enterprise-in-transition's leadership team must solve the complex non-linear optimization problem of having to simultaneously defend and grow the value of the investments in their enterprise. This, while faced with a wide range of short lived, time-dependent, potential solution sets, none of which is really optimal due to the conflicting objectives and interests of the multiple internal and external constituencies. Enterprise strategy is a long-term high stakes play. Outcomes are decisive. And in this arena, because of the power of the other players, temporary influence is typically more attainable than control. The enterprise-in-transition's leadership team faces a truly strategic dilemma. It remains under competitive attack in its shrinking core defense businesses while it is attempting to reallocate its limited assets to establish a "toe-hold" and expand into commercial markets. The enterprise's

- cost structures are inappropriate for commercial competition,
- marketing and sales processes are inappropriate,
- new product introduction and cycle times are too long,
- products are "over" designed and therefore too cost intensive,
- products are based on mature (proven) designs and lagging technologies,
- understanding of the new markets is immature.

Under such conditions, successful retasking requires an enormous amount of effort at every level. This must include the board members and the chief executive officer who must intercede with the regional and national (and in some cases international) forces to stimulate an environment conducive to the mission shift. The central government must provide incentives which will allow the firm team to compete for investment capital while incurring the costs and problems inherent in the strategic mission shift. Buy-in must also be obtained from the chief operating officer and his or her team[4] some of whom will have to be reskilled in an effort to align with a significant customer, and deliver results which far exceed the customer's initial expectations. The end results of this initial effort will both establish the credibility and the momentum for the enterprise as it transitions to the new environment. Every member of the enterprise must accept that survival depends, not on "just" delivering technology, but using the skills and capabilities of the enterprise to deliver technology solutions that absolutely delight their customers while providing competitive advantage to the strategic customers *in their own markets.*[5]

The Enterprise or Business Model

At the micro or operational level, the business strategy reflects the directions and intentions for nearer term survival. At this level, operating management attempts to win and maintain "share-of-mind" (and assets) from the senior management team. This is done while engaging and defeating the efforts of competitors on a field of play "defined" by legal and regulatory constraints imposed by external authorities (local, regional, national, and international.) Such external authorities may opt or be required to define the "norms" of acceptable behavior in terms that advantage "native" customers and/or home industries. This too, represents a very complex optimization problem, but tends to be more linear. Although significant, due to the multiple opportunities and the limited scale, individual operational level strategies tend to be somewhat less decisive to the longer term survival of the enterprise.

The Enterprise Strategy

There are no "magic bananas." The "right" strategy for an enterprise is unique to that enterprise and is not necessarily appropriate for others. When adopting a "me too" strategy, the absolute best the follower can achieve over time is parity. As a result, it's important that the enterprise-in-transition understand its significant competencies and how they can be leveraged to grow the business. Assuming that the strategy is directionally correct and appropriately funded, the vigor of execution and the synergy and internal dynamics of the team are key factors that determine the effectiveness of the core strategies. Within the context of mission shift, however there are some tenets of conventional wisdom (also see Hofer, 1975):

- The immediate mission is to survive and make an assault on emerging and transitioning industries and capture shares of new wealth created ("new" market capitalization).
- The collapse in the core defense businesses may be offset by serving non-defense government procurement requirements (Feldman, 1996).
- Existing products are not the engines for entry into significant "new" markets.[6]
- Joint development can be one of the more effective market entry strategies.
- Partnerships and alliances can compensate for reconfigured business value-chain weaknesses (i.e., channel, marketing, software technologies, etc.).
- Alliances can allow for sharing of risk (Cochrane *et al.*, 1996).
- Profitable "spotted puppies," otherwise known as non-core technologies should be nurtured and then spun-off as quickly as prudent to fund the diversification strategy.
- Licensing or sale of non-critical technologies can be a source of barter or cash.

The Business Value-Chain

The business value-chain (Porter, 1985; Shank, 1993; Hax and Majluf 1996) consists of the sum of the interdependent steps in conceptualizing product and/or service, converting it to substance, positioning the substance with a customer or customer set and delivering it in order for the customers to create or find value in its use. At a very high level, the business value-chain may include innovation of concept, sourcing of raw materials and components, engineering design and development, assembly, marketing, finance, distribution, installation, service and support. It is in the comparison of the revenues and costs associated with each step in the business value-chain, to the relevant revenues and costs of direct and cross-elastic solution providers, that defines the tactical business situation in which the enterprise-in-transition's offers will compete.

Short-term advantage can be created by understanding the costs of each value-chain activity, the linkages between and amongst the activities, and the cost drivers associated with each of the activities and linkages. It is here that the enterprise can reconfigure the "traditional" industry value-chain and precipitate an economic attack on pricing structures employed by the current industry players. Such actions, of course, are simply a first shot signaling a firm's intent to redefine the targeted market or industry sector. Before that shot is fired, repositioning of the enterprise business value-chain must be perceived internally as a matter of immediate survival for all the members of the firm. Further, in order to achieve its goal of becoming an architect of industry revolution, the enterprise-in-transition must:

- reconfigure the business value-chain to change the rules of the game
- anticipate and respond to technology changes faster than the competition
- achieve scale or scope[7] positions that provide better cost positions than current players
- reconfigure the solution (product or service) to create substantively more value for the customer **and** create a competitive discontinuity in the marketplace.

By expanding the Porter "value-chain" concept (illustrated on the left in Fig. 14.2) and identifying a distinct technology value-chain as a driver of the business, not a support activity, the technology intensive enterprise can simultaneously attack markets on two levels, technology and business efficiency. Whatever the specifics of the market attack strategies, the enterprise team must emerge from the business value-chain reconfiguration activity with better controls and improved understanding of the core strategies, the product value propositions, and their costs.

The management team must gain the experience and skill necessary for making the next difficult decisions in the next series of attacks as the assaults on the targeted markets and industry sectors continue. There must also be a better understanding and appreciation of the human costs and contributions to the business. Both the managers and the teams must emerge from the initial experience with a preoccupation for changing the traditional "industry rules of the game" without compromising the firm's sense of mission, relative perceived quality, and focus on delighting its strategic customers.

The Technology Value-Chain

Strategies must be driven by capabilities and competencies. Functional technical competencies, as well as core competencies, can hold keys to future competitive advantage (Ebeling and Snyder, 1992). As a result, both the business value-chain and the technology value-chain[8] can be leveraged to simplify the inherent systems level complexities of how the enterprise can add unique value for its strategic customers. The technology value-chain in Fig. 14.3 allows the enterprise management team to identify and align common cross-product technologies and skills required to support specific strategic thrusts within the framework of the enterprise vision. The enterprise can then identify the attributes necessary for "success" in the new industry sectors, comparing itself to its strategic competitors in those areas. Those attributes needed to support evolution of the enterprise can be flagged and compared to in-house competencies and capabilities. Those required can be retained, gaps can be filled via education or recruitment, and those no longer needed can be phased

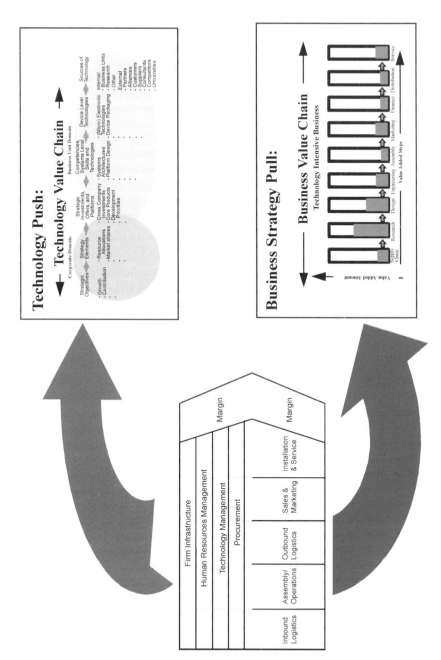

Fig. 14.2. Proposed evolution of the business value-chain concept

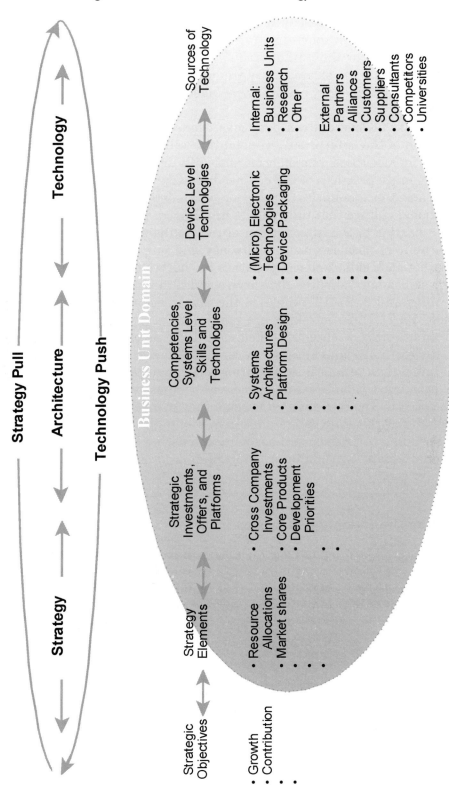

Fig. 14.3. Strategy, architecture and technology are all inter-linked in the technology value-chain

out. The technology value-chain in Fig. 14.3 expands the technology value-chain presented above in Fig. 14.2 to illustrate that either the push of technology or the pull of strategy can influence how technology is leveraged to meet the strategic requirements of the enterprise. Strategy, architecture and technology are all inter-linked in the technology intensive enterprise.

The majority of "current competencies" that don't necessarily support the evolution of the enterprise, are potential "killer" competencies. These will tend to include those competencies and skills unique to the procurement and maintenance practices of the defense industry. Such competencies can be so ingrained in the fabrics of the people, the defense business, and the organizational culture that they persist unchanged. In fact, they may well present a barrier to the introduction of new skills and capabilities. The result is built-in obsolescence, inefficiency and vulnerability (retasking requires that such skills and competencies be "harvested"). Over the longer term, the skills may "satisfice" and survive, but if that is allowed to happen, the viability and competitiveness of the enterprise will be jeopardized (Peterson, 1997). Different skills are required! For example, in a survey of 190 medium-sized defense firms, Feldman (1996) determined that market skills (hiring more marketers) were more important to successful diversification than technical skills (hiring more engineers). The enterprise-in-transition must balance its technical and business value-chain skills in order to succeed. The conventional wisdom that there can never be enough technical and engineering assets must give way to the understanding that technical skills are professional assets and must be balanced with other professional assets (i.e., planning, marketing and other business value-chain activities).[9]

Integrating the Planning Activities

Although discernible at a relatively high level, the processes and information requirements necessary to actually retask the enterprise are interrelated, inter-linked and must constantly be under analysis and revision.

Fortunately most of the information required to assert technology strategy already is in use by the firm to manage its ongoing activities. Fig. 14.4 illustrates the analysis and decision-making activities that have significant impact on the retasking effort. It also introduces the product technology roadmapping tool which can be used to influence executive decision making. The product technology roadmap reflects the analysis activities that verify and position the technology intensive product in markets. That positioning allows the dominant management team to ask the appropriate questions and, if appropriate, make investment decisions in real time. The analyses and potential decision alternatives reflected in the product technology roadmap include:

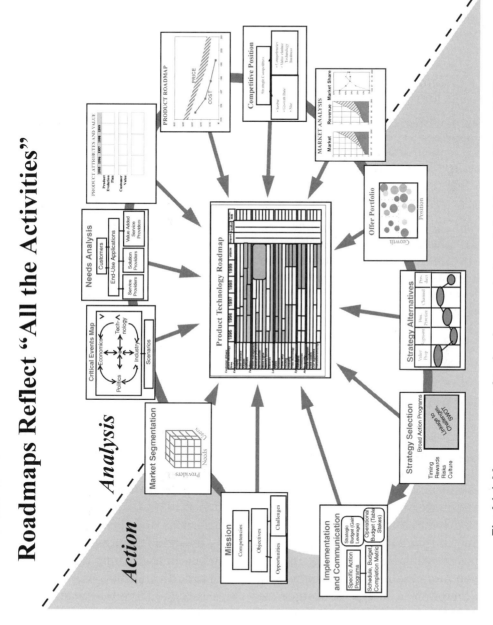

Fig. 14.4. Management of technology analysis and decision-making activities

Analyses

- Critical events map including environmental scan and "identification" of coming critical events and development of scenarios projecting their impacts on the firm
- Needs analysis including identification of specific needs of the targeted customer sets
- Definition of the offered product's attributes and value proposition
- Projection of prices and costs on the firm's planned product specific learning curve
- Identification of competitors value positioning and anticipated market plays
- Sizing of the targeted markets, their growth rates and projected "take-rates"

Actions

- Acceptable positioning of planned offers in both "cash generated and used" and "growth" portfolios
- Identification of how the offers impact the firm and what the strategy alternatives are (mix and match)
- Selection of the appropriate strategies to retask the firm
- Creation of specific action plans and operating budgets to support the selected strategies
- Revisit strategy alternatives in context of the redefined mission
- Revisit segmentation to assure appropriate targeted markets
- Iterate (re-enter analyses, i.e., step 1 above) . . .

Effective Implementation and Follow-up

The strategy processes require the "dominant managers" to manage by the strategic thrusts, not to the numbers. This means that the unskilled mid-level one-minute denominator manager **can no longer afford to** focus only on the numbers. The strategy processes require that non-quantified prerequisites be addressed and not overlooked. The management team must drive to a consensus concerning the priorities and directions for the business. In addition, they must individually assume the responsibility for communicating the strategies and their rationale to their functional teams. The management team members must personally monitor and enforce implementation of the strategies and the strategy elements within both their individual and team spheres of influence.

Modern strategists seem to consider the execution phase as a different aspect of the field of management (Hay and Williamson, 1991). It is typically assumed that the strategies providing the best overall time-phased returns are the appropriate strategies to execute. The execution steps are mostly abandoned and left to the tactical manager in each of the various specialized functions. Relatively speaking, and with some exceptions (e.g., Bourgeois and Brodwin, 1984), the transition from planning to execution has not attracted significant attention. In fact, the field of strategy and strategic planning has often "involved separating thinking from doing" (Mintzberg, 1994).

The simple reality is, as Arnaldo Hax of MIT's Sloan School says, **"Strategy without control is sheer poetry."** The senior management team is responsible for implementation. Within the enterprise, the management team must maintain buy-in and consensus concerning the core strategies, business directions, priorities and time-phasing for the business objectives. In addition, they must individually assume the responsibility for communicating the strategies and their rationale to their functional teams. The senior management team members must also think in terms broader than the mere "tyranny of the numbers." They must personally contribute to the delight of their customers as well as monitor and enforce implementation of the strategies and the strategy elements within both their individual and team spheres of influence. One means to address and manage these requirements is the adoption of a framework that allows the evaluation and recommendation of actions to the dominant management team. The framework facilitates the pulling together of the necessary information and integrating it into strategy recommendations that can be used to manage the firm. Fig. 14.5 takes the analysis and decision-making activities (illustrated in Fig. 14.4) and links them in a flow process diagram to the key measures of value.[10] The result is a high level framework that illustrates the inter-working of the key activities and links them to specific measures of firm performance. The resulting framework can then be customized to reflect the unique capabilities and opportunities for the enterprise-in-transition.

In the framework, the analytical activities interlink and drive the decision activities. All decisions are in turn linked to the mission and to specific financial metrics to provide the senior management team with traditional financial translations of the technology decisions. Such a framework allows closer alignment of business and technology value-chain activities to the implementation of the retasking strategy. By linking the considerations and activities to operating income and other measures of value (senior management compensation) the framework allows the visualization of action programs and alternatives on the same bases as the revised mission, the strategic thrusts and the market-driven product portfolios. It is only then, by aligning the appropriate business vision and strategies, effectively redefining the elements of the technology value-chain, reinventing the steps and focus within the business value-chain, and obsessing at every level of the enterprise-in-transition can such companies become the hoped for success, and in turn, engines of local, regional and national economic recovery and growth.

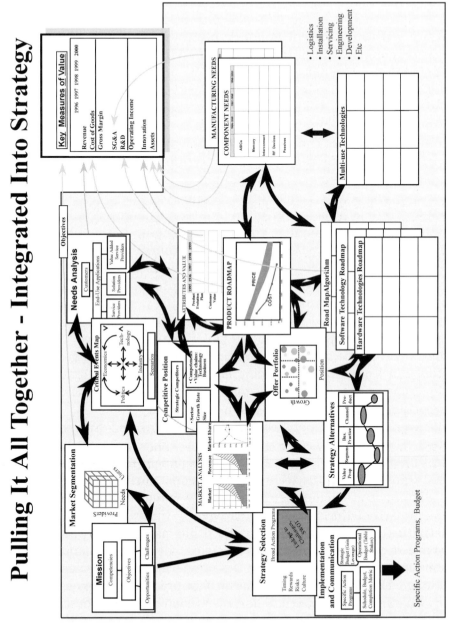

Fig. 14.5. Technology centered framework for strategy selection and implementation

Summary

The United States has apparently passed the high watermark of government investment in "defense-related" research and technology development. Recent changes in national policy have begun to shift the primary burden of innovation to universities and private businesses. The resulting impact on national defense capabilities is unclear. However the policy changes appear to have negated the National Command Authorities' traditional war strategy of immediate engagement (to disrupt "enemy" actions) and ramp-up to supply mission capability packages at an appropriate scale to wage a war of strategic attrition.

Current government bureaucracy, accounting and reporting requirements, and performance requirements, appear to significantly limit the effectiveness of strategies for dual-use product and technologies. During this time of transition from national defense-centric policies, the question becomes one of not whether the government should intervene by providing economic incentives to preserve some strategic capabilities, but where and how. Logic suggests that government could be most effective by migrating onerous military procurement and reporting practices to civilian market equivalents, thereby reinforcing competitiveness based on business and technology value-chain efficiencies necessary for global competition; by incenting the creation of new knowledge intensive markets, thereby preserving leadership in C^4I (Command, Control, Communications, Computers and Intelligence); by incenting university and private sector co-operation in revolutionary innovation; and influencing corporate appreciation for longer term returns on technology investments.

The authors believe that private market mechanisms should be used to the extent possible to support innovation and technological evolution. However, history shows that since World War II government defense research policy has helped create hundreds of thousands of jobs and trillions of dollars in new industries and innovation. The national defense and economic implications of such a shift away from historical national policy has precipitated consolidation in the traditional defense industries, especially at the meta-supplier level. Weapon platforms and supporting systems can probably be expected to remain mostly unaffected, that's probably where the surviving development budgets will be invested. It is at the subsystem and defense supply commodity level where the impacts of onerous requirements are felt most dramatically.

Pressures for profits and the effects of global competition are forcing small and mid-level suppliers to follow strategies of diversification and conversion to civilian applications. The depth and density of the defense industry will soon no longer be sufficient to support sustained war operations—there is no such thing as a free lunch. Survival of enterprises transitioning from defense supplier to free market competitor requires a reinventing of the enterprise. It requires vision, tough-minded management capable of effective planning and tough decision-making. Operationally, the retasking process requires that both technology and innovation be integrated as a source of potential advantage in both technology value-chain and business value-chain activities. Most importantly, survival requires enterprise missionaries with the right skills, competencies and good fortune that will allow them to accelerate and successfully complete the transition from "defense-supplier-entering-a-death-spiral" to diversified "engine of economic growth."

NOTES

1. In the technology value-chain, the corporation links its unique competencies and intellectual skills to add value to basic and enabling technologies to provide the products and systems that align with very specific corporate strategy elements and objectives.
2. In the business value-chain, the enterprise attempts to leverage technology to add value and efficiency to the creation and commercialization of products and services that align directly with segmented customers' needs.
3. See March and Simon (1958) for the original use of this term.
4. The role of the team is to become missionaries of the retasking, internally and externally. They must carry the message and convert the undecided. This must be achieved while making the targeted customer set absolutely giddy with delight.
5. One example frequently cited is Lau Technologies of Action, MA, as a case study for a defense supplier on the ropes which managed this process in the early 1990s (Brokaw, 1995). Lau Technologies is continuing its transition, and recently raised capital for its core business by spinning off a major portion of a "spotted puppy" in an initial public offering (IPO).
6. This is not to imply that current products may not be successful contribution sources while diversifying. Datron Systems' mobile satellite tracking antennae is an example of successful direct adaptation of military technology to civilian use. The antennae fixes on a television broadcast satellite and allows reception in large vehicles (trucks, mobile homes, etc.) while in motion (CNN April 29, 1997).
7. See definition of "manufacturing scope" in Coates et al. (1997).
8. The technology value-chain concept model was initially proposed by AT&T in 1988 by Richard Albright, now Director of the Office of Technology Assessment at Bell Laboratories, then at AT&T Corporate Strategy and Development, to address "value added" by technology to produce and services by the technology competencies of the various AT&T business units. Although more detailed than the core competency model advocated by Dr. Vijay Govindarajan of the Amos Tuck School, both build upon enterprise skills and intellectual abilities and move from technology through human value-added. The Govindarajan model is:

    Core competencies = Technological assets + Physical assets + Human assets (knowledge and learning)

    The Albright model also addresses functional technology competencies that generate business value-chain advantages for the business units. The technology value-chain establishes a linear model, originally a three by three matrix with multiple sub-elements that moves from basic device level technologies, which are microelectronics type components, to systems level enabling skills and technologies, which represent human value-added, and finally to products and systems offerings that are what the business units actually sell.

In 1992, the initial Albright technology value-chain model was further evolved by Gerald Zielinski and John Peterson of AT&T Network Systems/Bell Laboratories to include enterprise strategic objectives, the strategic elements supporting those objectives, technology and product investment priorities, and potential "best" sources of technology. The resulting modification allows the technology value-chain to map to both specific strategic thrusts and to product technology roadmaps and be used by the enterprise to identify the common cross-product technologies and skills required to support current and future products/applications.

9. Not to belabor the point, but teams as divergent as the large Japanese companies and U.S. military special operations teams support significant cross-training and rotational leadership experience (i.e., reskilling) to strengthen their "teams" (for more details on the Japanese experience, see Kawamura and Kurokawa, 1997).

10. For a more detailed introduction and explanation of the framework, see Eriksen *et al.* (1998)

REFERENCES

Bertelli, D. "Battle Over Defense Conversion." *Business & Society Review*, No. 91, pp. 11–14, Fall 1994.

Bourgeois, L. J., III, and Brodwin, D. R. "Strategic Implementation: Five Approaches to an Elusive Phenomenon." *Strategic Management Journal*, Vol. 5, pp. 241–264, 1984.

Brokaw, L. "Lau Technologies." *INC.*, 17n18, pp. 88–92 December 1995.

Coates, J. F., Mahaffie, J. B., and Hines, A. *2025, Scenarios of US and Global Society Reshaped by Science and Technology*. Oak Hill Press: Greensboro, 1997.

Collis, D. J. "How Valuable Are Organizational Capabilities?" *Strategic Management Journal*, Vol. 15(S2), 143–152, 1994.

Cochrane, J., Temple, J. and Peterson, J. "Shapes and Shadows of Things To Come, Information Strategy." *The Executives Journal*, Spring 1996.

Ebeling, W. and Synder, A. (Braxton Associates) "Targeting a Company's Real Core Competencies." *The Journal of Business Strategy*, pp. 26–32, November–December 1992.

Economist, An Ex-swordsman Ploughs Into the Peace Business." *Economist*, Vol. 336(7933), p. 59, 1997.

Eriksen, L. T., Wofford, K. O. and Peterson, J. W. "On Recreating the Enterprise Technology Management Infrastructure." In Lefebvre, L., Mason, R. and Khail, T., eds., *Management of Technology, Sustainable Development and Eco-Efficiency*. Oxford: Elsevier Science Ltd, 1998.

Feldman, J. "Diversification after the Cold War: Results of the National Defense

Economy Survey." Working Paper 112, Rutgers University Center for Urban Policy Research, 1996.

Gansler, J. *Defense Conversion*. Cambridge: MIT Press, 1995.

Hax, A. C. and Majluf, N. S. *The Strategy Concept and Process: A Pragmatic Approach*. Upper Saddle River, NJ: Prentice Hall, 1996.

Hay, M. and Williamson, P. "Strategic Staircases: Planning the Capabilities Required for Success." *Long Range Planning*, Vol. 24(4), p. 3693, 1991.

Hofer, C. W. "Toward a Contingency Theory of Business Strategy." *Academy of Management Journal*, Vol. 8(4), pp. 784–810, 1975.

Kawamura, K. and Kurokawa, S. *Japanese Technology Management: Implications for US Companies*. NCMS Workshop, May 1997.

March, J. G. and Simon, H. A. *Organizations*, New York: John Wiley and Sons, 1958.

Mintzberg, H. "The Fall and Rise of Strategic Planning." *Harvard Business Review*, Vol. 72(1), pp. 107–114, 1994.

National Center for Manufacturing Science (NCMS) "Management of Technology Feasibility Study." Ann Arbor, MI, 1996.

Peterson, J. "One Corporation's Strategy and Technology Alignment: A Case Study." In Szakonyi, R., ed., Boston: Auerbach Publications, pp. 63–83, 1997.

Porter, M. E. *Competitive Advantage: Creating and Sustaining Superior Performance*. New York: Free Press, 1985.

Prahalad, C. K. and Hamel, G. "The Core Competence of the Corporation." *Harvard Business Review*, Vol. 68(3), pp. 79–91, 1990.

Quinn, J. B. *Intelligent Enterprise*. New York. The Free Press, 1992.

Samad, S. A. "The Challenge of Change." *Cost Engineering*, Vol. 38(8), pp. 17–18, 1996.

Shank, J. K. and Govindararajan, V. *Strategic Cost Management*, The Free Press: New York, 1993.

Stalk, G., Evans, P. and Shulman, L. E. "Competing on Capabilities: The New Rules of Corporate Strategy." *Harvard Business Review*, Vol. 70(2), pp. 57–69, 1992.

Szakonyi, R. ed., *Technology Management 1997 One Corporation's Strategy and Technology Alignment: A Case Study, John Peterson*. Auerbach Publications, Boston, 1997.

Thouati, M. G., Radnor, M. and Levin, D. Z. "Corporate Growth Engines: Driving to Sustainable Strategic Advantage." *The International Journal of Technology Management*, forthcoming.

15

DIVERSIFICATION STRATEGIES FOR SECOND-TIER DEFENSE FIRMS IN THE INFRA-RED SENSOR INDUSTRY

GERALD I. SUSMAN
MARK A. BLODGETT

RATIONALE FOR STUDYING DIVERSIFICATION STRATEGIES

Defense-related firms in the late 1990s can increase their sales in at least three different ways. First, they can further penetrate markets in the defense sector in which they already sell products. Second, they can develop new products for defense and/or commercial sectors in which they currently do not sell products. Third, they can acquire or merge with other companies which already sell products in the defense and/or commercial sectors. In this third case, the combined sales of the new larger firm are not greater than the separate sales of the two firms before their merger, but the new larger firm is assumed to be more profitable from cost reductions due to consolidation of facilities and elimination of administrative overhead. Presumably, the first two strategies can be at least as profitable as the third one.

We are interested in the second of the three strategies for increasing sales, which is characterized as diversification via internal growth. Companies generally diversify because they have sufficient capital for expansion and expect to earn higher future returns in new product-markets than in the product-markets they already serve. Some companies diversify into new product-markets strictly to balance cyclical demand even if they do not expect future returns in the new product-markets to be higher than in current ones; however, such cases will not be explored here. Companies generally diversify into related product-markets because they expect synergies with the company's existing R&D, manufacturing, and marketing capabilities, and consequently increased profits and/or lower costs.

We are interested primarily in diversification into the commercial sector because the prospect for commercial sales growth is strong in many industries in which defense firms traditionally have had a presence, e.g., electronics, materials. Commercial diversification is especially relevant for small to medium-sized firms for which consolidation within the defense industry is not an option (other than to sell out to a larger defense-related firm) and/or for which deeper penetration of existing defense markets will be difficult. Market saturation from numerous similarly sized competitors or intense competition from newly consolidated larger defense firms that may have vertically integrated backward into the markets of smaller firms are both likely scenarios. Export of defense-related products is another possibility for increased market penetration, but we believe that other sales growth strategies can provide the nation with better and more immediate economic returns for its expenditures on defense. Effective commercial diversification can better leverage returns to the country's expenditure of defense dollars.

We believe that commercial diversification is a growth strategy that defense-related firms should be encouraged to pursue. It will increase the return to the American economy for the $45 billion invested in defense procurement annually in developing new technologies and intellectual resources. While investment by non-DoD federal agencies also might contribute to the nation's competitiveness, commercial diversification can provide greater returns to the nation's competitiveness and quality of life than defense expenditures contribute now.

We believe that diversification encourages firms more readily to find commercial applications for technologies which they developed originally for defense products. These firms already have the expertise from prior defense work to use to develop products for the commercial sector. Technology transfer is internal in a diversified company, is more focused, and occurs more naturally than through third-party brokers who transfer technology between government agencies and commercial firms. Diversified firms gain economies of scale and scope through use of common facilities, knowledge and skills to develop and produce similar products for defense and commercial markets. Designers within the same company can be encouraged and rewarded to design products and components for dual-use applications of technology. Such behavior is unlikely to be encouraged by less direct methods.

We will leave to others the public policy debate regarding how or in what ways the federal government can or should encourage defense-related firms to diversify into the commercial sector. The purpose of this chapter is to identify the strategic options to diversification that defense-related companies can pursue, what benefits and risks are associated with each option, and what barriers to implementation exist. We believe that awareness and understanding of diversification strategies can encourage firms to consider pursuing such strategies, if they have not considered them before or, if they had, were wary of pursuing them.

OBJECTIVE OF THIS STUDY

The objective of this study is to map the range of critical strategic choices that firms in the infra-red (IR) industry can make. The major critical strategic choice of interest in this chapter is whether to sell products primarily in the defense or commercial sector. Aside from this choice, firms also make choices to specialize by a single function, e.g., R&D or production, or perform multiple functions, or whether the firm will design and market products with a single application or multiple applications. Some firms make their initial choices, never revise them, and prosper. Other firms seek to grow by entering new markets, adding new functions, or developing new applications. The transition from a single function to multiple functions is "vertical integration," while the transition from a single application to multiple applications or from defense to commercial sectors is "horizontal diversification" (Ansoff, 1965). While our primary interest is in firms that revise their strategic choices, we get important clues as to what firms that do not revise them are like and the challenges they might face if they were to revise them in the future.

IR firms that initially choose to enter the defense sector often make related choices on the number of functions performed or applications developed. For example, firms that concentrate on defense sales are more likely to concentrate on a single function, e.g., R&D, or a single product application. By studying the range of critical strategic choices, we can better understand the related structure, processes, and culture that are aligned with these choices and the challenges facing firms that change strategic direction. As this study is cross-sectional rather than longitudinal, each firm is a snapshot in time. Consequently, we may not capture the drama involved when a firm changes direction. Nevertheless, some of the firms that we studied made the transition recently and their senior managers have observations and insights to offer about their transition.

We chose the IR industry because infra-red sensors are used commonly in weapon systems and sub-systems. The industry is an excellent candidate for studying diversification strategies in general, and diversification from defense into commercial businesses, in particular. The reasons are that infra-red sensor components and systems are found in a variety of defense and commercial products. The prospects for future growth in commercial sales appear bright, such as in nondestructive testing, geothermal exploration, fire detection, etc. The firms in the IR industry cover many SIC codes, each of which has both defense and commercial sales. The five SIC codes most represented in our sample have a substantial percentage of defense sales, i.e., neither lop-sided toward defense or commercial sales:

- optical instruments and lenses (SIC 3827), 29 percent;
- process control instruments (SIC 3823), 23 percent;
- electronic components (SIC 3679), 19 percent;
- engineering services (SIC 8711), 18 percent; and
- environmental controls (SIC 3822), 13 percent (cf Alic *et al.*, 1992).

Finally, no single IR market is large enough or growing fast enough to meet the growth objectives set by many end-use product producers (EUPP). Consequently these firms are likely to seek growth by diversification, and thus provide ample opportunities to study how diversification strategies are formulated and implemented.

We believe that the lessons learned from the IR industry are applicable to other industries which have a moderate percentage of defense-related firms, particularly those industries that involve electronics. Infra-red sensors remain critical to weapons that help maintain DoD "force preservation" policy objectives or allow the U.S. military to dominate the battlefield or "battlespace" of the future (Defense Research and Engineering, 1997). Night vision, for example, extends and enhances an infantry soldier's personal capabilities in conditions of limited visibility. The generalizations in this study may have to be tempered, however, when applied to industries that were hit harder by the downturn in the defense procurement budget of the late 1980s and early 1990s, and whose products may not have as bright prospects for future commercial or defense sales growth. Most IR firms in this study diversified in order to sustain or accelerate growth, whether by plan or opportunism. Few did so as a result of financial distress. Nevertheless, the manner in which these firms diversified is instructive for other firms. Also, some of the firms in the study were founded and have remained commercial firms throughout their entire history. The challenge that these firms faced as they added functions or developed new applications offer valuable insights for defense firms undertaking similar initiations.

The IR industry also has lessons to offer regarding prospects for dual-use applications. As IR firms have diversified along any of the three dimensions discussed above, they have gained economies of scale and/or scope by using common facilities, common machines, common product platforms or common people and skills. If the concept of dual-use is broadened to include cross-utilization of any these resources, then dual-use benefits are achievable that can lead to reductions in the cost of both defense and commercial products and enhanced force reconstitution, when and if it is needed. Dual-use prospects are especially promising in regards to IR component producers. Our evidence suggests that common IR components are being used increasingly in both defense and commercial products. IR component producers generally are remaining specialized at the component level, selling to a variety of defense and commercial EUPPs, investing in process innovation, and concentrating on cost reduction. While most of the value-added in this industry occurs in systems design and integration, many of the EUPPs are using one or more of the resources mentioned above across defense and commercial applications with accompanying dual-use benefits.

DESCRIPTION OF THE FIRMS STUDIED AND INDUSTRY CHARACTERISTICS

The analysis and conclusions drawn from this study are based on published information about the IR industry overall and about 120 individual firms. Detailed telephone interviews were conducted with forty firms. IR firms are typically small to medium-sized with annual revenues ranging from $1–150 million. Number of employees range from 12–450.

The industry is organized into three layers. EUPPs tend to specialize in developing and producing a limited number of different product applications, usually one to three applications (see Fig. 15.1). There are eight widely recognized applications in the IR industry. These applications are night vision, missile guidance, predictive maintenance, gas analysis, fire/flame detection, process/quality control, non-destructive testing, and research and development. Each application can be divided further by customer markets. For example, night vision has customers in military services, law enforcement, border patrol, fire departments and forestry services, marine navigation, search and rescue, wildlife groups, sportsmen/hunters. Predictive maintenance has customers in insurance firms, roof inspection firms, electrical circuitry inspectors, and plant maintenance operators. Customers use an application in similar ways, but the performance demands and level of product sophistication may vary considerably.

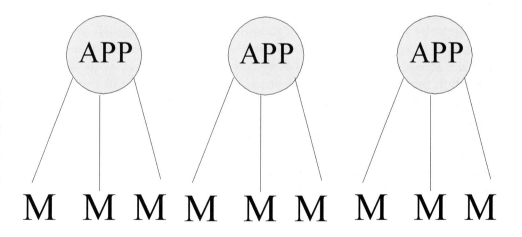

APP = APPLICATION
M = CUSTOMER MARKET

Fig. 15.1. EUPP specialization by applications and customer markets

EUPPs tend to compete within a limited number of application clusters. They compete primarily on product customization and quick response to customer requests. The latter is the primary basis on which EUPPs within the same cluster try to differentiate themselves. The EUPPs that compete within each application cluster know each other and each other's customers well. Smaller firms are reluctant to compete with each other directly, often choosing to stake out a modified niche instead.

Eltec Instruments is a small electronics firm that performs custom design and production of IR sensing systems in various markets for customers who want rapid turnaround. The firm consciously adopted a strategy of filling product lines not served by other established firms. When asked about security of their market position, the response was that "no one else wanted to be here," and the firm was confident they would have little direct competition, though they also acknowledged the difficulty of rapid growth.

Component producers compete primarily on product reliability and price. They are able to achieve economies of scale by selling similar or identical components to a wide variety of EUPPs. Few component producers desire or can forward integrate. Competition or retaliation by established EUPPs is a strong deterrent.

Fireye, a manufacturer of boiler safety controls, is part of a larger conglomerate corporation. In addition to strict corporate controls to prevent internal competition between divisions, the firm is not organized to conduct the market scanning necessary to "prove a market" before trying to forward integrate. There were numerous concepts within the firm that were believed to have significant market potential but that could not be supported during an introductory phase.

One component producer indicated that the exposure to proprietary knowledge of EUPP customers through close cooperation during custom design work made the thought of forward integration a breach of faith. An attempt to forward integrate would quickly drive away many of their most valued customers through fear of disclosure. EUPPs, in turn, frequently share as little information as possible about their customers and do not allow component producers or anyone else to open their products for servicing.

Laser Power, a lens manufacturer, described the situation as one in which the EUPPs had a captive audience. They were the sole service providers for their products and could mark-up replacement parts substantially while buying them for the same price they paid for original parts in the equipment. While this guaranteed proper servicing, it also prevented component manufacturers from capturing any of the additional profit from the market for replacement parts.

It also prevents any direct need for contact between the component manufacturer and the final customer since all information must flow through the EUPP. R&D firms also seldom forward integrate, partly based on conscious preferences of their managers. Most of the end-use product innovation originates from these firms, but these firms are more likely to license their innovations to EUPPs rather than commercialize them.

THE THREE DIMENSIONS

Single Function versus Multiple Function Firm

Production

Component manufacturers are essentially single function firms which concentrate on selling similar components to different EUPP customers. These companies invest virtually no money in product-related R&D, but may invest significant amounts of money in process-related R&D.

Dexter Research Center, an infra-red thermopile detector manufacturer, indicated that there is a high need for automation and scrap reduction. Recognizing that high volume production was essential for effective competitiveness, they reported their main R&D focus to be process innovation and efficiency.

The marketing function may be underdeveloped, especially for firms that sell most or all of their components to only a few EUPPs via long-term contracts. Throughout the industry, there was little evidence of active pursuit of any kind of strategic alliance relationships or long-term cooperation between firms.

Some IR components may be unique to end-use IR products, while others are applicable equally to IR or non-IR products. The major IR unique components are sensors, signal processing systems, lenses, and filters (see Fig. 15.2). Component producers generally sell sensors and signal processing systems as an integrated set to EUPPs. Sensors consist of heat sensitive material of varying types which convert infra-red energy to electronic signals. Sensors may be stand-alone components or clustered in focal plane arrays. The more complex the array, the more sophisticated the signal processing system must be since its function is to analyze electronic signals from the sensors and convert them into a format that is interpretable by an output device, e.g., video monitor. Component producers of lenses and filters sell their products to IR EUPPs as well as to firms that manufacture other types of opto-electronic equipment, e.g., lasers. Few component producers forward integrate to become EUPPs, either because of the formidable challenge of competing against established EUPPs or because of fear of retaliation by their existing EUPP customers.

Component producers focus primarily on cost reduction and increased volume for similar components. They expand their product lines by increasing the performance and complexity of the components offered in each product line. Components of a particular level of sophistication generally can be used in defense or commercial products of commensurate sophistication. Component requirements change little across the applications that demand

GENERIC INFRA-RED SENSING DEVICE

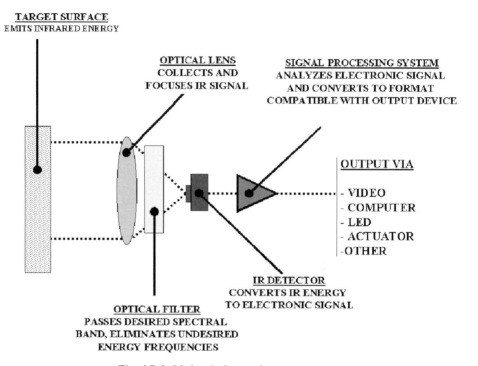

Fig. 15.2. Major infra-red components

similar levels of performance. Sales growth is achieved by discovering new uses for existing components, subject to minor modifications to suit new customers. The prospects for dual-use benefits are thus high.

Optek, Inc. is a producer of all types of opto-electronic and magnetic sensors, and feel they are very good at using "old technology in a new package." Concentrating on building their commercial market, they have placed little emphasis on developing new technology and have directed efforts at a more developmental approach to their business.

Research and Development

Firms that specialize in R&D also are single function firms. These firms hire a large number of Ph.D.s and have a culture that is not unlike that of a university. They develop expertise in responding to government requests for proposals and winning contracts to conduct specialized research on leading-edge issues. These firms must be disciplined in responding selectively to RFPs that allow them to develop a coherent and consistent base of knowledge and skills, thus developing expertise in a specific research area. Some of the R&D firms expressed concern that they otherwise might diffuse their energies and develop knowledge and skills with little potential for synergy.

Irvine Sensors, primarily a research and development firm, discussed an idea called "the divergent skill set." This situation occurred when the R&D efforts of a firm were not closely linked over time and the firm could not effectively assimilate its intellectual capital into a competitive competency. If not careful, other firms can gather more complete information about specific areas of expertise and thus be more capable of providing information to customers. Furthermore, if a firm that had acquired divergent skills tried to move toward commercialization, they would not have the complete skill set necessary in one area to try to compete with existing production firms.

Most R&D firms remain specialized in R&D, taking the products no further than the prototype stage. They have little interest in adding manufacturing and marketing functions, and instead prefer to license their technology to others to manufacture and sell.

Brimrose Corporation, another R&D firm, stated that they carefully selected research projects based on potential for commercialization. Though very interested in the profitability of a particular research project, the firm fully intends to stay away from manufacturing as a competency. Each decision the firm makes is based on this strategic outlook, and through licensing of proprietary technology, Brimrose intends to continue its efforts in the research and development areas.

An employee who is interested in commercializing a product probably would have to leave the firm rather than be able to develop it within the company. Adding production and/or marketing functions to a specialized R&D firm may divert its attention from what it does best. Some R&D firms that have added other functions have done poorly.

Spire Corporation is a research and development firm with a long history of government contracts, primarily under the SBIR arrangement. Though the firm would like to grow the commercial side of its business, it feels limited by product development issues. A large number of their products have high commercial value, but have only been developed to the point of proving a concept. A manager explained that the SBIRs paid to get the ideas 75 percent of the way to marketability, but that the last 25 percent, producibility issues, were not funded. Since commercial customers were frequently unwilling to fund completion of the research, the firm had extreme difficulty carrying the product further.

The number of single function firms that transition into multiple function firms is limited because of the difficulty of the transition. The transition is equally difficult to make for component producers as well as for R&D specialized firms. General management skills often are underdeveloped in sales and marketing, and single function firms often underestimate the challenge of managing a multiple function firm and managing the transition effectively.

General Microwave, an electronics firm with a strong tie to government contracting, felt the defense cutbacks in the mid-1980s and early 1990s as a motivation to try to diversify. Acquiring an unrelated business, the firm tried to integrate the new unit into its business for approximately five years. The attempt was unsuccessful, and the business unit was divested with the decision to "stick to what the firm knew how to do."

Many firms, in evaluating commercial markets, are quite aware of the difficulty of diversifying into such markets and are extremely reluctant to try. The Barnes Engineering Division of EDO Corporation manufactures satellites and space-related equipment. In this market, products are low volume and very highly priced. A senior manager explained that when a product can cost a billion dollars there is no room for design or manufacturing mistakes that will interfere with the sale. The firm, therefore, has "stayed very close to home" in its choice of product lines and pursuit of new customers.

Multiple Functions

We use the term end-use product producer (EUPP) and multiple function firm interchangeably in this chapter, although, in rare cases, a firm might develop, market, but not manufacture a product. EUPPs have learned how to integrate at least two of three basic functions effectively, i.e., research and development, production, and marketing. They sell their products to end-users. Their core competence is system design and integration, which is based on knowledge and skill rather than technology. These firms contribute a unique configuration of components and subsystems that accomplish a new end-use that is valued by a customer. In fact, most of the components used in their products do not embody leading-edge technology.

Detection Systems Incorporated, a supplier to EUPPs in the safety and security market, said that though products must continue to evolve, the need for price, functionality and reliability was far more important. Responsiveness to customization needs was what sold products, not reliance on leading-edge technology. In other markets, customers simply cannot afford leading-edge technology and therefore don't buy it.

The most successful EUPPs are able to integrate their design capabilities with marketing capabilities that involve reaching out to new customers and understanding how a potential application can meet their customer's needs.

The Optoelectronics division of Hughes Corporation discussed an idea called "domain expertise" in which the producer must intimately understand not only the technological needs of the customer, but the business of the individual customers as well. In the case of defense sales, for example, the need to be able to understand and discuss fleet tactics was essential for any expectation of winning a particular Naval contract.

Graseby Infrared, a sensor manufacturer, took the unusual initiative (for a component manufacturer) of closely partnering with both the EUPPs and final users so the firm could more clearly understand the needs of the customers. This approach was described as "reaching down the food chain" and was seen as a way to gain competitive advantage over competitors. EUPPs that produce and sell a single product application might grow by expanding their product lines, or, more interestingly, they might do so by seeking new applications for their products.

FLIR Systems, a producer for a wide variety of infra-red cameras, viewers, and equipment, was initially founded to sell airborne IR systems to the border patrol. The firm developed a successful product, and decided that selling aircraft mounted IR systems to the military was a logical next step. FLIR Systems successfully developed a wide variety of devices that could be used for both viewing and weapons guidance specifically designed for military utilization. The corporation next decided that their competency in IR and airborne platforms was not interdependent and developed IR cameras and viewers that could be hand-held. These systems were also very successful and sold to a variety of new market applications including law enforcement, predictive maintenance, and process control.

If EUPPs began primarily as defense or commercial manufacturers, they can seek new applications within their original sector or cross into the other sector. The shift from the defense to commercial sectors is at least as difficult as the transition from an EUPP that sells a single application product line to one that sells product lines with different applications. Different applications need to be designed to use common components, and marketing needs to be aggressive in order to understand different customer preferences.

Single versus Multiple Product Applications

This dimension applies to EUPPs because, by definition, an application is put to an end-use or purpose by a customer. EUPPs may produce a single line of products that is used for a particular purpose or they may produce multiple lines of products, each of which is used for a different purpose. Single application firms produce and sell products that are used to do Y. Some of these firms may prefer to remain single application firms. For example, firm X has a reliable customer base and reasonably steady profits. Firms like X are generally relatively young and focus primarily on commercial sales. They have no interest in growing, especially if they occupy a niche that few or no other firms wish to enter. This is partly a question of senior management temperament and the firm's culture. The marketing capability of these firms is weak, but this is not a problem if they have little need to scan their environment for opportunities. As long as their niche remains secure, they are comfortable. They can be vulnerable, of course, if this condition is altered.

The term "multiple application firm" refers to an EUPP that develops more than one product line, each of which has a different application. FLIR Systems, from the discussion above, is a good example of a firm with such a strategy. For example, the company designs and manufactures products for a wide variety of night vision and industrial applications. Its night vision products are used by law enforcement for drug interdiction, search and rescue, border and marine patrol, and environmental protection. Growth is a conscious objective of the firm with choices made regarding what markets to enter on the basis of growth prospects and utilization of existing resources. A prerequisite for multiple applications is product standardization of the core product, with modular peripheral components that can be easily interchanged to create new applications. FLIR Systems has actively pursued standardization as a means of streamlining its production process.

A limiting factor to growth is availability of knowledge and skill required to understand customers for different product applications. The most direct contact a firm would have with its customers is through its sales force. Persons with such customer knowledge and skill are increasingly scarce in the IR industry. In addition, they are expensive, and the requisite knowledge and skills take time to develop on the job which further limits a firm's ability to grow. Unlike hiring consultants to solve specific technical problems, marketing personnel need to be regular employees of the firm. The requisite knowledge and skills take time to develop on the job, which further limit the speed with which firms can grow.

Defense versus Commercial Sectors

The differences between defense and commercial companies are well documented (Gansler, 1995). Design, production, and marketing differ when selling to a commercial customer versus selling to a single large defense customer which specifies the product design and places higher priority on performance than cost. While many firms may flourish by producing and selling exclusively in one sector or the other, this section will concentrate on firms that produce and sell in both sectors, and on their transition between sectors.

Figure 15.3 shows the strategic paths that a defense-related firm might take. A defense single-function (DSF) firm might diversify into the commercial sector by taking Path A. Component producers might find this path relatively easy to take if they can sell essentially the same product to firms in both sectors. R&D firms, however, might find this path somewhat difficult because many commercial customers seem unwilling to pay for the latter's services. The same single-function defense firm instead might choose to stay in the defense sector, but forward integrate and become an end-use product producer (Path B), thereby enhancing revenue by adding more value for its defense customers. A commercial single-function (CSF) firm also might forward integrate and become an end-use product producer (Path C). Defense and commercial firms face a similar challenge to add and integrate new functions into their organizations and serve new end-use customers.

A defense end-use product (DEUP) producer might diversify into the commercial sector by taking Path D. This path can be relatively easier to take if the end-use product serves essentially the same purpose in defense and commercial applications. The path becomes more difficult, the greater the product must be modified for commercial use. The same defense end-use product producer might instead choose to stay in the defense sector and

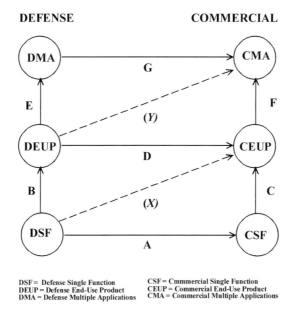

DSF = Defense Single Function
DEUP = Defense End-Use Product
DMA = Defense Multiple Applications

CSF = Cmmmercial Single Function
CEUP = Commercial End-Use Product
CMA = Commercial Multiple Applications

Fig. 15.3. Strategic position of firms and potential trajectories

develop new defense-related product applications (Path E). This option is limited because the customer, not the producer, plays the primary role in defining and requesting new product application. A commercial end-use product (CEUP) producer also might decide to increase sales by developing and marketing new commercial product applications (Path F). Finally, a defense multiple-application (DMA) firm could seek to increase sales by developing commercial multiple-applications (CMA) of its end-use products (Path G). A defense firm that is capable of developing new defense applications might be able to transfer this capability to the commercial sector. However, such a firm would still need to develop and integrate commercial marketing skills with its applications development skills.

One can infer from Figure 15.3 that vertical paths within a sector (Paths B, C, E, and F) and horizontal paths across sectors (Paths A, D, and G) are feasible means to generate new sales. Higher level paths require the firm to add, modify, or integrate more functions or applications than lower level paths. Consequently, lower level vertical paths (B, C) are easier to take than higher level vertical paths (E, F), and lower level horizontal paths are easier to take than higher level horizontal paths, e.g., A<D<G. One also can infer that diagonal paths (Y and Z) which cross sectors and add functions or applications simultaneously are very high risk strategies and unlikely to succeed. Rather than moving diagonally, however, a defense-related firm could succeed by proceeding step-wise either by moving vertically or horizontally first, then, after mastering the challenges associated with this move, take a horizontal move towards the commercial sector (if not yet taken) or take a vertical move.

ISSUES

Motivation to Diversify Across Defense and Commercial Sectors

Most IR firms diversified in order to expand sales from an already established sales base, whether defense or commercial. Few firms diversified because of a sharp drop in sales in the defense or commercial sector, though many did note that they foresaw such a drop and expanded commercial sales in anticipation of defense cutbacks. The direction of the expansion was sometimes based on a conscientious and thorough scan of market opportunities, but often it was driven by feedback and suggestions from commercial customers.

The Barnes Engineering Division of EDO Corporation scanned the commercial space arena and saw the telecommunications sector as a rising market. Concentrating on this market, the firm has developed invaluable leadership positions that have enabled it to remain highly competitive among other space producers.

Graseby Infrared reported that strategic thinking is an everyday part of each managers activities. Environmental scanning is constant and habitual in an effort to stay abreast of opportunities and potential threats.

Compix Inc. is a small firm that manufactures IR equipment for use by forestry services in its fire-fighting efforts. The market is very small, and, because there are no related competitors, the firm must sell complete systems of hardware, programming and engineering. The needs of each customer are so specific, it is difficult to anticipate them and design or build ahead. The firm, then, has adopted a reactive stance to customers' demands rather than a proactive one possible for other firms.

ISI Group, a small custom design firm specializing in rapid turn-around and customer responsiveness, has adopted several new variations of IR technology as a direct response to customer demand. Uncooled focal plane arrays, image analysis software, and color capability were all incorporated to meet specific needs of a customer.

It is hard to say whether firms that scanned for market opportunities performed better than firms that responded primarily to customer feedback. In the former case, however, more attention was devoted specifically to leveraging existing technology, and knowledge and skills. Some single application EUPPs and R&D firms chose to remain primarily in the defense sector because they believed that they had the knowledge required for successful government contracting. Although these firms believed that IR sensor sales to DoD would be growing in the next few years, their underdeveloped production skills (R&D firms) and/or marketing skills made entry into the commercial sector very challenging. There also were firms that chose to remain primarily in the commercial sector because they perceived that the paperwork associated with testing and documentation required by DoD was very burdensome. They also believed that DoD contracts were an unreliable source of revenue because they could be terminated at DoD's convenience.

Some concerns expressed by firms regarding the production and sale of both defense and commercial products are instructive. The financial, technical, and intellectual resources required to work on defense-related products was considered to be so great that they limit or detract the firm from competing simultaneously in the commercial sector. For example, contracts for defense-related products require the operation and maintenance of obsolete or mature technology for extended periods. This ties up capital and human resources, which then are not available for commercial use, especially when quick response to opportunities is essential.

Cal Sensors, one of the two leading players in the world market for InGaAs (indium gallium arsenide) sensors, stated the specific intent of remaining an industry leader. From the firm's perspective, military contracts are long life commitments to existing technology. The company intentionally avoids any pursuit of government work because managers see it as a distraction that ties up the resources necessary to remain competitive at the leading-edge. It is easier to keep up with the flexibility of a commercially oriented organization because there is a greater ability to respond to developments and take advantage of opportunities quickly.

Assets that are tied up in developing and making defense-related products pose considerable opportunity costs to firms that see commercial growth prospects to be very bright. Also, infrequent and expensive defense contract bids in pursuit of "home-runs" can pose greater financial risk to a firm than can a series of frequent and moderately expensive commercial ventures that seek to hit "singles or doubles".

Epitaxx Optoelectronic Devices, a component firm that manufactures several varieties of sensors, stated that commercial sales were a more reliable source of revenue because they move more quickly and had a higher turnover. The government contracts were slow and could be canceled at any time, making them far less preferable.

The technical skills required for defense contracts can be so highly developed that they are greatly underutilized when applied to commercial products. This is especially the case for high-end EUPPs or firms that specialize in R&D, but wish to commercialize a prototype product that they developed. Their skills are too sophisticated and expensive for application to most commercial products. Commercial customers are unwilling to pay the price for such skills, and only a few firms were willing to hire less expensive employees to carry out the less technically demanding work. Marketing skills, which are virtually unneeded for defense contracts, are absolutely essential for selling commercial products. Commercial customers for IR products need extensive training in how to use them as well as requiring after-sale service. A trained sales force that can give demonstrations and clearly understand the technical needs of customers is essential.

Obtaining Scarce Human Resources

A major limiting factor in diversifying across defense and commercial sectors is the knowledge and skill needed to compete in the commercial sector. Some of the knowledge and skills are transferable and useful across the two sectors. For example, some EUPPs continue to compete for defense R&D contracts in order to develop knowledge and skills that are useful for developing future commercial products. This can be very useful as long as these firms select contracts carefully to maximize the knowledge and skill they hope to gain from such contracts. Firms that have experience in developing very reliable products, especially those that require extensive testing and documentation, have found this experience to be useful for developing and manufacturing commercial products that have similar testing and documentation requirements, e.g., medical products.

Advanced Photonix, a firm that manufactures photo-diodes and does extensive government contracting, indicated that the company felt it was easier to penetrate a new market based on a quality capability rather than on new technology. One avenue for such a strategy is to pursue markets in the medical industry where the concern for reliability and quality was equal to that of defense customers.

Persons with the knowledge and skills required to develop new product applications and service new customers are scarce. Advanced Photonix stated that human capabilities of the firm were the first consideration made when evaluating a new product line. The Barnes Engineering Division of EDO Corporation said that human capabilities were the most limiting factor concerning where the firm could go. Inframetrics stated that drastic expansion of current production would be a significant challenge due to the need for acquisition of skilled personnel and training time required.

Industry growth is straining the capacity of educational institutions to train persons fast enough, smaller firms cannot afford to hire permanent employees until their market prospects are clearer, and some knowledge and skills are used too infrequently to justify hiring workers on a full-time basis, e.g., software writing. Consequently, many firms outsource for these required knowledge and skills or form partnerships with other firms that have the knowledge and skills they need.

Consultants are used extensively in the IR industry primarily to solve technical production problems and occasionally research and management problems. Most component manufacturers are facing increased pressure to expand production, improve yields and lower their costs in order to remain competitive. A guess at the motivation for EG&G Judson's recent acquisition of Graseby Infrared is to gain the competence in sensor manufacture. Firms appear to hire consultants less frequently to solve marketing problems. These firms are too small and technical, and related problems arise too infrequently to hire employees with such knowledge and skills on a full-time basis. The few instances in which consultants were hired for marketing purposes were to train customers to understand a firm's product and how to use it. The knowledge of a particular customer's needs requires continuous and intense contact with that customer. This is an area in which consultants are less useful and is a limiting factor in an EUPP's ability to develop applications for new sets of customers. Some firms that are interested in expanding into new markets have acquired other firms that have the requisite market knowledge.

Segregated versus Common Resources

Most firms in the IR sensor industry did not have segregated facilities. Component producers did not need to segregate manufacturing because the same components went into defense and commercial products. EUPPs that sell similar high-end customized products in small quantities to both sectors can develop and produce them in the same facilities. In some cases, stripped down versions of high-end products can be sold to commercial customers. In these cases, the only difference was that production of defense-related products required paperwork and documentation that was performed by management. One firm commented that the attention to quality required for defense production carries over into commercial production, even for products that differ in more than minor ways.

Some firms reported that they would segregate defense and commercial production if they could afford to do so, but they simply didn't have the facilities or finances to accomplish this. Firms that cannot afford to hire a separate less skilled work force to make commercial products will have to utilize their more expensive skilled defense workers for commercial production. Also, one firm with a strong defense background was concerned that habitual use of regulations regarding milspec standards might spread and "contaminate" commercial production. The firm didn't want to use milspecs unnecessarily for commercial production because of the higher costs involved. The most common form of segregation, when it exists, is separate marketing functions that are oriented to defense and commercial customers, respectively.

CONCLUSIONS

As discussed earlier, component producers are unlikely to forward integrate. They fear retaliation from their EUPP customers and face a formidable challenge in transitioning successfully into a multi-function company. Commercial EUPPs and large defense firms are unlikely to backward integrate into IR components. Both types of firms stand to benefit from the economies of scale that component producers achieve by producing components for both sectors. Thus, IR component producers and EUPPs will likely remain in non-overlapping, but complementary segments of the IR industry for the foreseeable future. Consequently, IR component producers and EUPPs that choose to diversify into the commercial sector will take different paths in that direction.

Component producers appear to face reasonably strong demand from growth in a wide variety of commercial IR applications. As a result, DoD need not be concerned about retaining domestic sources of IR components. Enough component producers will remain in the industry to make IR components for both defense and commercial markets. Defense-related component producers need make only minor modifications in their organization structure or culture to serve the commercial sector effectively. Their relationship with customers is similar in both sectors. However, they will face constant pressure on profit margins, which will influence how components are produced. IR components are essentially commodities. There will be little opportunity to raise prices, and strong interest in lowering costs via automation in order to maintain reasonable profit margins. Competition among component producers will lead to constant or lower prices.

While the variety of IR applications and related products will continue to grow, most EUPPs will specialize in only a few applications unless they can develop a common design platform for different products, which appears unlikely in the near future. Even so, meeting customer needs for each application requires substantial time and attention. Marketing capabilities will continue to remain a scarce resource and limit company growth beyond a few applications. EUPPs will find the transition between defense to commercial sectors more difficult than will component producers. Interestingly, defense-related EUPPs that successfully transition to the commercial sector are unlikely to remain permanently in the defense sector. The heavy investment in marketing and the organizational and cultural changes required to compete successfully in the commercial sector will lead diversifying firms to view the defense sector as increasingly less compatible with their newly developed capabilities. Consequently, the EUPP's defense business will either be sold or shrink to a small portion of their total business. The large defense contractors should be able to meet the limited demand for specialized defense products with in-house capabilities or from R&D oriented firms that remain specialized in defense.

REFERENCES

Alic, J. A., Branscomb, L. M., Brooks, H., Carter, A. B., Epstein, G. L. *Beyond Spinoff: Military and Commercial Technologies in a Changing World*, Boston: Harvard Business School Press, 1992.

Ansoff, H. I. *Corporate Strategy: An Analytic Approach to Business Policy for Growth and Expansion*, New York: McGraw-Hill, 1965

Defense Research and Engineering, *Defense Technology Area Plan DTOs Sensors, Electronics and Battlespace Environment*, 1997.

Gansler, J. S. *Defense Conversion: Transforming the Arsenal of Democracy*, Cambridge, MA: The MIT Press, 1995.

16

DEFENSE CONVERSION IN INFORMATION TECHNOLOGY SERVICE INDUSTRIES

DAVID J. BERTEAU

INTRODUCTION

Defense conversion policies and the studies that support them usually focus on companies that develop and manufacture hardware, both the prime contractors (original equipment manufacturers, or OEMs, paid directly by the government) and those that are subcontractors (paid by the prime contractor). These companies include the largest and best known defense contractors (Lockheed Martin, Boeing, Raytheon, United Technologies, and General Dynamics), and the programs which they work on are the most visible and widely-known defense budget items (the F-22 fighter, the F/A-18E/F attack aircraft, the B-2 bomber, and the new attack submarine).

It should be clear, however, that an approach to defense conversion policy which focuses only on hardware manufacturers ignores a large and growing segment of the defense private sector supplier community. That segment is the services industries, and it includes a wide array of industry types, ranging from small businesses that provide custodial or food services to massive systems integration corporations. While the Defense Department buys services from the full range of service companies, those firms that supply technology-driven services warrant a closer evaluation from a defense conversion perspective.

If we accept the view that public policy, at a minimum, is interested in encouraging defense-dependent companies to investigate options to expand their business in non-defense markets, then any review of the impact of public policy initiatives must look at the effects of those policies on defense-dependent services industries. This chapter begins to establish a framework for that review, identifies several significant issues, and recommends some guidelines for solutions.

HIGH-TECH SERVICE COMPANIES—POTENTIAL DUAL-USE TECHNOLOGY PROVIDERS

Of particular interest from a defense conversion perspective are those high-tech service companies that were relatively defense-dependent but which now serve business areas where government and commercial technologies most closely connect. These business areas include engineering and systems integration, information systems and services, communications, and software development. Such companies today provide the best examples of dual-use technology, providing development and applications that are available for both government and commercial markets. High-tech defense-dependent service firms find it easier to sell to both public and private customers than their defense hardware counterparts, and such companies are therefore more likely to find new or expanded markets when they try to diversify.

Such companies differ from both manufacturing and low-tech service companies in that they depend primarily on the skills of their workers for their success. Workers are more readily shifted from project to project than are facilities and equipment. The mobility of the American workforce is therefore coupled with the relative ease of redeploying assets, making service companies more likely to enter into dual-use markets. The kinds of government policies that will accord the most benefit are those that enable these companies to hire, train, retain, and reward their skilled workforce. This means that government actions that reduce the flexibility with which companies can use their workers will inhibit defense diversification success.

In addition, in areas like information services, communications, and software development, high-tech services companies already serve both sides of the defense conversion boundary, in government and commercial markets. Software and services are more readily adaptable to multiple uses and customers. Since this phenomenon works best when demand has been growing in both the public and private sectors, it is useful to examine recent business trends.

MARKET TRENDS

The past decade has seen dynamic growth in information technology services markets, both within the government and in commercial markets. Federal government expenditures have been growing at nearly 5 percent per year and are projected to grow at a similar rate for the next decade. Reported budget estimates for federal information technology expenditures are $30 billion per year, although some experts believe that additional IT expenditures are masked within federal agencies' operating budgets.

State and local government IT expenditures have become larger than federal outlays in recent years ($37 billion in 1997), and they are projected to grow at a faster rate over the next 5–10 years.[1] State and local governments now make up more than 10 per cent of the total IT market for the U.S.

Even so, the total combined government IT market (federal, state, and local) is barely one sixth of the total domestic market for information technology. Commercial and international markets have been expanding at a rate that is twice as fast as government. Paul Strassmann (1997) estimates that U.S. corporations spent more than $500 billion on information systems in 1995, or nearly 1.5 times the amount of total profit. These markets are also growing at a faster rate than either the federal or the state and local marketplaces.

As a result of these market shifts, a number of companies that grew originally as federal information technology suppliers have been branching into new markets, including state & local government, commercial, and international markets. These companies include BDM, Computer Sciences Corporation (CSC), PRC, BTG, EDS, and Science Applications International Corporation (SAIC). For example, Computer Sciences Corporation has become the 12th largest U.S. defense contractor (*Defense News*, 1997) with more than $1 billion in DoD revenue. Yet CSC is less dependent on defense today than in the past. Even though its defense revenue has grown by 30 percent in recent years, the defense portion of CSC's total revenue has dropped from 22 percent to 19 percent over the same period[2] because CSC's non-defense revenue is growing even faster. A similar pattern is evident among most of the other top defense information technology firms.

Many of these companies have expanded into commercial markets by applying classic diversification strategies. Their approach to this diversification is often through internal development, technology transfer, and by acquiring or merging with other companies already in target markets. Where they differ from traditional diversification, though, and pursue dual-use technology business is in re-applying their newly acquired capabilities back to the government marketplace. For example, SAIC acquired a number of transportation planning and technology companies and is now using those technologies to develop high-tech border crossings for the federal government. SAIC has also acquired Bellcore, the Bell Communications Research firm formerly owned by the regional Bell operating companies. Similarly, BTG justified its intention to acquire Micros-to-Mainframes as a way to "expand its business with private sector companies."[3]

On the other side of the market, some companies which were born in the commercial business are now pursuing government markets. For example,

SAP America, the U.S. arm of the German software firm SAP AG, has begun a public sector marketing effort to sell its highly touted financial and management systems software. SAP joins companies like Oracle and PeopleSoft in creating federal versions of their commercial products (Federal Computer Week, August 1997: 52). Still, there are many high-tech service companies like Perot Systems that assiduously avoid any contracts with the federal government, in large part to avoid the rules, accounting system requirements, and profit controls that accompany such business.

ISSUES FACING DEFENSE AND THE FEDERAL GOVERNMENT

Despite the success that some companies have had in expanding their business from defense and other federal agencies into commercial markets, there are several issues which inhibit further success. These issues include the differences between government and commercial requirements, government accounting and contracting rules, human resources issues, and funding for research and development.

Issue 1—Government Difficulties in Establishing Requirements

The government's requirements process makes it harder to buy information technology services from suppliers that are providing similar services to the private sector. In part, this is because the federal government sets requirements that do not exist in commercial business. Sometimes these requirements are driven by law, as when the government calls for a portion of subcontracts to be with small and disadvantaged businesses.

Often, though, these requirements go well beyond those required by law. For example, when the Defense Department announced its intention to buy an off-the-shelf system for managing procurement information, the Standard Procurement System, the announcement required features which no purely commercial system would include. Among these were the requirement that the vendor would provide its programming code to the Defense Department, the need to comply with a form of computer security that was largely absent from the commercial world, and the demand that the company incorporate changes in future laws on a fixed-price basis (despite the difficulty in predicting what laws Congress will pass in the future). No commercial off-the-shelf system would have such features, yet DoD's contract maintained the fiction that it was procuring such a commercial system.

Issue 2—Accounting Rules Reduce Access to Commercial Capability

Many companies that provide services to both the public and private sectors find it advantageous to segregate the business units by customer, because of the government's rules on, among other things, cost accounting, allow-ability of costs, and cost and pricing data. For example, SAIC has organized its commercial IT outsourcing business into a sector that provides services only to commercial and international customers. One of the consequences of this segregation is that personnel with desirable skills are not available for government work, including the best outsourcing management and technical personnel. This is solely because of the difficulties in business that would be caused by the accounting systems. Commercial IT contracts are based on the value they provide to the customer, and their prices are driven by the marketplace. On the other hand, the federal government's accounting rules control costs and prices, within the limits of the contract, and provide little insight as to the relationship between those costs and the corresponding value to the government.

Issue 3—Reduced Access to Skilled Workers

In order for the Department of Defense to have access to the best workforce, particularly in areas of high-technology service industries, it will need to be able to purchase those services from commercial companies. In some cases, the federal government has difficulty gaining such access. To gauge the extent to which this could be a problem, one need only look at the situation today, as the government faces challenges in upgrading its computer programs for the changes in date required by the year 2000. The issue is one of available personnel to assess, update, install, and test programming code for successful functioning post-2000. Such skills are already in short supply, and the federal government does not have adequate in-house staff and must rely on the private sector. Yet the federal government is ill-equipped to compete with the commercial sector for such skills, in part because it takes longer to negotiate contracts, in part because the government is unwilling to pay market rates, and in part because of the other issues already described above.

   Yet, even as the government comes to rely more on private companies with commercial services for its information technology needs, there will be an increasing need for the government to recruit and employ personnel with the skills needed to be a smart buyer of such services. The most likely source of these personnel will be from the very companies that provide such services, but the myriad pre- and post-employment restrictions that face employees in the federal government today make it virtually impossible for such personnel to move back and forth between government and industry. This will make it harder for the government to hire the skilled workforce it will need.

Issue 4—Government Research and Development Costs

In the past, when the Defense Department bought information technology services from private vendors, it would pay for the research and development

(R&D) costs of its unique information systems. Much of the cost, for example, of the WWMCCS (World-Wide Military Command and Control System) Improvement Program paid for software development in DoD's peculiar programming language, Ada. These costs were borne by the R&D accounts of the Defense Department, and the contractors were reimbursed based on their costs, plus an added fee. Similarly, the Reserve Component Automation System, or RCAS, was a program whose development costs were paid by DoD. These and many other software programs were procured to meet unique DoD needs and had little if any commercial or dual-use value.

This type of approach no longer applies in the dual-use information technology market of today, as DoD buys IT services with funds appropriated for Operations and Maintenance (O&M), rather than paying for R&D and procurement with funds appropriated by Congress specifically for those purposes. In fact, the federal government is prohibited from reimbursing R&D costs from any funding other than that provided directly for that purpose. Thus, in theory, DoD and the rest of the federal government will no longer be able to provide R&D support for IT services it wants to buy in a more commercial fashion.

In a true dual-use market, contractors fund their own research and development costs out of profit dollars. As long as the government does not levy unique requirements on the contractors, this does not pose a problem, as each contractor makes its R&D investment decisions based on expected returns from both commercial and government markets. In cases like the Standard Procurement System noted above, though, DoD required the contractor to perform additional development on an existing system but did not reimburse the contractor for such development costs. In such cases, the government will end up paying the costs of development, in the transaction costs of the services. These transactions will not support dual-use technology, though, because they will only serve government-unique requirements.

If, on the other hand, DoD procures IT services in a commercial fashion, then companies can invest the necessary R&D to support unique government requirements, provided that those requirements are stated in performance terms that can translate into equivalent commercial demands. In that way, firms can determine the levels of investment and risk that they are willing to undertake in order to meet those performance requirements, and the dual-use commercial market will help determine the price and return on that investment.

Benefits of Dual-Use Information Technology Companies

The issues summarized above are not insurmountable, but without a clearly directed effort to resolve them, they make it difficult for the federal government to take advantage of the benefits that dual-use information technology companies can offer DoD and the federal government. Some of these advantages are outlined below.

The most significant benefit for the federal government is access to the latest developments in technology. In many information technology applications, the commercial sector is advancing the state-of-the-art at a rate far beyond the pace of government-unique applications. Commercial companies are expending more resources than the government can afford, their development life cycle is shorter, and their return on investment is higher than if DoD were providing funding. DoD can only keep up with such rapidly changing technology by procuring services from these commercial providers. This creates the market for dual-use technology.

In addition, the commercial marketplace drives standardization and compatibility in ways that the federal government can take advantage of, and even help shape, but cannot replace. These standards are developed and incorporated into commercial practice far more rapidly than the historical standards development process provided for, because companies that ignore these rapidly evolving compatibility standards are left behind in the marketplace. DoD benefits from the reduced costs, increased reliability, and enhanced performance provided by commercial compatibility in software applications, communications, and information systems.

As the DoD weapons system and equipment inventory continues to age, the government has increasing requirements for the modernization and upgrade of existing equipment, as well as for maintenance of that equipment. The Defense Department has historically performed much of this work with government employees in government facilities, often with apparently low marginal costs but with total costs that are unknown but potentially high. (Federal government accounting and management information systems track expenditures very carefully, in accordance with the categories for which funds were appropriated by Congress. Actual costs for specific activities and tasks are much harder to pin down.)

The only alternative available to DoD in the past to such in-house work has been by contracting with the OEMs for modernization, upgrades, and repair. In some cases, this reliance has created the potential for non-competitive situations to arise, in which DoD becomes dependent on a single company for specific maintenance services. In such situations, DoD could pay excessive prices to these OEMs, because it would have no alternative.

Increasingly, the requirements for such upgrades have been in communications and electronics equipment, interfaces, and software. The hardware itself can last for decades (some of today's military aircraft are older than the pilots who fly them, for example), but the communications and electronics equipment, the signal processors, and the internal computers and software become obsolete in a few years. Expanding the reliance on dual-use IT service providers can give DoD an alternative to reliance on OEMs for upgrades and modernization, thus increasing competition and reducing costs. When coupled with the benefit of access to the latest technology (see above), these benefits multiply.

However, some question whether the government can rely on receiving priority delivery of such services if they are provided by dual-use companies with commercial as well as DoD customers. Historically, defense contracts have been subject to laws like the Defense Production Act, under which DoD can force priority delivery from its contractors. When the government procures services from commercial providers, these contractual protections may not automatically be created. Short of returning to the lengthy process of separate government requirements fulfilled by separate contracts, and thus abandoning the benefits of dual-use technology providers, what other avenues of protection would be available to DoD in time of crisis?

OUTSOURCING AND PRIVATIZATION

Perhaps the best avenue of protection is for DoD to contract directly with companies for the performance of entire functions. This is increasingly the path being taken by successful, globally competitive firms in many IT areas. Such a path will require the federal government to adopt a different set of roles than it historically has undertaken.

In large, complex information technology programs, there is a need for some entity to tie together all the hardware, software, networks, and sub-systems that make up the total system. That entity is referred to as the system integrator. In the past, the federal government often filled the role of system integrator itself. This role may have been acceptable, given the large percentage of government-unique features of such systems. In the future, using dual-use information technology solutions may lead to needed changes in systems integration. The federal government will not have the necessary resources (in terms of skilled personnel, tools, familiarity with the myriad numbers and types of systems and emerging technologies, etc.) to perform such integration. In addition, the commercial world has learned that, if one can define and measure the performance outcomes that are required, it is cheaper and faster to outsource IT functions to companies that can provide turn-key operations in both government and commercial markets. Under such a scenario, DoD can indeed contract for guaranteed performance, thus providing both for protection in time of crisis and for better performance at lower cost.

GUIDELINES FOR SOLUTIONS

When coupled with the benefits described earlier, outsourcing opportunities create the business case for increased reliance on and use of commercial dual-use information technology companies by the government, especially the Defense Department. The benefits to the government from buying commercial information technology solutions can be the catalyst for marshaling the resources needed to address and resolve the issues outlined above. Many of those issues can only be addressed by changes in the government's acquisition process. The emergence and sustenance of dual-use markets, technologies, and skills are integral parts of acquisition reform as it has been legislated and is currently being pursued by DoD and the federal government.

Acquisition reform has long been a goal of those who would improve the operations of the federal government.[3] Since 1993, however, an unprecedented set of changes in law and regulation have opened up new opportunities for commercial and dual-use companies to sell services to the federal government. The Federal Acquisition Streamlining Act of 1994 (FASA) and the Federal Acquisition Reform Act of 1995 (FARA), coupled with the associated changes in the Federal Acquisition Regulation (FAR), are aimed at removing many barriers for commercial vendors.

Barriers remain, however, in contracting practices and especially in financial areas. In essence, the government does not yet know how to contract well for IT services from dual-use technology integrators. The federal government still has difficulties defining what performance it seeks, but it cannot settle alternatively for simply buying the best capability available today. In order to offset that difficulty, the government still has a strong tendency to define what the supplier should do, since it cannot define the performance the government expects. This input-oriented form of contracting must be abandoned in order for DoD to benefit from dual-use commercial technology.

Government contracting officers and contract audit officials need to adopt and practice the new regulations for commercial procurements. Often, they still raise questions about the allowability of costs, still attempt to impose limits on profit, and still require that contracts include rules requiring that vendors keep cost-and-pricing data records not normally used in commercial business. Whether these actions are based on out-of-date rules not yet rescinded or on a cultural idea that remains strong, they serve to prevent companies from successful commercial–government business integration by making them continue to segregate their government and commercial operations.

Sometimes it is more than rules and their applications that constrain integration. Limitations in the skills, experience, and expertise of contracting officers in the federal government inhibit successful commercial-government integration. The government needs better trained contracting officers, with more commercial experience. Little of this inadequacy should be blamed on the officials themselves, but it reflects the slowness of the changes in regulations and in training for those changes.

These limitations manifest themselves in another way. Since the government buyers cannot define the actual price or measure the actual value of the services they seek to buy, they may focus instead on the price per hour of the various types of labor needed to perform the job. In part, this is due to government accounting systems that remain input-oriented and that cannot calculate actual costs of specific functions or activities. This approach of contracting for services through negotiating the lowest hourly rates, which might seem to protect taxpayers from exorbitant fees, in fact may cost the government more money, when lower-paid, less-experienced staff take disproportionately longer to perform the same work or even perform it less well. By contracting for total performance rather than hourly rates, the government will receive more value at less cost.

CONCLUSIONS

The changes in the marketplace for information technology have produced successful defense conversion, or, more appropriately, defense diversification results for a significant number of information technology companies. These results have created and supported the potential for true dual-use technology development and applications.

These successes can also translate into substantial benefits for the federal government in general and the Defense Department in particular, but a number of barriers remain that limit these benefits. Unless the government develops and pursues policies that address these barriers and resolves a number of other issues, the benefits will remain elusive, and the development of dual-use technology will be constrained. Since commercial applications will continue to move forward independent of any government procurements, based on the size and growth patterns in those markets, the ultimate loser in this scenario will only be the federal government and the taxpayers.

NOTES

1. Government Technology, State and Local Government Market Overview, December 1997.
2. *FY 1997 Annual Report*, Computer Sciences Corporation.
3. Reform commissions, beginning with the Truman Committee hearings in 1942–43, have focused on the need for the federal government to rely more on commercial-type practices in procuring its goods and services. See, for instance, the 1986 reports of the President's Blue Ribbon Commission of Defense Management.

REFERENCES

Defense News, p. 6 (chart), July 21, 1997.
Federal Computer Week, August 11, 1997, p. 52.
Strassmann, P. *The Squandered Computer*. New Canaan, CT: Information Economics Press, 1997.*Washington Post*, p. C10, September 3, 1997.

17

ARE DEFENSE AND NON-DEFENSE MANUFACTURING PRACTICES ALL THAT DIFFERENT?

MARYELLEN R. KELLEY AND TODD A. WATKINS

INTRODUCTION

There are tens of thousands of firms in the defense manufacturing base. With defense procurement outlays in FY 1997–98 60 percent lower in real terms than a decade before, much ongoing policy discussion centers on the conversion and diversification of defense contractors into commercial markets. Yet, beyond case study investigations of leading firms, there has been little attempt to systematically investigate how defense manufacturers' practices compare to commercial ones. Our ongoing research is the first systematic comparison in 35 years of defense contractors to establishments that do no defense contracting.

We evaluate the differences among the defense and strictly commercial industrial bases and the implications for the prospects for dual-use manufacturing and diversification by addressing the following questions: How separated are commercial and military production? How do defense and non-defense contractors compare in their use of advanced flexible manufacturing technologies? How do the collaborative practices of the broader production networks of defense manufacturers operate in comparison with those networks of strictly non-defense establishments? Wherever possible, we answer these questions through statistical comparisons of defense contractors to manufacturers with no contract ties to DoD but operating in the same industries and relying on the same underlying process technologies. In addition to our own 1991 survey of a representative sample of establishments in the machining-intensive durable goods sector (MDG), we use supplementary information from specialized government data sources and case studies describing the practices of selected defense contractors.

The defense industries in 1997–98 are rapidly moving targets in both structure and size. Caution is clearly in order concerning the applicability of conclusions from our 1991 data and case analyses done over the past few years. Nevertheless, we believe that our findings—which are contrary to long-held conventional wisdom—may refocus debate over the possibilities and challenges of defense diversification and dual-use manufacturing. To us, the policy questions are not about the extent to which they are possible: even during the peak years of defense procurement budgets, we show that much defense manufacturing was already considerably integrated with commercial manufacturing. The questions should be, instead, more along the lines of how to maintain the strengths of the defense industrial network— since World War II the principle instrument of U.S. technology policy—in an era of dramatically lower defense spending.

CONVENTIONAL WISDOM: DEFENSE INDUSTRIES AS A CLOSED, DISTORTED SYSTEM

High-tech products made for the military such as tanks, missiles, satellites, submarines, or fighter aircraft have some similar features. Each is a complex product customized to the requirements of a single customer. The manufacturing processes can involve exotic materials, sophisticated technology, and specialized engineering expertise. The customer also has the political power to restrict the sale or use of products to other potential customers. A company that makes a new high-tech weapon for the DoD cannot sell that weapon to another customer (i.e., another government) without the Defense Department's permission. The DoD still forbids commercial use or sale of some of the components of these systems. These and related peculiarities provide the basis for the assumption that there has been very little integration between a production system that satisfies defense procurement and one designed for commercial transactions (Gansler, 1995; Markusen and Yudken, 1992; Melman, 1974; Weidenbaum, 1992). More generally, the conventional wisdom holds that the defense industrial base is a closed, oligopolistic production system grossly distorted in its use of technology, batch sizes, and its dependency on a single customer. An analysis of the defense industry by Murray Weidenbaum (1992, p.131), former Chairman of President Reagan's Council of Economic Advisors, is illustrative: "Truly, military and civilian decision-making differ so substantially that they are almost worlds apart."

In the remainder of this chapter, we examine two propositions about the character of defense-dependent companies and industries. Beliefs in these propositions discourage companies from embarking on their own diversification campaigns and underlie skepticism about dual-use manufacturing and the extension of industrial technology policies to the diversification of defense-related resources into civil and commercial applications.

(1) The unique market structure and regulatory environment of the defense industry compel companies to isolate their defense production. The engineers, managers and workers employed in these operations have little experience with commercial customers. Conversion of this base would require re-organization, considerable re-training, and in many cases, entirely new management. Comments by Alic, *et al.* (1992: 142), among the leading proponents of dual-use manufacturing, are typical of this widely-held belief that firms "conduct military business in divisions that are managed separately from commercial operations, often with separate work forces, production and research facilities, accounting practices, engineering design philosophies, and corporate culture."

(2) The absence of competitive pressures implies that government contractors will refrain from making investments in new productivity- and flexibility-enhancing technologies and organizational practices to the same extent as producers in commercial markets. As a result, compared to strictly commercial enterprises, defense contractors' manufacturing processes are technologically and organizationally behind and are limited in their flexibility to diversify into commercial markets. Retooling and reorganizing will be necessary to achieve the integration of military production capabilities into the commercial industrial base. Again, Weidenbaum (1992: 146): "Under the circumstances [Federal Procurement Regulations] it is not surprising that the major military contractors have been reluctant to make substantial new investments in their factories and production equipment."

These propositions about the singularity of the defense industry are largely unexamined and hypothetical. Defense contractors' practices have been subject to considerable public scrutiny, and numerous studies have focused on the special problems of the military enterprise. Yet we cannot identify a single empirically rigorous study since Peck and Scherer (1962) 35 years ago in which the market structure and manufacturing and organizational practices of defense-dependent establishments are systematically compared to those of manufacturers of commercial goods. As Gansler (1995: 29) notes, "It is at the plant level itself, which is the most important area as far as individual employees are concerned, that surprisingly little information is available."

Moreover, manufacturing of such complex products as aircraft and other weapons systems requires capabilities that no single company has by itself. Yet the larger set of facilities engaged in manufacturing some part of these products have hardly been studied at all. Our main criticisms of the literature are the absence of any systematic effort to compare practices of defense contractors to their commercial counterparts, and the failure to include in those comparisons the vast supplier network. Yet, that network's prospects for dual-use manufacturing and its flexibility in diversification are arguably as important a policy and economic issue as the top-tier prime contractors. Only with such comparisons is it possible to sort out the real differences that divide the defense and commercial industrial spheres. In this chapter, we undertake such an examination of defense and commercial production and the larger networks that support that production in the machining intensive durable goods sector (MDG).

THE EXTENT OF DEFENSE MANUFACTURING

From 1980 to 1987, purchases by the U.S. Department of Defense were the single largest contributor to the growth in U.S. manufacturing. At the 1987 peak of the Reagan defense build-up, DoD accounted for almost 12 percent of the sales of durable goods manufactured in this country (Alic, et al., 1992). Yet, the concentration of Pentagon spending leads many analysts to conclude that only a small number of large companies benefited from this increase in defense spending. A frequently cited statistic in support of this conclusion is data compiled each year by the Pentagon: the largest 100 defense contractors receive about 60 percent (58.4 in FY 1996, 61.5 in FY 1993) of the value of the total defense prime contract awards in excess of $25,000.[1] Current concentration by this measure is down considerably since the early 1960s through late 1970s when it hovered around 70 percent.[2] Nonetheless, this domination by the major defense contractors has led to a long-standing and widely held belief in the difficulties the defense industry would have in competing in commercial markets.

Yet, when considering the implications for diversification and dual-use manufacturing, it is important to consider as well the vast supplier networks upon which these firms rely. Major weapons systems manufacturers stand at the top of a production chain similar in structure to manufacturing systems for complex commercial products. Within their own facilities, these prime contractors largely confine production activities to final assembly. In commercial manufacturing industries for such complex products as automobiles, trucks, and civilian aircraft, the final producers do not manufacture all of the components of these products. Subcontractors make the parts and components as well as the specialized equipment, e.g., robots or machine tools, which the final producers use in assembly operations. For defense products, the establishments that have either prime or subcontracts make up the defense industrial base. The size and proportion of all DoD purchases that go towards subcontracts provide convenient indicators of the importance of this broader industrial base, indicators missing from previous analyses, which focused on the largest prime contractors.

More importantly, the characteristics of this broad supplier network have important implications for the competitive abilities of the national defense industrial base to undertake dual-use manufacturing and diversification. For example, Gansler (1995) and Alic, *et al.* (1992), among others, have speculated that the prospects for dual-use manufacturing may be higher at the component level than at final assembly.

Unfortunately, the only systematic information on subcontracts that DOD collects from prime contractors concerns subcontracts to small enterprises.[3] Various analyses that do exist of weapons systems' costs suggest that subcontractors are responsible for a substantial share of defense manufacturing.

In published sources and in our own interviews with manufacturing managers at major prime contractors, we find that the dependence on subcontractors ranges from 60 percent to more than 70 percent of prime contractors' costs. At Pratt & Whitney, a manufacturer of aircraft engines, "approximately 60 percent of the dollar value of its engines is materials and components purchased from suppliers."[4] Similarly, at Allied Signal, which makes sub-systems, purchases of materials and components accounted for 60 percent of total costs.[5] For GE Aircraft Engines, subcontracts consume two-thirds of the overall cost of producing a military aircraft engine. At what is now Lockheed Martin's Fort Worth F-16 manufacturing plant, managers involved in supplier development activities estimate that subcontracts consume more than 70 percent of the cost of the aircraft. These levels of pass-through (as reported by major prime contractors) are consistently higher than the 50–60 percent range reported in the Rand Corporation's 1965 study of the subcontracting cost of selected weapons systems (Hall and Johnson, 1965).

In 1991, we surveyed 1124 manufacturing establishments, of which 973 were still in manufacturing and engaged in the precision machining process. The sample is the cohort of plants in the MDG sector, first surveyed by Kelley and Brooks (1991) in 1986–87. That sample was selected by stratifying all establishments identified in Dun and Bradstreet Company's plant universe (of 1984) as belonging to the most machining-intensive industries into the following five employment size categories: fewer than 20, 20 to 49, 50 to 99, 100 to 249, and 250 or more workers. An equal number of plants was selected from each stratum, resulting in proportionally greater sampling from the larger plant sizes (and statistically weighted accordingly to give corrected overall population estimates). The overall effective response rate to the combined telephone and mail survey in 1987 was 89.3 percent. For the 1991 survey, the effective response rate is 91 percent.

The original survey was completed on a national sample of plants engaged in the machining production process.[6] Each of the 21 industries selected for inclusion in the sampling frame accounts for at least 1 percent of all employment in machining occupations in all of the manufacturing sector. Moreover, for each industry, machining employment constitutes at least 10 percent of all production employment in the industry. We call this set of industries the machining-intensive durable goods (MDG) sector.[7] The manufacturing of high-tech military hardware, in the form of aircraft, satellites, and missiles, is concentrated in this sector. Collectively, the 21 industries selected by these criteria account for more than half (51.3 percent) of all durable goods purchased for defense[8] and more than one-fourth of all U.S. manufacturing employment.

Because the survey's technological focus (to enable comparing similar production processes across establishments) was on machining, the sampling frame does not include a number of important defense industries, notably communications equipment, electronics, computers, tanks and shipbuilding. Conversion and dual-use issues may differ in these sectors, so our results for MDG manufacturers may not be generalizable to the entire defense base. Tanks and shipbuilding are among the most defense dependent of all industries, and technological change is notoriously rapid in communications, electronics and computers. Thus, we hesitate to speculate about that half of the defense base.

Nonetheless, our sample industries do represent a large and important fraction of defense manufacturing. Of the top 50 (4-digit) defense industries at the peak of the defense buildup in 1987, according to Alic et al. (1992), our sampling frame encompasses 26, including 16 of the top 20 as ranked by defense share of total industry sales. As a result, defense dependency in our sample industries may be overstated compared to the average defense industry. This makes the degree of dual-use manufacturing we find already occurring all the more remarkable.

The Department of Defense is the final customer (through prime contracts or subcontracts) for an enormous number of production facilities in the United States. From our 1991 survey of a sample of establishments, we estimate that 48.8 (± 3.1) percent of all plants in the MDG sector were defense contractors (Table 17.1). That amounts to nearly 40,000 facilities.[9] Our sample estimate is consistent with Census Bureau data on the extent of the defense industrial base in this sector. In 1988, the Census of Manufactures conducted a special survey on technology and defense manufacturing. With data from the Census survey,[10] we estimate that 49.7 (± 1.0) percent of establishments with 20 or more employees in the MDG sector in 1988 were either defense prime contractors (selling directly to one of the federal defense agencies) or were subcontractors to defense prime contractors. Despite declines in defense spending in real terms between 1988 and 1991, we find no statistical evidence of a decline in the share of the overall manufacturing base in the MDG sector that served the Department of Defense as of 1991.

Table 17.1
Distributions of Defense Contractors' Sales in 1990 to Military and Commercial Customers
by Type of Contractor
(Sales in $ millions)

Defense Contractor Type	All Sales	% of All Sales by Type of Contractor	Defense Sales	% of Defense Sales by Type of Contractor	Commercial Sales	% of Commercial Sales by Type of Contractor
Only Prime Contracts	$16,848.83	6.7%	$765.61	0.8%	$16,083.21	9.9%
Prime & Subcontracts	$153,254.91	60.6%	$70,338.89	77.0%	$82,918.02	51.3%
Only Subcontracts	$92,968.87	32.8%	$20,200.09	22.1%	$82,768.58	38.8%
All Defense Contractors	$253,072.63		$91,902.59		$161,769.82	
% of All Sales		100.0%		36.1%		63.8%

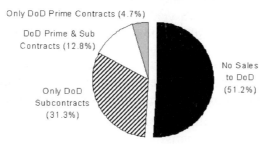

Distribution of MDG Sector Plants by
Defense Contracting Status

Only DoD Prime Contracts (4.7%)
DoD Prime & Sub Contracts (12.8%)
Only DoD Subcontracts (31.3%)
No Sales to DoD (51.2%)

The continued defense drawdown since 1991 may have reduced the breadth of the defense supplier base. However, a 1997 GAO study reported on a random sample of small California aerospace businesses that supplied goods or services to large military aircraft programs. That survey showed that between 1992 and 1995, 94 percent were still in business while 3 percent had either merged or been acquired (GAO, 1997).

Even if the base consolidated since our survey year, in U.S. manufacturing, there remains a vast hidden defense industrial base consisting of a large number of subcontractors with no direct dealings with the Pentagon. Allied Signal, for example, among the top 25 defense prime contractors, had about 3000 direct suppliers in early 1997 (Minahan, 1997). As Table 17.1 shows for the MDG sector, 64.1 percent of plants with any defense-related sales did not sell directly to DoD in 1990, but rather served only as subcontractors or suppliers to defense prime contractors. Our survey data indicate considerable pass-through from prime contractors to this subcontracting base. Spending on subcontracts alone accounted for 41 percent of all defense-related sales and shipments in the MDG sector during 1990. Moreover, more than half (54 percent) of the value of shipments to prime contractors from subcontracts in the sector came from lower tier suppliers, i.e., those that had no prime contracts with a federal defense agency.

Conclusions about the uniqueness of the defense industrial base that rely solely on information about prime contractors miss the influence of DoD on the tens of thousands of subcontractors that make equipment for the military. Related policy prescriptions overlook whether that broader industrial base is flexible enough to support diversification and undertake dual-use manufacturing.

THE EXTENT OF COMMERCIAL AND MILITARY INTEGRATION IN PRODUCTION

Even the largest defense contractors belong to companies that depend on commercial sales for the greater part of their total revenues. At the corporate level, Alic, *et al.* (1992) found that, among the 100 largest defense prime contractors, even during the height of the 1980s defense buildup, the 67 publicly traded firms derived only 9 percent of their total sales from defense prime contracts over the five-year period ending in 1988. Moreover, only 9 of those 67 firms were highly defense-dependent, with 50 percent or more of their sales going to DoD during those peak years of the build-up.[11] Yet, because many of these companies have set up separate divisions for their defense business, little or no interchange is assumed to take place between the defense and commercial sides from the top to the bottom of the enterprise.

A separate division within a corporation indicates a separate chain of command for managers responsible for defense production. However, such an organizational structure may not imply a physical separation between the people and machines actually involved in defense and commercial manufacturing operations. For example, corporations commonly employ what is called a matrix reporting structure, in which groups with the same functional responsibilities have dual reporting responsibilities: to a product division, and to the director of a functional area, such as the chief of manufacturing operations. In such a matrix structure, the alleged segregation of defense work from commercial work may simply be an artifact of reporting lines of authority. This does not imply that the organization has literally constructed separate work groups or facilities for the two divisions.

Even so, the segregation from commercial operations of facilities and production equipment used exclusively for the manufacture of military products is frequently described in the academic and business press as if it were the established practice of most defense contractors.[12] This separation is also thought to extend from the headquarters to the shop floor. According to Markusen and Yudken (1992), special accounting rules, technical requirements, and the like are responsible for a "wall of separation" that divides production for the military from commercial manufacturing. Such high profile examples as Lockheed Martin's Skunk Works (the incubator for the U-2 and SR-71 "Blackbird" spy planes and later the F117A Stealth fighter), or General Dynamics' Fort Worth F-16 manufacturing facility (now also Lockheed Martin's), have helped perpetuate the view that defense production largely takes place in facilities where no commercial products are made.

Drawing on our 1991 survey data of manufacturing establishments, we attempt to measure the extent of segregation between defense and non-defense production through an assessment of the following:

- What proportion of the defense industrial base in the MDG sector operates specialized facilities dedicated to the manufacture of military hardware?
- What proportion of total defense output in the sector is produced in segregated facilities?
- Do prime contractors (especially those that are part of large companies) have a greater propensity to operate defense-dedicated production facilities than subcontractors?

Our first indicator of segregation is the percent of total 1990 shipments from each plant sent directly to a federal defense agency (including any branch of the U.S. Armed Forces, the Defense Logistics Agency, depots of the services, and the Department of Energy) or to a prime contractor of one of these agencies.

As Table 17.1 shows, nearly two-thirds of all the output generated from defense contractors in the MDG sector in 1990 went to commercial customers. Overall, defense shipments contributed only 36.1 percent of the total value of 1990 shipments from plants that make military equipment in the U.S. MDG sector.

We find that, contrary to the conventional wisdom, the typical defense contractor in this sector is not very defense dependent, even at the establishment level at which we collected data. The median defense share was only 15 percent among plants with any 1990 defense sales in the MDG sector. The vast majority (80.4 percent) of establishments integrated commercial and military production in the same facility, selling more than half of their 1990 output to commercial customers. As Fig. 17.1 shows, only 21.4 percent of plants with prime contracts had more than 50 percent

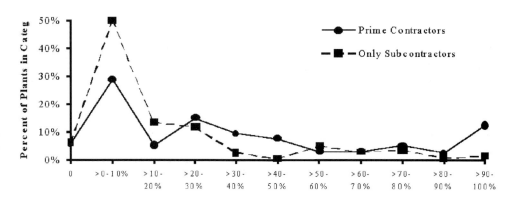

Plant's Dependence on Sales to Defense Customers

Fig. 17.1. Defense contractors' dependency on sales to DoD in 1990 by contracting tier: all prime contractors vs. subcontractors.

of their sales going to DoD in 1990. For the lower tier contractors, only 18.5 percent shipped more than 50 percent of their 1990 output to defense prime contractors.

Because a large number of these establishments are small, we also investigated whether, despite the numbers of integrated establishments, most of the value of defense-related manufacturing was done in defense dependent facilities. In Fig. 17.2 we cumulatively add 1990 shipments by establishments that reported any shipments to defense agencies or prime contractors to defense agencies at all, ordered by the degree of the plant's dependency on sales to defense customers. As Fig. 17.2 shows, we estimate that more than half of the value of defense-related work in this sector came from plants that did the majority of their work for non-defense customers. Moreover, less than one-third (32.7 percent) of the value of total shipments of military goods from the sector in 1990 came from highly segregated facilities (with 80 percent or more of their output going to defense).

Another way we attempted to account for the variation in size and organizational strategies of companies in the defense industrial base was to look at the differences between multi-plant firms and single-plant firms. Multi-plant companies have the option to place all of their defense orders in one facility and commercial work in another. If multi-plant corporations adopt such a segregation strategy, we should find a higher incidence of dedicated facilities among branch plants doing defense work than among single-plant enterprises. In Fig. 17.3, we see that there is no statistical difference between these two types of companies in the proportions of facilities which are highly specialized in making defense products.[13] For large, multi-plant firms, and small single-plant enterprises alike, fewer than one in five of the plants that do defense work sell more than 50 percent of their output to DoD or a prime contractor.

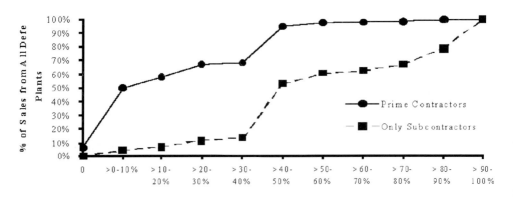

Fig. 17.2. Cumulative distributions of 1990 sales by defense contractors to defense & commercial customers.

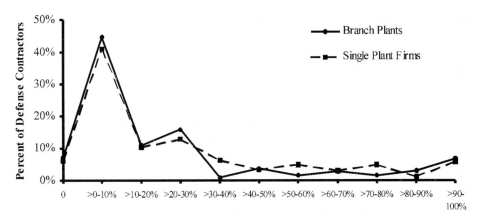

Plant's Dependence on Sales to Defense Customers

Fig. 17.3. Defense contractors' dependency on sales to DoD in 1990 single-plant firms vs. branch plants.

Although larger firms are not more defense dependent, on average, than smaller firms are, we do find that facilities dedicated to defense production are somewhat more common among those branch plants of large corporations that receive prime contracts. As we show in Fig. 17.4, which looks only at branch plants of multi-plant firms, prime contracting defense plants belonging to multi-unit firms are significantly more dependent on sales to DoD, on average, than branch plants that only have subcontracting ties to DoD. For example, a larger fraction of prime contractors (22.3 percent) than subcontractors (12.1 percent) depend on DoD (or other prime contractors) for 50 percent or more of their sales. These differences are statistically significant ($p<0.05$). If branch plant prime contractors are more

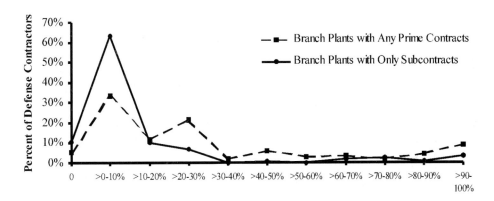

Plant's Dependence on Sales to Defense Customers

Fig. 17.4. Defense contractors' dependency on sales to DoD in 1990 branch plants of multi-plant firms: prime contractors vs. second-tier subcontractors.

likely than subcontractors to do assembly work, then this finding is consistent with the view that prospects for dual-use may be higher on the component level than the system/assembly level. Regardless, facilities that serve both commercial and military customers are the overwhelming norm across both branch plant categories.

In short, at the level of the plant, we find considerable integration between the commercial and military industrial spheres in the MDG sector. Large multi-plant firms that do defense prime contracting tend to be slightly more dependent on average than subcontractors. But overall, we find that defense production in the MDG sector (whether directly for DoD or indirectly through subcontracts) more often than not takes place in facilities in which the majority of production is for commercial customers.

In order to satisfy ourselves that the integration of production for military and commercial customers suggested by our statistical data reflects practices on the shop floor, we undertook a number of more detailed case studies. We selected cases from respondents to our survey, from a review of previous research, and from interviews with managers of major defense contractors. Our case investigations were designed to identify if separate equipment, production workers, or engineers are assigned to military production in plants that are engaged in manufacturing for both DoD and commercial customers.

High Profile Cases of Commercial-Military Integration

General Electric has been among the top 10 DoD prime contractors for more than 40 years.[14] OTA (U.S. Congress, 1992: 202) describes GE as the quintessential case of a company that fully integrates its commercial and military production: "GE Aircraft Engines is the leading example; . . . [it] combines all aspects of its military and commercial business except for marketing, while still complying with DoD requirements." The two sides of the business share management, R&D facilities, and manufacturing. Despite a huge dollar volume in defense sales, GE Aircraft Engines derives more than half of its revenues from commercial engines. Sometimes GE even sells the same engine to both defense and non-defense customers. In a joint venture with the French firm Snecma, they produce CFM56-2 jet engines, the technical core of which powers the B-1 bomber, for DC-8 commercial airplanes as well as the Air Force's KC-135R tanker aircraft. Similarly, Pratt & Whitney sells the PW-2037 engine for both commercial and military use.

Hughes Space and Communications Company, is a leading supplier of commercial and military satellites and also consistently among the top 10 defense prime contractors. Not only does Hughes produce commercial and military components in the same facilities, but it also integrates its design processes as well. According to Albert Wheelon, former CEO of Hughes, "The design and fabrication of spacecraft subsystems is centered in the engineering and manufacturing divisions. In order to capture the benefits of scale and retain the flexibility to interchange parts and manpower when needed, these two divisions serve all programs, regardless of the structure of the individual customer's contract. One implication of this organizational design is that technical manpower in the engineering and manufacturing divisions is entirely interchangeable among projects" (Alic *et al.*, 1992, p. 179).

Another example is Vought Aircraft. Vought produces major aircraft structural subsections for both commercial and defense-related customers. e.g. tail sections for Boeing's 747, 757 and 767; engine nacelles for Canadair's CL-601RJ regional jet; wings for the new Gulfstream V (G-5) corporate jet; tail sections, aerial refueling receptacles, and engine nacelles for McDonnell Douglas' C-17 Globemaster III military transport; and until recently, the complex B-2 Stealth intermediate wing section, about one-third of the total B-2 airframe structure by weight. In August 1994, Vought Aircraft was acquired by the Northrop Grumman Corporation, itself created when Northrop acquired Grumman in March 1994. Vought employs approximately 5000 people in its facilities near Dallas, Texas, and its annual revenues, according to Northrop Grumman, near $600 million.[15] Vought was made the home of Northrop Grumman's Commercial Aircraft Division. As discussed in detail in Watkins (1997), during our visits and dozens of interviews, managers and shop-floor personnel there clearly demonstrated how cross-functional teams are the organizational norm at Vought, making it a truly dual-use operation. Integrated, centralized functional groups such as engineering, machining and fabrication, quality assurance, supplier management and so on, serve all programs, both military and commercial, with the same people and procedures. Vought operates under an "integrated product/process development" (IPPD) philosophy with what their human resource managers call a "strong matrix" organizational structure. In addition to reporting to a functional group, one axis of the matrix, people also report to (and are co-located with) multi-functional product or process teams, the other axis, that have full responsibility for integrating and managing all aspects and the whole life-cycle of each commercial or military program, from development through delivery and post-production support.

Alic *et al.* (1992) report similar integration at the Lord Corp., a leading supplier of rubber-to-metal adhesives and computerized vibration-control equipment. Lord uses a single division and the same engineering group to work on the Boeing 737, 757, and 767 aircraft as well as the Black Hawk helicopter and the Osprey tilt-wing transport. The Castings and Forging Division of Wyman Gordon Co. employs the same people, processes, and equipment in supplying special alloy castings to GE Aircraft Engines, Pratt & Whitney, Boeing, and McDonnell Douglas. Hewlett-Packard's Microwave semiconductor division integrates military and civilian production as well.

Commercial–Military Integration in Subsystem and Component Manufacturers

We selected a number of smaller defense contractors that vary in their degree of dependence on sales to DoD or prime contractors. In every case, we find that these subsystem and component manufacturers operate completely integrated facilities, using the same people and equipment for both commercial and military products. We offer three examples to illustrate how these production sites handle differences in production requirements (if any) between their commercial and military customers.

Tecknit

In our first example, we might expect a high potential for segregating military work, because approximately 50 to 60 percent of the firm's business is defense-related. Tecknit, founded in 1958 as Technical Wire Products, Inc., designs and produces materials and components for electromagnetic interference (EMI) shielding, grounding, and static discharge. The firm employs about 300 people in manufacturing and sales facilities in the U.S. and U.K., the majority of whom work in the main plant in Cranford, New Jersey. We visited the Cranford facility.

The company's original product line was seamless knitted-wire-mesh rings and gaskets, manufactured for both military and commercial markets on equipment of their own proprietary design. Thus, the core capability of the firm was a process technology designed and built in-house. However, rather than remaining a wire knitting specialist, the company has re-focused on technologies that provide solutions to problems from electromagnetic interference.

Tecknit's product line now also includes a wide array of products with similar functions: patented conductive elastic polymers (similar to rubber), conductive adhesives, paints, and greases, as well as shielding screens, coated windows, and air vent panels. Their products are used in electrical equipment or components that either emit or suffer interference from electromagnetic radiation in the power, radio or microwave ranges of the spectrum (e.g., personal computers, power supplies, aircraft navigation equipment). Tecknit's largest customers include Westinghouse, Rockwell, Raytheon, Boeing, Hughes, Magnavox, and Texas Instruments. In addition, Tecknit sells to telecommunications equipment firms and computer companies, including Apple, DEC, Dell, Siemens and IBM.

Although the company offers a standard range of EMI products, Tecknit operates primarily as a job shop. Its production is low volume and labor intensive. Much of the assembly work (for example, the tasks of bonding elastomer gaskets to machined aluminum frames) is still done by hand.

We find no differences in the manufacturing process for defense and non-defense products. According to the manufacturing manager we interviewed, and our own observations, there is no special labeling on products made for defense contractors that would distinguish them from products made for commercial customers. The technology, manufacturing equipment, process flow, labor, and engineering are indistinguishable from one customer to another. The only differences in requirements occur in the documentation during final inspection. DoD has reporting requirements in tracking materials and in documenting inspecting and testing procedures that are not demanded by other customers.

Electroid

This case is more typical of defense contractors in terms of dependence on defense work. The share of sales going to defense prime contractors has never exceeded 20 percent. Again, we find complete integration between military and non-military manufacturing. The Electroid Company is a specialty manufacturer of high performance motion-control devices. Their core business is in electro-mechanical clutches, brakes, and solenoids. With current manufacturing facilities in Springfield, New Jersey, Electroid is a privately owned division of Valcor Engineering Corporation. Electroid employs about 100 people at this facility.

When most of us think about clutches and brakes, our automobiles come to mind. Yet Electroid has a clear corporate strategy that completely avoids the automobile industry, which management considers a low-performance, high-volume (and low-profit margin) market. Instead, Electroid supplies medium- to high-performance electro-mechanical stop/start motion-control equipment in low volumes. Their products are used in industrial machinery and a number of aerospace applications, largely in motors or actuators for the purpose of engaging or disengaging mechanical power, or stopping or locking moving parts. The company supplies such devices for packaging machinery, photocopiers, industrial robots, the Apache helicopter, and the turret on the M1 tank. NASA's Space Shuttles have Electroid's fail-safe brakes to lock the doors in position. Some of Electroid's major customers include Allied Signal, Boeing, General Electric, Scientific Atlanta, and Westinghouse.

By defining their technological niche as motion control, Electroid naturally serves a broad cross-section of industries. Most of their orders come from makers of industrial machinery, but the company has been a defense contractor since the beginning of the Reagan build-up in 1981. For its products, Electroid identifies military requirements to be as stringent as those in commercial applications in aircraft and nuclear power plants. That group of customers is collectively known at Electroid as "NAM" for "nuclear, aerospace, and military." NAM accounts for 15 to 20 percent of total sales. According to the vice president for manufacturing, the designs, materials, tolerances, inspection and reporting requirements for Electroid's NAM work are more exacting than the demands of their other customers.

Nonetheless, this manufacturing facility remains completely integrated between defense and non-defense work. The manufacturing process on the shop floor uses the same production workers, the same manufacturing equipment, and the same engineers for both commercial and defense jobs. A worker could spend one hour on a NAM job and then the next hour on an industrial machinery part. No equipment or any employee in the plant is dedicated to military (or NAM) production. The only feature of the production process that identifies military products is also used to distinguish all NAM work from products made for other customers. Work for NAM customers is placed in blue plastic tote bins and pieces for non-NAM customers in tan-colored bins. The main reason for this color-coding is to alert the employee at each work station to follow the written instructions that accompany the NAM item. For NAM products, whether for military or commercial customers, detailed specifications dictate the tasks to be performed (and checked) at each stage in the production process.

Delroyd Worm Gear

This case is another example showing that the process and standards for making some commercial and military products are indistinguishable. Delroyd Worm Gear manufactures large speed-reducing worm gears for use in high torque applications. The company is a division of IMO Industries and employs fewer than 100 people at the facility we visited. Their products are used, for example, in conveyors, printing presses, oil drilling pumps, and cranes. They make gears for canal locks, including some used on the Panama Canal, and for aircraft hanger doors. Customers can order products from the company's catalogue, but custom orders are also accepted.

Delroyd has been a defense contractor since World War II. The DoD accounts for 5 to 10 percent of their business, largely buying replacement parts for Naval ships.

There is absolutely no separation of manufacturing work for the Navy from other work. No machines or workers are specially set aside. Nor are there any special testing or inspection requirements associated with defense contracts. The Navy simply places an order, identifying the replacement part it wants Delroyd to build.

Worm gears are a "mature" product. Technological changes occur very slowly. One production manager we interviewed said that new materials are the only major product improvement in the past 40 years. Today, worm gears are made with less expensive, more durable materials than two generations ago. The only defense-related peculiarity is that government purchase orders for replacement parts generally specify the same materials as the original order. For replacement parts on older ships, this can mean the Navy purchases products made with inferior (and more expensive) materials than used in new gears.

COMPARING NETWORKING AND TECHNOLOGY INVESTMENT PRACTICES

In addition to finding that most defense contractors are already considerably diversified into commercial markets, we also find evidence that they may be better suited than their strictly commercial brethren for dealing with changing markets and dynamic competition. Organizational and technological flexibility are important elements of any diversification strategy. By several key indicators, defense contractors have adopted more flexible manufacturing technologies and organizational practices.

Our findings lie in stark contrast to prevailing views on the defense industrial base. Appealing to economic theory, Markusen and Yudken (1992) argue that the normal conditions of exchange between buyers and sellers do not operate in the defense industry. As a consequence, defense contractors do not face the kinds of competitive pressures to innovate or to minimize costs that (at least in theory) affect companies operating in strictly commercial markets. In a similar vein, Demski and Magee (1992) identify a number of unusual features of military product markets that can be expected to further distort firms' behavior: administered prices, uncertainty, and a single buyer (DoD) with considerable regulatory power over its suppliers. Rogerson (1992) argues further that the cost-based pricing rules for defense contracts actually provide a perverse incentive for suppliers to under-invest in technology and to subcontract out less, employing more direct labor, than would be expected of enterprises operating in commercial product markets. This is because standard accounting procedures usually allocate overhead in proportion to direct labor.

Failure to invest to improve productivity has long been identified as a possible source of high costs among defense contractors. Indeed, as early as 1976, a major Pentagon review of procurement practices concluded that defense contractors used only 42 percent as much capital equipment and facilities per dollar of sales as durable goods manufacturers overall (U.S. Department of Defense, 1976). In 1980, the House Armed Services Committee drew similar conclusions about the lack of investment in new manufacturing technologies by defense contractors (U.S. Congress, 1980). The Pentagon undertook a variety of initiatives, such as the Industrial Modernization Incentives Program (begun in 1982) and reforms of contract pricing practices introduced through the Competitiveness in Contracting Act of 1984. These initiatives provided incentives to keep costs down (with the awarding of more fixed-price contracts) and assisted those contractors wishing to undertake productivity-enhancing investments in new technology. If these reforms have had any effect on defense contractors, in our 1991 data we should expect to see the gap narrowing between their propensity to invest in new technology compared to their counterparts operating in strictly commercial markets.

More recent reforms have also been along the same vein. The Federal

Acquisition Streamlining Act of 1994 provides clear statutory preferences for commercial products and "best value" contracting, and also raised the threshold for simplified acquisition processes to $100,000 from $25,000. Similarly, the Secretary of Defense in 1994 directed DoD programs where possible to use commercial specifications.

More broadly, firms face a general problem of imperfect information for learning about and effectively adopting new technologies. Accumulated knowledge and expertise is important for assessing potential benefits from adopting a new technology. Differences among firms in their access to such expertise explain, in part, why some firms are more likely than others to adopt a new technology or innovate themselves (Arrow, 1962; Cohen and Levinthal, 1990; Dosi, 1988; Kelley, 1993; Nelson and Winter 1977, 1982; Rosenberg 1972, 1982; Watkins, 1991). Thus, to the extent that these government policies provide a more supportive and information-rich environment for long-term investments in new technologies than companies ordinarily have in strictly commercial customer–buyer relations, we may even expect to find a higher level of investment among defense contractors.

The issue is not whether the product market and the competitive conditions in which defense contractors operate differ from some hypothetical ideal, but whether government policies through the defense procurement system have (positively or negatively) affected the propensity of private manufacturers to invest in flexible and productivity-enhancing technologies. Flexibility may be particularly important in the context of defense diversification. In this section, we draw on our 1991 survey data once again to address the differences in investment in new flexible and productivity-enhancing manufacturing technologies between defense contractors and their counterparts in the MDG sector.

Defense Contractors' Leadership in Flexible Manufacturing Technologies

Programmable automated machine tools (PA) are a particularly important and recent manufacturing innovation. PA, or computer-integrated manufacturing, refers to information technology applications in which computer software and microelectronic-control devices are used to direct and monitor such ordinary production operations as machining, welding, testing, and inspecting. What distinguishes PA from previous generations of productivity-enhancing technology is that the instructions controlling the operation of machines are incorporated into software rather than hardware. As a result, PA is a very flexible innovation that can be used to reduce the costs of product diversification and of both large-volume and small-batch production, even in the smallest companies and in a wide variety of industries.

A large literature addresses the cost, performance and flexibility advantages of PA (e.g., Ayres and Miller, 1983; Dosi, 1988; Freeman and

Perez, 1986; Hirschhorn, 1984; Kaplinsky, 1984; Kelley and Brooks, 1991; Piore and Sabel, 1984). PA blurs the distinction between the economics of Fordist mass production and of small batch production (Cohen and Zysman, 1987). Batch production on conventional machine tools involves general-purpose machines hand operated by skilled workers. Unit production costs tend to be high because hand operations are time consuming and set-up costs are spread over small numbers of units. Mass production, on the other hand, involves high fixed costs for dedicated machines. Unit costs are low because set-up and equipment expenses can be amortized over large output volume. However, retooling for new products can be time consuming and costly.

In the diversification context, what is most relevant about PA is that it added "economies of scope" to the manufacturing lexicon. Cohen and Zysman (1987) believe PA's potential for improving both static and dynamic flexibility is key to economic growth and competitiveness. Static flexibility allows for switching production among a number of different products. PA reduces the time it takes to switch from one product to another, and at the same time can increase utilization rates by limiting set-up costs. Dynamic flexibility provides the ability not just to produce more than one product on a single line, but to enable production to evolve quickly with changes in either product or production technology. Dynamic flexibility, in this view, is critical for the timely realization of new ideas. Hence, both static and dynamic flexibility are relevant for diversification.

Going further, Piore and Sable (1984) and more recently Harrison (1994) see the introduction of flexible specialization as having fundamentally altered the nature of competition in manufacturing industries. If this is true, defense contractors may be comparatively well situated for competitive success in what Harrison (1994) calls "the age of flexibility."

Our survey results confirm a statistically significant difference (p<0.0001) in PA adoption related to defense contracting: 66 percent of plants with any defense prime- or sub-contracts have PA machine tools, compared to 50 percent of plants serving exclusively non-defense markets (i.e. with no sales or shipments to defense agencies or defense prime contractors). Moreover, defense contractors that adopt this technology employ a much higher fraction of PA tools in their total machine tool stock than do establishments engaged in the same manufacturing processes but with no contractual ties to DoD.

Defense contractors have been more innovative in their application of information technology to other production processes as well. For six common uses of computers in manufacturing, we compared the adoption rates of defense and strictly non-defense plants. In addition to PA, the applications include computer-aided design (CAD), computer-aided manufacturing process control systems (CAM—used to plan and monitor inventory, work-in-process and materials flow), computer-aided materials planning, and the use of programmable automation in other production processes. For every one of these technologies, we find higher adoption rates (p<0.0001) among defense contractors than in plants serving exclusively commercial markets.

Figure 17.5 graphs the ratios of adoption rates of these technologies between defense contractors compared to strictly commercial establishments, for various size firms. In the figure a ratio of 1.0 would mean that the same proportion of defense and strictly non-defense establishments adopted the technology, while a 2.0 would mean defense establishments adopt at twice the rate of those operating strictly in non-defense markets. The adoption advantages are particularly important (ranging from about 1.5 to more than 2.5) for small firms, with fewer than 50 employees.

Although it is difficult to single out a particular cause for these differences, we believe that government policy initiatives and programs directed at the defense industrial base are at least partly responsible for the large technological gap we find between defense contractors and other U.S. manufacturing establishments in the MDG sector. From 1982 to 1992, the Industrial Modernization and Incentives Program of the DoD provided technical assistance to contractors in assessing the applicability of advanced manufacturing technologies to defense contractors' operations. Through its manufacturing technologies (ManTech) program, DoD has supported the development of advanced technologies and improvements in process technologies among defense suppliers. DoD spent between $150 to $200 million annually throughout the 1980s on ManTech alone, exceeding the level of spending by all state governments on technical assistance programs aimed at manufacturing firms during the same period (Shapira, 1990; U.S. Congress, Office of Technology Assessment, 1990). Although these programs

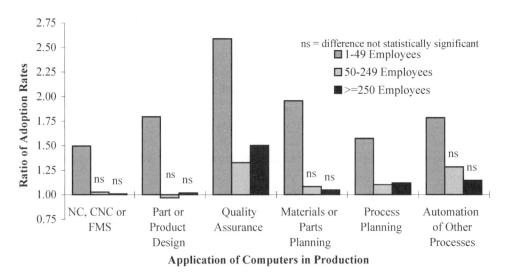

Fig. 17.5. Ratio of technology use in defense and non defense manufacturing plants, by firm employment.

directly assisted a relatively small number of defense contractors, DoD also sponsors annual conferences and workshops on new developments in manufacturing practices. These sessions highlight the lessons learned from the experiences of early adopters of advanced manufacturing technologies, providing the kind of learning opportunity for the larger defense industrial community that Von Hippel (1988) and Kelley and Arora (1996) identify as important institutional mechanisms for diffusing new technologies.

Defense Contractors' Advantages in Collaborative Networking Practices

Indeed, as we discuss in this section, we find considerably more collaborative networks of information sharing and supportive relationships surrounding defense contractors in the MDG sector than surrounding establishments with no defense-related sales or shipments. Drawing on theories of the economic value of collaborative production networks, in our survey we gathered 43 measures about the history of external relationships each establishment had. These economic exchange relationships were with key external organizations, including: technology vendors, subcontractors, competitors, customers and other sources of technical information such as government agencies and colleges and universities. For each of the 43 measures we performed statistical tests to determine if the particular type of relationship, such as "collaborated in developing new products with subcontractor" was more likely among firms with or without defense ties. For continuous variables, such as "the number of years you have been doing business with this technology vendor," we tested whether one group of plants had more durable (longer) and more intensive interchanges with their external partners. Again, we split the sample by whether or not the establishments had any sales or shipments to the defense agencies or defense prime contractors.

   Which group engaged in more collaborative networking relationships with their customers, suppliers, competitors and other external organizations? Whether the comparison is of defense prime contractors to plants with strictly non-defense work or between defense subcontractors and other plants, the overall pattern is striking. For defense prime contractors, on 19 of the 43 separate measures we find stronger or more prevalent collaborative external links than for plants with no defense sales. Similarly for defense subcontractors, 19 of the 43 measures are statistically greater than for other plants. Among both defense prime and subcontractors, a higher proportion of plants have close ties to their customers, competitors, sub-tier contractors, or technology vendors. Moreover, these relationships are more durable and intense, on average, for defense contractors than they are for other plants.

   Figures 17.6 and 17.7 summarize these findings. These figures include only those external connections that were more prevalent or stronger in

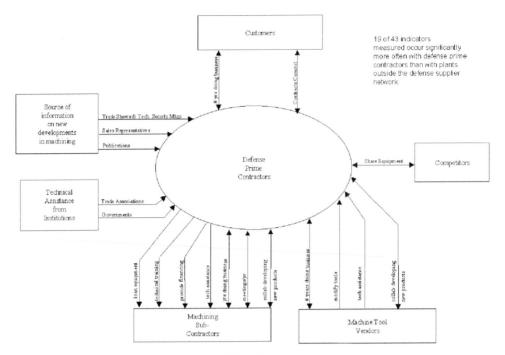

Fig. 17.6.

plants tied to defense compared to strictly non-defense plants. For each measure shown, the differences between plants inside and those outside defense production are statistically significant at p<0.05. The arrows indicate the direction of the connection.

In Fig. 17.6, we see that the differences between defense prime contractors and non-defense plants are particularly strong when comparing each group's vertical relationships to their subcontractors and technology vendors. Out of ten different indicators of close ties to machining subcontractors, seven are significantly more collaborative for defense prime contractors than for plants in strictly non-defense markets. Defense prime contractors far more frequently say they provided technical assistance, loaned equipment or machinery, and provided financing, and technical training to their subcontractors in 1989 or 1990 than did non-defense plants. In addition, defense primes have a much more intensive relationship with subcontractors, meeting with the technical staff of their subs more than two-and-a-half times as frequently in 1990 as managers from non-defense plants report about contacts with their important subcontractors. With respect to links with technology vendors, we find that four of the seven measures are significantly greater for defense primes than for non-defense plants.

Note also the relative stability of the relationships that defense primes have to their key partners. Prime contractors have been doing business with their largest customer, most important subcontractor and technology

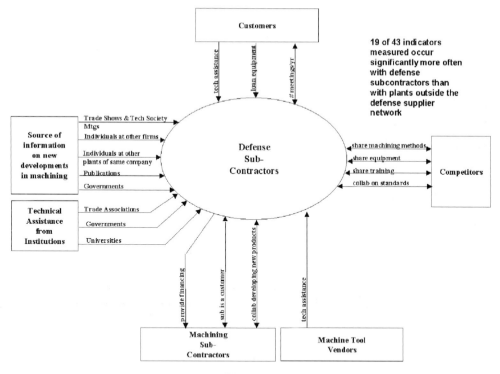

Fig. 17.7.

vendor for a significantly longer period of time than non-defense plants. On average, defense prime contractors have been supplying their largest customer for more than 16 years, which is in the same range (15 to 20 years) recently reported to be typical of subcontractors belonging to *keiretsu* in Japan's metalworking sector (U.S. Congress, Office of Technology Assessment, 1990: 135).

While prime contractors have relatively stronger collaborative ties to subcontractors and technology vendors than do non-defense plants, defense subcontractors have comparatively closer relationships with competitors. Figure 17.7 shows that a higher proportion of defense subcontractors have lateral collaborative ties to competitors and are better connected to sources of information and technical assistance outside of their exchange relationships than plants that have no defense contract work. For four out of six indicators of links with competitors, we find a significantly higher proportion of defense subcontractors reported recent collaborative experiences than were identified by strictly non-defense plants. Defense subcontractors are more apt to share information on methods of using machining tools and to share equipment with their competitors. Defense subcontractors are also more likely to engage in joint training activities and to collaborate with one another on standards. Moreover, defense subs appear to be better connected to external sources of information. They report using five of eight outside sources of information about new developments in machining technology significantly more often than do strictly non-defense plants. And defense subs are also at least 60 percent more likely to have received technical assistance in 1989–90 from trade associations, government programs and institutions of higher education.

Defense subcontractors are not more likely than strictly non-defense plants to have long-term contracts with their customers. Nor do defense subs have a longer history of a business relationship with their largest customer. However, compared to other plants, the largest customer of a defense subcontractor is more likely to provide technical assistance and to loan equipment. On average, defense subcontractors also have more intensive (frequent) interactions with the technical staff of their largest customer than typically occurs with the customers of strictly non-defense plants.

Finally, strictly non-defense plants have a higher incidence of collaborative ties to their customers, suppliers, competitors and other external organizations than do plants inside the defense network in only 4 of the 86 statistical comparisons we made. This includes only one of 43 comparisons to defense subcontractors, and three of that same set of 43 comparisons to prime contractors. Compared to prime contractors, non-defense plants are more likely to depend on colleges or universities for information on new developments in machining technology, and to collaborate with competitors and technology vendors in developing new products. However, on these same three indicators, a larger proportion of DoD subcontractors has these ties, but the differences are statistically insignificant.[16]

While stronger collaborative networking by defense contractors might be seen as inhibiting diversification—hindering breaking out of the defense industrial network—note again that most defense contractors already do more non-defense than defense work. To the contrary, as we show elsewhere (Kelley and Watkins, 1992, Kelley and Cook, 1997), this collaborative networking enables defense contractors to learn more quickly about flexible information technology applications than enterprises outside the network. Moreover, learning advantages are not confined to transactions specific to the Pentagon, but benefit the non-military operations of the networked enterprises as well.

CONCLUSIONS

Defense spending reaches a broad segment of manufacturing facilities in the MDG sector, affecting one-half of all establishments. Contrary to conventional wisdom, commercial-military integration is not only feasible, but is largely the normal practice in this sector. Our analyses indicate that the vast majority of defense contractors in the MDG sector manufacture military products in the same facilities in which they produce items for commercial customers. Most of the output from these facilities actually goes to commercial customers. For these already dual-use facilities, we see few technical or organizational barriers to converting these facilities to further serve non-defense markets. More targeted policies should be aimed at the small number of defense prime contractors and subcontractors, by our count fewer than one in ten in this sector, that are heavily dependent on DoD funds.

Moreover, the strictly commercial industrial base in the MDG sector lags behind the defense industrial base in using advanced flexible manufacturing technologies and in practicing collaborative production networking. Our research indicates that Defense Department policies and programs have supported the widespread adoption of these productivity-enhancing technologies. The DoD has supported a series of initiatives designed to provide technical assistance and incentives for defense contractors to improve their (and their suppliers') manufacturing processes. Our findings suggest that these initiatives have positively influenced the practices of a broad range of contractors.

In the MDG sector, DoD has provided a more supportive and information rich environment for long-term investments and the transfer of technology than we find among strictly commercial customer–supplier relations. The policy challenge is to preserve, within the constraints of much reduced defense procurement spending, the benefits of the relationships that have developed within this defense-contracting network.

ACKNOWLEDGMENT

Support for this research was provided by grants from the National Science Foundation (award no. SES-8911141) and the former Office of Technology Assessment of the U.S. Congress.

Neither sponsoring agency is responsible for the opinions expressed. Parts of the data and analyses herein we have previously reported in more abbreviated form in Kelley and Watkins, 1995a and 1995b.

NOTES

1. U.S. Department of Defense, 100 Companies Receiving the Largest Dollar Volume of Prime Contract Awards–Fiscal Year 1996, DIOR/P01-96, Washington, D.C., US Government Printing Office, 1996. Also available at http://web1.whs.osd.mil/peidhome/ procstat/top100/top100.htm.
2. See Gansler (1980: 37) who charts this ratio from 1959–78.
3. For certain contracts, an enterprise is considered to be "small" if it employs fewer than 500 employees. In other contracts, cutoffs are 750 employees or 1000. For service and construction industries, a small enterprise is defined in terms of revenue, not employment. The DoD also tracks prime contracts to small firms. These two sources provide estimates of the share of all defense spending that "leaks out" from the large prime contractors. Using these data, the Congressional Office of Technology Assessment (1992) estimated that 35

to 37 percent of all defense purchases in the 1980s went to enterprises that met one or another criteria as "small." This estimate applies to all DoD purchases, including both services and manufactured goods. About one-third of all defense procurement in the mid-1980s went to the service sector (Blank and Rothschild, 1985).

4. "IMIP: Pratt & Whitney Dependable Engines," informational brochure, E. Hartford, CT: Pratt & Whitney, p. 10.

5. *Wall Street Journal*, p. A3, August 17, 1993.

6. Machining involves the use of precision tools to cut and shape metal and includes grinding, drilling, milling, planing, boring, and turning operations. It is a process found in many manufacturing industries. Based on the industry-occupational matrix for 1984 constructed by the Bureau of Labor Statistics of the U.S. Department of Labor, we identified industries employing workers in occupations requiring specialized skills in these tools.

7. The industries are: nonferrous foundries (SIC 336), cutlery, hand tools and hardware (SIC 342), heating equipment and plumbing fixtures (SIC 343), screw machine products (SIC 345), metal forgings and stampings (SIC 346), ordnance and accessories, not elsewhere classified (SIC 348), miscellaneous fabricated metal products (SIC 349), engines and turbines (SIC 351), farm and garden machinery and equipment (SIC 352), construction and related machinery (SIC 353), metalworking machinery and equipment (SIC 354), special industrial machinery, excluding metalworking (SIC 355), general industrial machinery and equipment (SIC 356), miscellaneous machinery, excluding electrical (SIC 359), electrical industrial apparatus (SIC 362), motor vehicles and equipment (SIC 371), aircraft and parts (SIC 372), guided missiles and space vehicles (SIC 376), engineering and scientific instruments (SIC 381), measuring and controlling instruments (SIC 382), jewelry, silverware, and plateware (SIC 391).

8. This figure based on estimates of direct and indirect effects of defense spending in 1990 reported in Industrial Output Effects of Planned Defense Spending 1990–94, Office of Policy Analysis, Economics and Statistics Administration, US Department of Commerce, Washington, D.C., February 1991.

9. In 1989, there were a total of 81,506 establishments in this sector (Source: County Business Patterns, 1989).

10. These statistics were calculated by the authors based on unpublished data supplied by the Bureau of the Census from its 1988 special survey of approximately 10,000 plants belonging to the set of industries we have designated as the MDG sector. For a description of the survey and the data, see: U.S. Department of Commerce, Bureau of the Census, 1989.

11. McDonnell Douglas, General Dynamics, Martin Marietta, Grumman, Loral, Oshkosh Truck, Avondale Industries, Dyncorp, and the United Industrial Group.

12. GAO, 1997; Gansler, 1995; *The Economist*, October 2, 1993; *Business Week*, September 6, 1993; Lundquist, 1992; Markusen and Yudken, 1992; Weidenbaum, 1992; Center for Strategic and International Studies, 1989, U.S. Congress, Office of Technology Assessment, 1989; Melman, 1974.

13. We performed several statistical tests (at $p=0.05$) to examine the relationship between the size of plant or size of firm and defense dependence, measured

by the percent of total shipments from the plant in 1990 that went directly to a defense agency or a prime contractor. We find no significant correlation between the degree of dependence on defense purchases and either size of the parent company or plant size (both as measured either by sales or employment). Moreover, Chi-square tests fail to show any significant differences in the distribution of plants among (plant or firm) employment size categories (1–49, 50–249, ≥250) and the extent of the establishment's dependence on defense sales, grouped by categories (0, 1–9%, 10–19%, 20–29%, etc.). We also performed ANOVA F-tests, categorizing plants as single-plant enterprises or branch plants of large multi-unit companies, to determine if there were any statistical differences between large and small firms in the sample plants' dependence on defense purchases. Again, we found no significant relationship.

14. See OTA (1992) and the various years of U.S. Deparatment of Defense, 100 Companies Receiving the Largest Dollar Volume of Prime Contract Awards–Fiscal Year 19XX, Washington, D.C., US Government Printing Office.

15. Northrop Grumman press release, quoted in *S&P Daily News*, August 31, 1994.

16. Note here that if we would assume no differences between the populations of defense contractors and non-defense plants, and that the measures tested are independent, then with a 95 percent confidence level we would expect about 1 in 20 tests on data from random samples to show a statistical difference, even when there is no difference. That is, we would expect about 5 percent false positives—type II errors. The 4 of 86 cases where other plants have statistically higher incidence rates could possibly, then, be explained by random chance. The 19 of 43 cases where primes have statistically stronger ties, and the 19 of 43 cases we find for subs, though, are not within any reasonable realm of chance.

REFERENCES

Alic, J. A., Branscomb, L. M., Brooks, H., Carter, A. B. and Epstein, G. L. *Beyond Spinoff: Military and Commercial Technologies in a Changing World*. Boston: Harvard Business School Press, 1992.

Arrow, K. "The Economic Implications of Learning by Doing." *Review of Economic Studies*, pp. 155–173, 1962.

Ayres, R. U. and Miller, S. M. *Robotics: Applications and Social Implications*. Cambridge, MA: Ballinger, 1983.

Blank, R. and Rothschild, E. "The Effect of United States Defence Spending on Employment and Output." *International Labour Review*, Vol. 124(6), pp. 677–697, November–December 1985.

Business Week "The Godfather of Stealth Won't Let This One Slip By," pp. 60–62, September 6, 1993.

Center for Strategic and International Studies. Defense Industrial Base Project. *Deterrence in Decay: The Future of the U.S. Defense Industrial Base*, Washington, D.C.: CSIS Panel Reports, May 1989.

Cohen, W. M. and Levinthal, D. A. "Absorptive Capacity: A New Perspective on

Learning and Innovation." *Administrative Science Quarterly*, Vol. 35, pp. 128–152, 1990.

Cohen, S. S. and Zysman, J. *The Myth of the Post-industrial Economy*. New York: Basic Books, c1987.

Demski, J. S. and Magee, R. P. "A Perspective on Accounting for Defense Contracts." *The Accounting Review*, Vol. 67, pp. 732–740, October 1992.

Dosi, G. "Sources, Procedures, and Microeconomic Effects of Innovation." *Journal of Economic Literature*, Vol. 26, pp. 1120–1171, 1988.

The Economist "From Guns to Gadgets," p. 17, October 2, 1993.

Freeman, C. and Perez, C. "The Diffusion of Technical Innovations and Changes in Techno-economic Paradigm." Science Policy Research Unit, University of Sussex, 1986.

Gansler, J. S. *The Defense Industry*. Cambridge: MIT Press, 1980.

Gansler, J. S. *Defense Conversion: Transforming the Arsenal of Democracy*. Cambridge: MIT Press, 1995.

Hall, G. R. and Johnson, R. E. "A Review of Air Force Procurement, 1962–1964." Report No. RM-4500-PR Santa Monica, CA: Rand Corporation, 1965.

Harrison, B. *Lean and Mean: The Changing Landscape of Corporate Power in the Age of Flexibility*. New York: Basic Books, 1994.

Hirschhorn, L. J. *Beyond Mechanization: Work and Technology in a Postindustrial Age*. Cambridge: MIT Press, 1984.

Kaplinsky, R. *Automation: The Technology and Society*. Harlow, England: Longman, 1984.

Kelley, M. R. "Organizational Resources and the Industrial Environment: The Importance of Firm Size and Inter-firm Linkages to the Adoption of Advanced Manufacturing Technology." *International Journal of Technology Management*, Vol. 8, pp. 36–68, November 1993.

Kelley, M. R. and Arora, A. "The Role of Institution Building in U.S. Industrial Modernization Programs." *Research Policy*, Vol. 25, pp. 265–279, March 1996.

Kelley, M. R. and Brooks, H. "External Learning Opportunities and the Diffusion of Process Innovations to Small Firms: The Case of Programmable Automation." *Technological Forecasting and Social Change*, Vol. 39, pp. 103–125, April 1991.

Kelley, M. R. and Brooks, H. "Diffusion of NC and CNC Machine Tool Technologies in Large and Small Firms." In Ayres, R. U., Haywood, W. and Tchijov, I., eds., *Computer-Integrated-Manufacturing, Volume III: Models, Case Studies, and Forecasts of Diffusion*. London and New York: Chapman & Hall, 1992.

Kelly, M. R. and Cook, C. R. "The Institutional Context and Manufacturing Performance: The Case of the U.S. Defense Industrial Network," Ninth Annual Conference on Socio-Economics, Montreal, July 5–7, 1997.

Kelley, M. R. and Watkins, T. A. "In From the Cold: Prospects for Conversion of the Defense Industrial Base." *Science*, Vol. 28 , pp. 525–532, April 1995a.

Kelley, M. R. and Watkins, T. A. "The Myth of the Specialized Defense Contractor." *Technology Review*, Vol. 98(3), pp. 52–58, April 1995b.

Kelley, M. R. and Watkins, T. A. *The Defense Industrial Network: A Legacy of the Cold War, Contractor Report*. US Congressional Office of Technology Assessment, Washington, DC, 1992.

Lundquist, J.T. "Shrinking Fast and Smart." *Harvard Business Review*, pp. 74–85, November–December 1992.

Markusen, A. and Yudken, J. *Dismantling the Cold War Economy*. New York: Basic, 1992.

Melman, S. *The Permanent War Economy*. New York: Simon and Schuster, 1974.

Minahan, T. "Purchasing Rebuilds to Battle Poor Quality." *Purchasing*, Vol. 122 (1), p. 53, January 16, 1997.

Nelson, R. R. and Winter, G. *An Evolutionary Theory of Economic Change*. Cambridge: Harvard University Press, 1982.

Nelson, R. R. and Winter, G. "In Search of a Useful Theory of Innovation." *Research Policy*, Vol. 6, pp. 36–76, 1977.

Peck, M. J. and Scherer, F. M. *The Weapons Acquisition Process: An Economic Analysis*. Boston: Harvard University Graduate School of Business Administration, 1962.

Piore, M. and Sabel. C. *The Second Industrial Divide*. New York: Basic Books, 1984.

Rogerson, W. "Overhead Allocation and Incentives for Cost Minimization in Defense Procurement." *The Accounting Review*, Vol. 67, pp. 671–690, October 1992.

Rosenberg, N. "Factors Affecting the Diffusion of Technology." *Explorations in Economic History*, Vol. 3, pp. 3–33, Fall 1972.

Rosenberg, N. *Inside the Black Box: Technology and Economics*. Cambridge: Cambridge University Press, 1982.

Shapira, P. *Modernizing Manufacturing: New Policies to Build Industrial Extension Services*. Washington, D.C.: Economic Policy Institute, 1990.

U.S. Congress, General Accounting Office (GAO) *Defense Industry — Trends in DOD Spending, Industrial Productivity and Competition*, GAO Reports PEMD-97-3. Washington, D.C.: U.S. Government Printing Office, 1997.

U.S. Congress, House of Representatives, Armed Services Committee. *The Ailing Defense Industrial Base*, H.R. Doc. No. 29. Washington, D.C.: U.S. Government Printing Office, 1980.

U.S. Congress, Office of Technology Assessment (OTA) *After the Cold War: Living with Lower Defense Spending*, OTA-ITE-524, Washington, D.C.: U.S. Government Printing Office, 1992.

U.S. Congress, Office of Technology Assessment (OTA) *Making Things Better: Competing in Manufacturing*, Report No. OTA-ITE-443. Washington, D.C.: U.S. Government Printing Office, 1990.

U.S. Congress, Office of Technology Assessment (OTA) *Holding the Edge: Maintaining the Defense Technology Base*, OTA-ISC-420. Washington, D.C.: U.S. Government Printing Office, 1989.

U.S. Department of Commerce, Bureau of the Census, *Current Industrial Reports: Manufacturing Technology 1988*, SMT(88)-1. Washington, D.C.: U.S. Government Printing Office, 1989.

U.S. Department of Defense, Office of the Assistant Secretary of Defense, Installations and Logistics, *Profit '76 Summary Report*. Washington, D.C.: U.S. Government Printing Office, 1976.

Von Hippel, E. *The Sources of Innovation*. New York: Oxford University Press, 1988.

Watkins, T. A. "Technological Communications Costs Model of R&D Consortia as Public Policy." *Research Policy*, Vol. 20, pp. 87–107, 1991.

Watkins, T. A. "Dual-Use Supplier Management and Strategic International

Sourcing in Aircraft Manufacturing." Lean Aircraft Initiative Case Study. Cambridge: Massachusetts Institute of Technology, October 1997.

Wall Street Journal "Allied Signal's Chairman Outlines Strategy for Growth", p. A3, August 17, 1993.

Weidenbaum, M. *Small Wars, Big Defense: Paying for the Military After the Cold War*. New York: Oxford University Press, 1992.

18

THE CONVERSION OF DEFENSE ENGINEERS' SKILLS: EXPLAINING SUCCESS AND FAILURE THROUGH CUSTOMER-BASED LEARNING, TEAMING, AND MANAGERIAL INTEGRATION AT BOEING-VERTOL

JONATHAN M. FELDMAN

INTRODUCTION

This chapter explores two military diversification projects by Boeing-Vertol in the post-Vietnam War period, the development of a trolley car, the Light Rail Vehicle (LRV) for Boston and San Francisco, and a subway car for the Chicago Transit Authority (CTA). Each of these projects was supplied by the same firm and industrial facility of Boeing-Vertol, a helicopter manufacturer based in greater Philadelphia. Unlike the trolley car project, the subway car was successful, still in use after more than twenty years. Both civilian projects made extensive use of defense engineers so that the ability to transfer engineers was not a sufficient condition for successful diversification.

Boeing Vertol's entry into civilian markets was facilitated by the similarities or relatedness between military and civilian markets and technological skills. On the one hand, the LRV failure illustrates the limits of claims that conversion is easy precisely because military and civilian projects can share common resources (cf. Kelley and Watkins, 1995a; Kelley and Watkins, 1995b; Rumelt, 1982). On the other hand, the CTA success was noteworthy because, in contrast to commercial aircraft and satellite products, it involved defense engineers building products which were not directly or immediately interchangeable with defense products. This contradicts the claims of some "naysayers" who argue that defense firms can only convert engineering resources in such limited kinds of product lines. Furthermore, the CTA success took place at precisely the time period when defense CEO Norman Augustine claimed that the history of conversion was "unblemished" by success (Adelman and Augustine, 1992; Lundquist, 1992; Weidenbaum, 1992). The CTA project also reveals the limits to Wall Street claims that conversion almost always fails (cf. Markusen, 1996; Markusen, 1997). This chapter is based on: interviews with key managers and engineers at Boeing-Vertol, the Chicago Transit Authority, San Francisco Municipal Railway (MUNI), and the Massachusetts Bay Transportation Authority (MBTA) in Boston; internal reports from MUNI and MBTA; and various secondary sources.

A THEORY OF HOW DEFENSE FIRMS' SKILLS CAN BE CONVERTED

I will now examine the mechanisms by which defense firms overcome barriers to serving civilian markets, explaining the understudied mechanisms by which defense firms "learn" and collaborate. In contrast to those who suggest that defense firms' conversion is easy or next to impossible, other scholars view military specialization as a potential barrier which defense firms seeking to expand civilian sales must overcome (Melman, 1974; Melman, 1983; Dumas, 1986; Markusen and Yudken, 1992). Empirical work on small and medium-sized defense firms shows that specialization can be overcome by:

- reducing overhead barriers,
- moving into civilian government markets centralized with few customers like defense markets, and
- through the acquisition of new skills through new hires or collaboration (Feldman, 1997a; on collaboration generally, see: Dodgson, 1994; Freeman, 1994; Sabel, 1994).

I argue that large defense firms also use these "avenues of diversification" to support entry into civilian markets.

Defense firms face other diversification challenges, some of which have little to do with military specialization and are shared with civilian counterparts. These additional factors shaping diversification outcomes include:

- the ability of innovators to secure needed financing (Veblen, 1965) which sometimes manifests itself as "corporate commitment" (Pages, 1992, 1995; Oden, 1997);
- the sophistication of the customer as user; the ability of users to communicate with producers (cf. Rosenberg, 1982; von Hippel, 1988);

- the ability of producers to integrate diverse engineering disciplines in new production routines (Merton, 1947; Bowen *et al.*, 1994).

These various challenges can be summarized as the need for various stakeholders such as suppliers, customers, engineers, project managers, and top managers to share their respective knowledge and thereby extend learning (cf. Malerba, 1992). But, the problem is not simply one of mutual learning because the division of labor among key stakeholders creates potential barriers to the exchange of information or its proper application. As a result, various forms of decision-making power and knowledge can be separated but need to be integrated (Melman, 1958, 1975; Child, 1984; Lazonick, 1990, 1992; Sayer and Walker, 1992; Bowen *et al.*, 1994). I will now explain how defense firms' innovative resources centered in engineers can be converted successfully.

Defense engineers' skills are converted through a "managerial equation" which involves both learning (or "knowledge") and power (or the ability to act on or enforce that knowledge). Learning depends on the application of what works in a specific product application, subsystem or technology, i.e., what I will call "acquired knowledge." There are two key forms of acquired knowledge that must be transferred to defense engineers. First, individual subsystems or components must be tested out. Second, all subsystems must be integrated together. The design of the components may be related to core skills acquired working on military projects. Yet, the knowledge of how whole systems fit together ("systems integration") cannot be automatically transferred from military projects. For example, the shift from military to civilian satellites may involve an easier transfer at the system level than that between helicopters and mass transit vehicles. Or, the systems integration tasks may simply be different (Melman, 1983: 252–259). Acquired knowledge is based on formal tests or accumulated knowledge based on trial and error (cf. Nelson and Winter, 1982; Vincenti, 1990), engaging key parties such as producing engineers, suppliers, product users or regulators. In order for diversification projects to succeed, acquired knowledge about an application or civilian product must be extended to engineering designs and practice.

Diversification projects can fail if there is a break in the transmission of acquired knowledge or if the power to apply that knowledge is lacking. In other words, for acquired knowledge to be extended to defense engineers, there must be a clear communication "channel" (cf. Fiske, 1982) between the information gleaned by acquired knowledge and directives issued to engineers. If acquired knowledge is not communicated clearly to the engineers, civilian projects can develop problems, creating difficulties that may jeopardize the technical and therefore market success of a project. This transfer can be blocked in several ways.

First, a proper test must exist of a component or the integration of different components (cf. Melman, 1983; Bowen *et al.*, 1994). Tests may be improper, as when prototypes of novel systems are not based on in-service products or when products are not tested long or hard enough. Sometimes users or producers fail to respond to tests that show a product failure, preventing the transmission of acquired knowledge.

Second, in cases where the customer as user has acquired knowledge and the producer does not, the customer must have a system for transferring that knowledge to engineers doing detailed design or production tasks (cf. Rosenberg, 1982). This can involve specifications (Vincenti, 1990; Hedrick, 1995) or design reviews between users and producers based on the history of a product in service. These specifications and reviews can help overcome the proclivity of defense engineers to overdesign a product, making it too complex, or produce designs which make a product hard to maintain (cf. Melman, 1983; Hedrick, 1995).

Third, some diversification projects may not gain full access to relevant learning concentrated in a firm's various engineering disciplines. In addition to the learning and communication system between user and producer, there is also an important system of learning that goes on independently of such interactions. When defense engineers work on various military applications they gain important knowledge that can sometimes be transferred to commercial projects. These generic skills include knowledge about various technologies or subsystems which is acquired by specific engineering disciplines. In any firm there exists a hierarchy as to which engineers have a better grasp of such generic knowledge. The ability to know who is best in various disciplines may itself be shaped by having broad, generalist knowledge of engineering problems (cf. Lazonick, 1990, 1992). In Japanese firms like Canon this knowledge is acquired as employees are exchanged from one part of the company to another and jobs rotated (Imai *et al.*, 1985: 356).

The transmission of acquired knowledge to engineers is not simply a process of "learning" or clear communication channels. It also can depend on power relations or monitoring between customer and producer or between project managers and their engineering subordinates. One definition of monitoring is "the determination by . . . transacting parties that the gains from learning be distributed according to the standards agreed between them, as interpreted by each" (Sabel, 1994). As applied here, monitoring refers to an oversight function in which customers convey instructions (representing customer-based acquired knowledge) to another party whose design routines are shaped by specialized military knowledge or a history of serving defense customers, i.e., the defense engineering team. The transmission of acquired knowledge can involve power relations when independent parties (which can include not only outside customers but internal team members with knowledge of the commercial application) must have design authority vis-á-vis defense engineers. The independent parties are the "monitors" who enforce acquired knowledge. In addition, managers of diversification projects, or those learning about production problems, must have power to respond to directives from the monitors. If customers have knowledge about what works and producing engineers do not, then customers may have to enforce adherence to proven methods on producing engineers. They may have to tell defense managers and their engineering team to follow clear guidelines, or formal specifications, as to how a product should be designed. These reviews can help avoid the introduction of military-inspired designs that would not work in a specific commercial application. This relationship is not simply based on exchanging information. User knowledge must be joined with power as monitors order defense engineers to design systems according to proven methods or defense engineers become dependent on some way on the party with greater user knowledge. This integration of design knowledge and decision-making power is facilitated when the customer purchases through a direct contract with the defense producer, as is common in civilian government mass transit agencies. Like the relations between defense contractors and the Pentagon (Melman, 1970a), knowledge and power can be joined in commercial applications through design reviews and specifications.

Diversification project managers may also have to monitor their defense engineers to insure that acquired knowledge (such as from customer specifications) is applied. Defense managers at Boeing headquarters in Seattle, Washington, have proposed a naysayer view that defense engineers can not learn from collaborations with civilian counterparts and can not be monitored to make sure they properly follow guidelines laid out by acquired knowledge. One top manager argued that higher authorities could not provide specifications or guidelines regulating how defense trained engineers designed commercial projects: "you tell an engineer that he's not going to make any changes on aircraft," but "he's going to make changes because he always wants to make it better. That's the nature of the beast. That's the way they're trained. You can't shut off changes" (Confidential interview, 1993a). He also argued that with digital computers, supervisors find it hard to know "exactly what that engineer is doing. The only time I can tell what an engineer is doing is when the digital data set comes out the other end and I sit there and look it over and understand geometrically what it means. But, changes are very insidious . . . Engineers are interested in improving the process" (Confidential interview, 1993a).

PRECONDITIONS FOR SUCCESS: MARKET ENTRY AND REORGANIZATION

During the Vietnam War, Boeing-Vertol's 13,500 employees produced helicopters for the military (Melman, 1983: 253). The company was the result of Boeing's acquisition of the Vertol Aircraft company in 1960 (Dancik, 1995). The major product line was the Chinook CH-47 helicopter whose "production dropped rapidly after the conflict peaked in 1968" (DeGrasse,

1987: 102). In the 1970s, as a response to military budget reductions and increased management interest in diversification, Boeing-Vertol won its bids for electric trolley and subway car orders for the domestic U.S. market. They decided to enter the transit business as a potential commercial market in part because of their engineering skills. The chief design manager for the CTA subway car project was Paul Dancik. He recalls that the company had dynamics and acoustic skills which could be applied to mass transit and the company had "terrific electronic engineers and structural designers." These skills emerged out of the need to respond to technical problems in helicopters. When the blades of a helicopter pass over its fuselage, this pushes air down on the craft. The downward pressure of air produces a "beat" that induces vibration and diverse sets of skills were required to reduce this vibration (Dancik, 1995).

While such "related skills" provided a basis for entering the mass transit market, Boeing-Vertol also needed to reorganize operations and established a new group to produce transit products. The Surface Transportation Group joined everyone "from the top boss down" in the same building. The CTA project and the LRV program's engineering staff both were located on the same floor in the Scott Paper Building, although in a different area (Dancik, 1995). Related skills gave Boeing-Vertol a "foot in the door," but such technological similarities were not always sufficient for either collocation with defense production or guiding defense engineers into civilian projects.

The transit projects could benefit from the new location of the building separated from high military cost centers without losing access to engineering talent within the firm. Such separation would, by definition, foreclose the burdens imposed by charges for many items which comprise the overhead in military firms. First, the project was able to establish new accounting systems, separate from the military. Second, transit projects would not be charged with costs for dedicated testing equipment. Nevertheless, the project could borrow any needed personnel required from the aerospace program as in a larger matrix program. For example, on acoustical tests, aerospace personnel were brought in "as needed" and not paid on a full time basis (Dancik, 1995).

The new organization not only provided space for physical separation from overheads. The new organization also embodied new practices which reduced the managerial layers contributing to overhead. The head manager of the transportation division gave project engineers like Mr. Dancik "more freedom" than he was accustomed to having on military programs. These programs were managed in a less hierarchical fashion and were not subject to as much control by top aerospace management at Boeing-Vertol. Like a "skunk works" there was less hierarchy. The head of the transit division was at the apex and Mr. Dancik was third in charge, only two management layers removed from him. On defense projects, Mr. Dancik would be fourth or fifth in the management hierarchy. The decrease in the layers of decision making meant that Mr. Dancik "had less people to report to." He didn't have to discuss things with as many people, which could "waste time." Part of the reduced need for technical monitoring and supervision was that fewer personnel were required to produce simpler transit as opposed to more complicated helicopter programs (Dancik, 1995).

In addition to physical separation and flattened hierarchies, costs in the new civilian group could be reduced by eliminating unnecessary military production requirements. These requirements are usually more stringent than in the commercial world which leads to higher costs (cf. Melman, 1974; Ullmann, 1995). More stringent requirements required by military specifications lead to more extensive tests of components and systems. These tests are the most significant aspect of military production that add costs. The military requires very stringent tests such as environmental testing, e.g., a component will be subject to extremes of heat and cold. There are also "salt spray," "fungus," fatigue and other tests. This adds costs to a product that wouldn't be found in the world of commercial production (Dancik, 1995).

While it is hard to develop a direct cause and effect relationship between the new organization and its ability to compete, the evidence suggests that such reorganization facilitated civilian market entry. Interviews by the author with transit officials in one state and its potential local defense supplier has established that very often defense firms' bids are significantly higher than lower cost civilian firms. A key bidder who lost attempted to win bids from a defense facility with high overhead costs and using facilities dedicated to military production (Confidential interview, 1993b). The physical separation of military and civilian production facilities does not prevent practices which may add costs to production, however. In the LRV project Boeing-Vertol used in-stock connectors on its trolley cars. These were hermetically sealed and more expensive than the standard rail road connector. Here, Boeing-Vertol was responding to the needs of its biggest customer, the Pentagon (Mackay, 1987).

Boeing-Vertol won its contract with the CTA in a December 1973 competition because it had the lowest bid. Yet, their major competition was another defense firm, Rohr. Rohr's bid was $32.8 million and Boeing's bid was for $29.3 million, i.e., there was a total difference of only $3.5 million (Keevil, 1995). Boeing-Vertol completed the design, development and delivery of 200 subway cars in 1976. Its total costs were about $80 million on the LRV and about $60 million on the CTA rail cars (Dancik, 1995). Boeing's bid weight was originally over the specification weight by 1500 pounds per car. Rohr's bid weight was also over the specification weight, but they were more "over weight" than Boeing. There were also disincentives (or penalties) for noise levels. Here again, Boeing performed better than Rohr. Boeing's design was cheaper and appeared to weigh less and was quieter (Keevil, 1995).

Even if Boeing's costs were lower than Rohr's, a set of relatively unique circumstances may have protected them from even more competitive suppliers. Several manufacturers became preoccupied with other sales which reduced competition, e.g., Pullman was building the R-46 subway car for New York City; Rohr was building 300 cars for Washington, D.C. and some cars for BART; Budd was busy building cars for AMTRAK. When the bid was originally advertised, there were thirteen prospective car builders. One of these companies went out of business. Bids were also expected from Japanese firms, but with the devaluation of the dollar the Japanese would not be able to compete on cost and so dropped out (Keevil, 1995). Under the Nixon Administration's "Phase One" economic edicts, recipients of federal funds from agencies like UMTA were required to spend money within the United States. In the LRV case, this prevented MBTA from buying proven cars from Düwag, a West German supplier, which it favored (Young, 1993: 30). Even if Boeing had a relatively sheltered market, one could argue using the so-called "infant industry" thesis that the company required protection in order to provide space to learn how to draw down its costs. The extent to which Boeing-Vertol was able to compete on cost is described in the section titled, "Aftermath of the CTA Program."

Boeing-Vertol and other aerospace firms like Rohr were also encouraged to participate in non-defense markets like mass transit because of federal transit policies. Federal policies aimed to help defense firms overcome dislocation caused by defense budget cutbacks and transition into mass transit markets. These incentives included potential subsidies from the federal government's Urban Mass Transportation Administration, known as UMTA (Melman, 1983) and increased surface transit expenditures. Boeing was also encouraged to use its design capacities for transit uses by a series of government sponsored technology transfer programs (Keevil, 1995; Neat, 1997). Boeing was involved in two of these projects which included separate programs: the State of the Art Car (SOAC) and the Advanced Concept Train. The latter was beyond the state of the art and addressed very advanced concepts for vehicles and subsystems. SOAC was run at the same time the CTA cars were being built (Keevil, 1995). One source argues that Boeing-Vertol's participation in UMTA programs (like a contract for the standard U.S. light rail vehicle) gave the company "the inside track" on the LRV contract (Radin, 1980: 13). Other reasons why Boeing may have won the LRV contract include their superior financial resources as an aerospace firm. The winning bidder would have "only 230 examples against which to set off design and tool costs . . . an attractive proposition only to a company wanting to break into the business and with plenty of cash to spend" (Young, 1993: 31).

THE LIGHT RAIL VEHICLE

A Product that Failed

Boeing entered the mass transit business with a contract in the early 1970s to produce 275 light rail cars. A chronology of key dates in the history of the LRV and CTA projects appears below (Appendix 1). The cars were produced for a joint purchase by the Boston area Massachusetts Bay Transportation Authority (or MBTA) which bought 175 cars and the San Francisco Municipal Railway (MUNI) which bought 100 cars. The LRV performed poorly in Boston and San Francisco. By the third year of the program, on June 27, 1979, the MBTA reported that of 175 cars ordered, only 30 were in service on the previous day (Melman, 1983: 255). Cars in both cities were "generally considered to be a failure in design and construction. . . . They are being retired as fast as possible by both owners." Boston has gotten rid of most, if not all, of these cars. These cars have been criticized as "unreliable, very expensive to maintain and operate and they're not a design which anyone would want to replicate" (Cihak, 1994).

An earlier study has claimed that MUNI's "LRVs have proven reliable in revenue service," leading San Francisco to purchase "31 of the cars Boston refused to accept" (DeGrasse, 1987: 104). Yet, transit officials in San Francisco and consultant studies have documented numerous problems with MUNI's LRV (Appendix 2; Lewis T. Klauder and Associates, 1985a and b). As of November of 1996, San Francisco MUNI used 121 out of 130 original cars delivered from Boeing-Vertol. Three cars were slated to be scrapped. San Francisco has been actively retiring its trolley cars and has turned to Breda, an Italian company, to replace the Boeing-Vertol cars (Miller, 1996).

Suppliers and Defense Engineering Routines as the Baseline of Acquired Knowledge

Acquired knowledge can be gleaned from producers, customers, or testing procedures. Yet, the Boeing-Vertol project did not make full use of the information from these sources. Turning to producers first, within the United States, the main models and specifications for trolleys was the Presidents' Conference Committee Cars first produced in 1935 (Bei, 1978: 18; Cihak, 1994). By the early 1970s, this technology was several decades old. American universities could not help producers develop new specifications because the academy had not trained engineers "for this long-dead industry" (Melman, 1983: 254). A more modern source of acquired knowledge lay in Europe. A small team made a brief sojourn to Europe to tour plants manufacturing rail cars (Melman, 1983: 254) and the LRV copied designs found in Europe and used articulations and motors found in European light rail vehicles. Yet, this transfer of European knowledge was superficial.

The LRV was new to U.S. firms and did not rely on long-term learning found in European manufacturing practices. Boeing-Vertol moved from making zero to making 175 cars for Boston and another 100 cars for MUNI quite rapidly. By industry standards, this was a rather ambitious undertaking. The much more experienced German firm Düwag, linked to the Dusseldorf car factory now owned by Siemens, estimated that on such an order they could only have made about 50 to 100 cars (Vychick, 1996). The progress payments which Boeing's customers were required to make helped speed LRV production faster than traditional car builders (Young, 1993: 32).

As a pioneer, Boeing attempted to learn from suppliers experienced in the details embodied in equipment, yet this strategy was fraught with problems. For example, it turned to foreign suppliers but their component supplies often failed: "the doors on the original cars, designed and built abroad, had 1,300 parts and failed repeatedly" and had to be redesigned (Melman, 1983: 255). In the division of labor to build the cars, Boeing-Vertol "established themselves as system designers and assemblers, leaving the principal functional work to specialized subcontractors." Yet, by relying upon subcontractors for the primary knowledge in how to build a car, Boeing-Vertol had no direct source of knowledge of the complications in integrating diverse components or assembling a car: "the engineering staff attempted to skip an otherwise long learning period and to dodge the manufacturing problems they might have encountered if they had tried to design and produce many of these components themselves" (Melman, 1983: 254).

In addition to suppliers, Boeing-Vertol relied heavily on design routines established on military projects for developing its American LRV. Eighty of the engineers working on the LRV, save one, came from the aerospace departments of the company (Melman, 1983: 254). In his account of the origins of the LRVs design problems, Seymour Melman argued that they were based on the application of design principles fashionable in the military world: "the Boeing-Vertol management/engineering team was prepared to accept 'sophisticated' designs in many components, when what was needed was rugged simplicity." The trolley car doors, reflecting aerospace needs, were not designed for repeated use: "trolley-car doors, which, unlike those on airplanes, are opened and closed thousands of times a week, impose special design criteria of simplicity and durability. . . . The doors on the original cars, designed and built abroad, had 1,300 parts and failed repeatedly. Boeing-Vertol did its own redesign, scaling down the complexity of the door mechanism to about 600 parts" (Melman, 1983: 254–255). A MUNI manager stated that the LRV was "plagued with intermittent failures due to its complexity of design" (Wynn, 1982: 6).

I will describe how learning from aerospace projects can be applied to transit projects without customer input, based in part on testing and production experience in military projects, e.g., component or subsystem technologies. There are serious limits to such "pathogenic learning" as will be described below. In other areas, whether influenced by military practices or not, the LRVs were simply designed according to poor engineering judgment. For example, the design did not allow the air-conditioning system to be quickly disconnected. Air-conditioners could not be serviced "without cutting and then resoldering piping, a long and costly process" (Melman, 1983: 255).

Testing Procedures

The overly complicated and poor design practices introduced by defense engineers were not the only problems which the LRV project would have to be wary of. In addition, the program as a whole was based on introducing novel, and therefore, more risky technologies. Boeing had agreed to an UMTA standard for cars that was novel: "in effect, Boeing agreed to produce a new generation of trolley cars" (DeGrasse, 1987). The car used basically new propulsion systems and body designs which produced "maximum risk," i.e., a greater risk of product failure (Cihak, 1994). The articulation section joining two sections of the rail car was always a problem in the Boeing-Vertol design. Boeing-Vertol's engineers modified or redesigned it three times. The MBTA claimed that the bearing-pad equipment on which the articulation pivots was inferior to a ball-bearing design "used on virtually all other makes of LRVs" (Radin, 1980: 11). Such design problems were linked to Boeing's corporate strategy of entering the market based on novel and therefore differentiated technology. Philip Craig, former head of MBTA's LRV project explained Boeing's strategy as follows: "Boeing made a decision that it was not in their commercial interest to license known proven designs" (Craig as quoted in Radin, 1980: 12).

If producers lack a baseline of acquired knowledge based on old projects, appropriate testing can reveal the limits of various designs. Yet, Boeing had adopted a strategy "for producing a new generation of trolleys without proper testing" (DeGrasse, 1987: 103). The LRV's testing program was insufficient on a number of grounds. The company practiced the testing protocol applied in military markets in which "full service testing of prototype equipment was dispensable" (Melman, 1983: 256–257). In Boston, the trolley cars which arrived in 1975 "were put through an eleven-week series of tests" which were completed on August 13th. According to reports at the time, these cars had bettered or equaled "all specified performance requirements, i.e., speed, acceleration, braking, energy consumption, ride quality, and noise level" (Fischler, 1979: 134–135). In the Fall of 1975, three vehicles were sent to the U.S. Department of Transportation test track in Pueblo, Colorado. Over a period of about six months, the LRVs were tested singly and in two-car trains (Bei, 1978: 19). None of these tests detected the extent of problems introduced by more extensive in-service use in Boston and San Francisco. From Boeing-Vertol's perspective, the LRV was tested "in accordance" with "what was requested in the specification" (Confidential interview, 1997b). Yet, the novelty of the specification meant that tests had to be much more rigorous than they actually were. Boeing "agreed to a compressed schedule" for producing the cars which prevented proper testing: "the schedule, coupled with penalties for late delivery, encouraged Boeing to meet schedule milestones by tearing out components found to be unreliable and modifying the cars as they proceeded with the production line." Boeing officials have conceded that many of their problems would have been avoided if they had "interrupted production, paid the scheduling penalties and completed design modifications prior to moving into full production" (DeGrasse, 1987: 103).

Customer Innovation and Testing: Boston and San Francisco

Although San Francisco's trolley cars were not a stellar success, their cars were more successful than Boston's. The several reasons for San Francisco's relative success underscore how the customer helped provide a baseline for acquired knowledge. First, MUNI was under less pressure to introduce cars than Boston and therefore tried to place demands on Boeing to make improvements in the LRV "at the Boeing factory before delivery, rather than on the property" (Young, 1993: 39). Boston faced "tremendous political pressure to get [the cars] in the street." Some link this pressure to Boston's failure to conduct extensive "prototyping" and "testing" as well as its acceptance of cars that it knew had problems (Miller, 1996; cf. Radin, 1980). According to Philip Craig, former head of MBTA's LRV project: "The administration knew there were problems when they accepted the first shipments of cars . . . but news clippings and notes coming down from the States House (to operating personnel) created the atmosphere that 'the State House wants more cars accepted'" (Craig as quoted in Radin, 1980: 13). San Francisco also faced pressure to use LRVs because other transit systems had broken down (Confidential interview, 1997a). Yet, MUNI delayed delivery of their fleet from Boeing-Vertol until modifications were implemented and proven in Boston (Miller, 1996).

Second, San Francisco's resulting delay in accepting the LRVs underscores how they used Boston's cars as prototypes and their procurement as a *testing system* for themselves. San Francisco effectively used Boston's LRVs as its own prototypes. Boeing's trolleys "delivered to San Francisco were modified based upon the difficulties experienced in Boston" (DeGrasse, 1987: 104). This appears to account for the greater ability of San Francisco to assimilate the cars. If San Francisco's continued use of the Boeing-Vertol trolley car is the criteria used to define success, then this criteria is misleading. Boston found an escape hatch not available to San Francisco. Boston sued Boeing-Vertol and got a big settlement. They did not take delivery of all the vehicles and wisely invested the funds from this settlement which were used to buy 95 new cars made by the Japanese firm Kinki Shayro without any federal assistance. Boston got a fifteen year variance from the federal government for early retirement of the cars (Miller, 1996). The MBTA made about 200 modifications to the LRVs in Boston even before MUNI received their first cars (Rogers, 1997b). It is unclear if political pressures to rapidly introduce the cars were greater in Boston or San Francisco, but San Francisco clearly learned from Boston's difficulties.

Third, San Francisco MUNI made major investments to repair the Boeing-Vertol trolley cars. After the cars were put in service in San Francisco, they only became reliable after $17 million was invested in revamping 131 trolley cars (Miller, 1996). Even after Boston's modifications, MUNI "ultimately implemented or planned" an additional 200 (Rogers, 1997b).

Barriers to Communication: Weak In-House Expertise and the Use of Consultants

In addition to formal producer tests, another source of acquired knowledge is the information collected by customers, the transit agencies as product users. On several occasions, however, communication between the customer and producer broke down. For example, in April of 1975, Boston's Acting Director of Maintenance encouraged "more face-to-face communication between the technical staff of Boeing-Vertol" and MBTA personnel. In May of 1976, however, the company argued that the benefits of its pilot car program were not being realized because of MBTA's decision not "to maintain or assist Boeing-Vertol in maintenance" of the LRV (Crugnola, 1979: 25, 34). This communication link between customer and producer broke down in the LRV project in several ways.

First, the test program of the transit agencies were as limited as that of the producers. Boston did not have a stringent test and requirement program. San Francisco's need to make several hundred modifications even after learning from Boston illustrated the limits of learning from Boston's "production prototypes."

Second, as potential monitors, the MBTA and MUNI provided limited guidance to the producer through specifications because of policy decisions and weak in-house design capacities. One reason is that the MBTA and UMTA did not put "much technical detail" into its specifications and UMTA favored open-ended performance specifications that they thought would be cheaper and attract aerospace innovations (Lewis T. Klauder and Associates, 1985a: 10; Young, 1993: 30–31). San Francisco was constrained in passing its design requirements along to the producer because it had at the time only one person who acted as a direct liaison with the manufacturer. As a result, it turned to Lewis T. Klauder and Associates as consultants who contributed "a lot" to "the design" specification of the car (Rogers, 1997a).

The use of consultants, however, created a buffer in communication between the user and producer, according to both transit industry observers and Boeing-Vertol engineers. The use of consultants created problems because they "can act like a stone wall." When the customer is "not knowledgeable and relies on the consultant, the consultant becomes the knowledgeable party." As a result, the consultant "makes the decisions that the customer should be making." If a customer lacks the specific knowledge required, then consultant recommendations will "run the show." In fact, "the customer may not even know what they're getting until the product shows up on their door step" (Keevil, 1995).

The different operating environments and multiple consultants used by Boston and San Francisco LRV programs prevented the development of a specification which clearly reflected acquired knowledge. There were five original members of the committee which developed the specification: Lewis T. Klauder, both properties, Kaiser Engineering, UMTA, MBTA in Boston and MUNI in San Francisco (Keevil, 1995; see also Massachusetts Bay Transportation Authority *et al.*, 1972). The multiple involvement of consultants helped break down the communication between customer requirements and supplier designs. As the customer did not take direct responsibility for oversight, the LRV program had both design and general contracting consultants. So "there ended up being consultants upon consultants . . . between the purchaser and the builder. That was highly problematical because the consultants didn't agree with each other." As a result of such disagreements, the monitoring system between suppliers and defense engineers was weakened as was the government's attempt to create a standardized car for two cities whose consultants disagreed with each other (Keevil, 1995). There was no authority to arbitrate the disagreements between consultants which created a major problem when the consultant groups took different approaches to the same problem. One observer concluded that "the best consultant works for the most knowledgeable customer" as such customers have exposure to what works as a product is used over a long period of time and is thereby able to develop the best performance specification (Keevil, 1997a and b).

The problem was not simply that the resulting communications buffer created by the consultants prevented defense engineers from doing things wrong, i.e., inappropriately applying defense design practices. It also prevented them from doing things right, i.e., applying what worked best in aerospace technology. As one engineer who worked on the LRV program at Boeing-Vertol explains, if consultants "specify the shape and weight of the car and everything else therein, there's not a lot that you can do with it. If they want to veto or vote down any of your suggestions on acoustics, noise transmission or anything else, you just don't have a chance. We had a terrible time with those consultants" (Bevan, 1997).

THE CTA PROJECT

Long-Term Learning

Although the LRV itself was "not a totally new concept" (Vuchic, 1996), it was novel to Boeing-Vertol. Yet, unlike the LRV, the CTA cars were built on long-term learning over many, many years. The LRV was an experimental program, prompted by UMTA subsidies. The CTA program was not. The long-term learning embodied in the CTA design helped make their specification a useful tool for transferring technology to Boeing-Vertol.

Mr. Paul Dancik was the key engineering manager responsible for the supply of transit cars to the CTA. His career was enmeshed in military production and aerospace culture (Dancik, 1995). He attended the College of Aeronautics where he got training in aeronautical design. He worked for Fairchild Aircraft in Hagerstown, Maryland before and after World War II. During the war, he was with the Eighth Air Force as a B-17 gunner. In a break from defense projects, he worked for Republic Aviation in New York on an ultra-modern, high speed propeller commercial airliner for Eastern Airlines. He was the design manager of the Boeing Sikorsky joint team venture to design the Comanche Helicopter (Dancik, 1995). He had some exposure to commercial aircraft work and, as noted earlier, his team of engineers had many of the skills related to those required in transit projects. Yet, despite such related skills and exposure to a "dual-use" industry of helicopter technology, there was a clear gap between the aerospace and transit worlds. The personnel who worked on the CTA project "were all aircraft oriented" (Dancik, 1995). One report notes that Boeing-Vertol "hired engineers and supervisory personnel from other firms in the transit industry. However, 75 to 85 percent of the engineers and supervisors working on the transit projects were shifted from Boeing's defense production." Somewhere between 95 and 100 percent of its production workers on the transit car program had worked previously on helicopters (DeGrasse, 1987: 105). The CTA program was dominated by aerospace engineers and external hires played a negligible role throughout the program. The reservoir of talent came strictly from Boeing-Vertol facilities in the greater Philadelphia area. Boeing's Seattle operations were preoccupied with their billion dollar plus commercial aircraft business (Dancik, 1995).

Mr. Dancik was the key engineer, but when assigned to the program he had "never worked on trains before." He "thought a truck delivered milk" and was not familiar with this term for the assembly that contains the wheels, the gear boxes and the brakes. In many areas, the skills which Dancik and his colleagues required on commercial and military helicopters were completely different from those required of rail cars. One area of important difference was the tolerances and materials standards required in helicopters. He realized that "if you tried to adapt aeronautical requirements to trains you could get into trouble because of the different tolerances and requirements" of planes and trains. In the helicopter field, engineers were more conscious of weight. The engineering staff "had to make sure that we used the right technology and applied it properly, but not get too sophisticated." The team adapted their aircraft skills and technologies by "trimming it down to suit the commercial requirement" (Dancik, 1995).

The sensitivity Mr. Dancik had towards the differences between aerospace and transit technologies may have played a role in helping the program avoid the overly complex and costly designs that plague unsuccessful diversification efforts. Yet, Mr. Dancik and his engineers clearly needed a baseline to guide them. The key project managers at the CTA, Frank Cihak and Walter Keevil, played this role and had as their primary tool the accumulated knowledge of transit designs which they and their suppliers had developed over a forty year period.

The predecessor model which informed the specification used by Boeing-Vertol was based on 144 cars built by the Budd company from 1969 to 1970. These cars had a number of problems which the new car designs were designed to eliminate. First, the trucks had problems with the wayside vibration and noise. This vibration extends to the train platform and affects the quality of the ride. At the time, these were hot topics in the transit world. As a result, there was a lot of federal money to study these problems and abatement methods at the Transportation Systems Center (TSC) in Cambridge, Massachusetts. This center was a former defense research group that got converted by conducting transit research. Second, the Budd cars had self-ventilated traction motors or motors which had internal fans to cool themselves. The absence of external fans created problems. During bad winter weather, the snow would get in the traction motors, melt into water and then freeze. This led the insulation in the motors to crack (Keevil, 1995). While Budd's 144 rail cars were not formal prototypes, the CTA learned from the limits of these cars so that they served as informal prototypes for newer models. The specification issued by the CTA to Boeing-Vertol embodied long-term learning within the agency and by long-time transit producers. Beginning in the 1950s, there was a gradual development in the design of subways cars which built directly upon improvements upon previous models. By 1964, there was a new family of cars developed so that all subways cars since that time have basically been related. The history of this family of cars informed the specification issued to Boeing. For its part, Budd as the previous supplier was very experienced, having been in the railroad car business since the mid-1930s. In sum, the CTA specification reflected long-term learning over many years (Keevil, 1995).

The subway program was largely based on applying the tried and true designs of the CTA to performance specifications for Boeing-Vertol. The CTA's Walter Keevil notes that in the 1970s, the aerospace firms tried to jump in "and solve all the problems of mass transportation" (Keevil, 1995).

Mr. Frank Cihak, who worked on the early phase of the program, observes that the purchasing program Chicago developed was not the result of trying "to enter a development program to develop technology" but simply resulted from purchasing cars to do a specific job. The Chicago Transit Authority "had a very strong interest in making sure these cars would be successful so we guided them very carefully when they got down to designing the car." The Authority recognized that "it was in our own best interest to have those cars operate as successfully as possible" (Cihak, 1995).

Chicago's policy was based on learning from prior successes. The Transit Authority took an active interest in issuing detailed specifications to its suppliers "because it was a successful way of procuring equipment." After the Authority developed a successful pattern for negotiating the acquisition of equipment, they had "little incentive to venture into unknown areas that would involve more risk" (Cihak, 1995). Mr. Keevil notes, however, that "the CTA has always had a preference for tried and tested equipment. We have never in a production order specified beyond the state-of-the-art or even the newest state-of-the-art." This tradition was based on the necessity of "having to run a railroad to carry passengers twenty-four hours a day" (Keevil, 1995). Unlike the federal government, Chicago could not afford to experiment, i.e., they were limited in their ability to take the risks associated with a major demonstration program of new technology.

In sum, "there were . . . successful examples of the same car that had been built in recent years and were in existence." At the system level, defense designers did not face the task "of making up something new." Designers could use previous designs of 1100 Chicago cars already in operation to guide their design of the subway car. The Authority had 3000 buses supplied by six different manufacturers. This gave Chicago experience in dealing with a diversity of suppliers: "we had lots of experience with lots of different manufacturers—good and bad"(Cihak, 1994).

The form of the specification issued by the CTA differed from that found in the LRV. The LRV "started from ground zero." The LRV was designed in response to a performance specification. There can be numerous design solutions that meet a performance specification. The CTA project used a detailed or hardware specification. Performance specifications are very broad and subject to a lot of interpretation. For example, a performance specification may be a requirement for a device that can write on paper. With performance specifications, "if a designer doesn't know what you want and know what you expect, then he may design something that doesn't meet your needs." In contrast, a detailed specification would be for "a ball point pen, with blue ink in it that is four inches long and a quarter inch in diameter that will write a line 800 feet long . . ." So with a detailed specification, the customer will use their knowledge as an owner of transit properties and instruct the designer to "the best extent possible" what he or she wants (Cihak, 1994; see also Vincenti, 1990: 98–101). Sometimes a detailed specification is called a hardware specification. The CTA's Walter Keevil explains that their specification was a combination of a hardware and performance specification. But, whether the specification was one governing details of the hardware or performance, it was written "with many specific details" (Keevil, 1995). Mr. Dancik was in constant communication with CTA's Walter Keevil, the author of the detailed specification (Dancik, 1995).

Mr. Cihak was the chief equipment engineer for the Chicago Transit Authority (CTA) in 1970 and was responsible for developing the specifications and conducting the procurement of buses and rail cars. He explains how Chicago's detailed specifications helped constrain and guide Boeing-Vertol's engineers:

> The Chicago order had a very tight, detailed specification. It was largely not a performance specification. It was a detailed specification: the parameters [governing the design] of the cars to be built were very constrained. They had to fit and operate with existing cars so there was little latitude for a designer to get into trouble.

The specification laid out the detailed requirements for the design of a car including the size of the wires, the diameter of the wheels, as well as the length and width of the car and doors. Such specifications are presented "in fine detail by drawings, or references or other standards." Such level of detail helped prevent Boeing-Vertol's engineers from making innovations that would not correspond to the requirements of the product in use (Cihak, 1994).

Limiting and Controlling Defense Design Practices

The CTA specification and intervention by managers like Walter Keevil helped convert Boeing-Vertol's engineering resources. Kevil participated in about fifty design reviews in which CTA staff joined their counterparts at Boeing-Vertol in discussions regarding the defense firm's subway car plans. In some cases he was joined by a staff of three or four but these assistants made no changes without his knowledge or agreement and Keevil was always present at the review (Keevil, 1997a and b). This suggests, however, that Chicago's design review staff was larger than San Francisco's. In some cases, Keevil would ask Boeing why it used tolerances that were narrower than would be required. As a contributing author to the CTA specification, his suggestions helped bring Boeing's designs into conformity with commercial and customer requirements. As a customer, the CTA also specified materials that they wanted used on the rail car. So, if Boeing-Vertol, as an aerospace firm, was accustomed to using more expensive materials, the specification provided a way to overcome that problem (Dancik, 1995).

The CTA's specified requirements influenced the parameters on Boeing-Vertol's designs. That is, the design knowledge established within the CTA was transferred to Boeing-Vertol through the specification. This knowledge often comprised a parameter on what would work under CTA conditions rather than actual details of how to implement a specific requirement for performance.

Mr. Keevil explained that one example of a fairly tight specification was in the car body section. It required that the "maximum stresses between bolsters under horizontal loads shall be 30 percent less than outside the bolsters to the ends of the car." The weight requirements indirectly influenced design choices by encouraging Boeing-Vertol to develop lighter weight cars. The need to reduce weight encouraged the company to turn to aerospace-based technology that could help them produce a lighter car. Therefore, Boeing "came up with a light weight structure that met our specifications for weight and stress levels" (Keevil, 1995).

The process by which the CTA's requirements were passed on to Boeing-Vertol did not reflect the simple transmission of information as suggested by many models of "learning." It also was enforced by the CTA's power vis-à-vis Boeing-Vertol. Mr. Keevil explains how CTA's intervention prevented the introduction of many aerospace designs which could have produced the post-Vietnam diversification failures associated with the aerospace manufactured trolley cars, people movers and buses. On some design questions, Mr. Dancik at Boeing-Vertol:

> wanted to do things that were just flat out not acceptable for a railroad car . . . I remember being on the phone with him discussing things for hours. It finally got to the point where we had to tell him, "no you cannot do that." In most cases, he listened to what we said and abided by our suggestions (Keevil, 1995).

In the partnership between Boeing-Vertol and CTA, Mr. Dancik was "willing to listen to CTA" even though on some issues, "he fought us tooth and nail." Mr. Keevil says this relationship resembled teaming, "because it was always a case of give and take on both sides between Boeing and CTA." The tension between CTA sensitivity to transit requirements and Boeing-Vertol's attempts to introduce new techniques produced a partnership that usually tapped the best of both worlds (Keevil, 1995).

In many areas, the specification required that CTA grant prior approval before it would release a component for use in a design. They required the supplier to submit samples, drawings, engineering data and test plans: "there must have been 300 different items that had to be approved before you could release them for manufacture." Mr. Dancik created a matrix of all these approval items by category based on items required by the specification. This matrix included a schedule and was summarized on one piece of paper. Both CTA and Boeing used this as a check list and they "both sang from the same piece of paper." Mr. Dancik submitted this matrix to CTA. This submittal confirmed the requirements which the consumer placed on the supplier (Dancik, 1995).

CTA's contract guidelines also placed cost constraints on Boeing-Vertol which governed the price of the product. This form of regulation involved an exchange of power based on "the power to price" (cf. Melman, 1970a; Gorgol, 1972). Boeing as an aerospace firm was accustomed to cost-plus contracts in the 1970s, but Chicago negotiated a "firm fixed-price contract per unit with a relatively tight delivery schedule." Chicago would only make final payment for cars after they were delivered and accepted (Cihak, 1994). Boeing also received incentives, "progress payments" based only on a rigid schedule with milestones. Progress payments were coming into the mass transit industry at about this time. One milestone was completion of a car body shell. The progress payment would be a payment which was a given percentage of the car value for each car completed. Another milestone was the delivery of the propulsion system to the factory where the car was going to be made. There were about seven or eight milestones (Cihak, 1994).

Learning from the Specification: A Form of Internal Training

When Mr. Dancik first visited the CTA, for about one out of every three questions he asked, the transit authority told him the answer could be found in the specification. He became embarrassed so he "spent a whole week reading every word in that specification." The CTA had "a good, detailed hardware specification." Mr. Dancik made sure his group engineers or lead design engineers for the truck, electrical work, car body, equipment, air conditioning, etc. "knew that specification as well as the guy who wrote it." The specification was a double-sided manual about an inch thick. The specification included general requirements and test requirements for components and the vehicle, materials and equipment needed, as well as drawings. The drawings specified the dimensions of the vehicle, e.g., the height of the floor above the railroad track and other aspects (Dancik, 1995).

Elimination of Buffers in Communication by Direct Communication between User and Engineering Disciplines

While there were elements of power in the relationship between CTA and Boeing-Vertol, the relationship rested on the interactive process in which there was clear communication between the defense engineers and the customer as a "design authority." The relationship was one of reciprocity rather than a unilinear transmission of information. As we have seen, the LRV program consultants actually acted as buffers to the communicative process between producer and user. By bringing its Boeing-Vertol CTA engineers to the bid meeting and eliminating the buffer of consultants as middle men or managers, Boeing-Vertol opened up a direct channel between the disciplines represented by its engineers and the customer (Dancik, 1995).

Before the bid, the CTA held a series of technical meetings with various potential bidders, companies like Pullman, Budd and Boeing-Vertol. They discussed questions about the various subsystems and other details. Mr. Dancik noted that at such meetings the supplier's program manager would be in attendance and maybe another engineer. At one such meeting, Mr. Dancik invited five Boeing-Vertol engineers, each responsible for a given discipline. In this way, the engineer responsible for a discipline got first-hand information about the customer's requirements. The information would not simply be filtered second-hand through Mr. Dancik. Mr. Dancik told the CTA, "I want those men to hear from you what you want—your interpretation of the specification." Mr. Dancik brought his assistant to take notes. This direct teaming between supplier and customer allowed each discipline engineer to ask specific questions and follow-up questions that might be needed to understand how the customer interpreted the specification (Dancik, 1995). Mr. Dancik was asked by someone in Chicago on the CTA project what he thought of full-time consultants. He said that "if we need someone between us to tell us what language you wrote, we're going to be in deep trouble." Mr. Dancik advised that consultants should only be used for discrete tasks. If someone at CTA had a question, then they could phone a Boeing engineer charged with a discipline task directly. In this way, the CTA could get their answer "from the horse's mouth" (Dancik, 1995). In the LRV program, the specification writing was funded by the federal government which encouraged a standardized vehicle involving multiple customers and novel aerospace producers (Bei, 1978; Lewis T. Klauder and Associates, 1985a; Young, 1993). The CTA was largely protected from any federal incentive or policy which might have avoided the use of tried and true specifications (Keevil, 1995).

Managerial Job Rotation and Integration

Beyond the CTA's intervention, management practices within Boeing-Vertol also contributed to the success of the project. Mr. Dancik says that a project could still fail even with a good specification if "it didn't have the proper organization and the proper people for it." One issue is how the specification is interpreted (Dancik, 1995). As noted earlier, managerial job rotation turns specialists into generalists. In the conversion context, engineers who rise through the ranks can gain broader understanding of various disciplines and how they should be integrated into production. Such integration is important for avoiding the failures associated with the LRV, in which parts proven separately fail to work properly once put together. Engineers who climb job ladders or rotate through various disciplines not only gain skills in how to put the engineering puzzle together, but also gain wider exposure to those among the reservoir or matrix of engineering talent who are the "best and brightest." Engineers who have wide exposure from age, experience or migration up and down job ladders gain knowledge about the abilities of their colleagues.

In Boeing's LRV program, a key project engineer was versed primarily in dynamics, a "technology discipline." He was a specialist in technical issues related to vibration in a helicopter. The problem with putting an engineer only versed in one theoretical discipline in charge of a project is that he or she will not be able to discriminate between better and worse designs. Some project engineers are simply "bean counters" who make sure a project is within its budget and "drawings are released in time whether they are good or bad drawings." But, such engineers need to understand hardware if they want to avoid problems, e.g., failures associated with approving a faulty drawing (Dancik, 1995).

Although Mr. Dancik had studied the specific technologies applied to the rail car (many of which were common to helicopters), he was not an expert in them but was able to discuss them and understand a wide variety of technology experts. He had worked in structures at Fairchild Aircraft and Republic Aviation. He had worked in flight controls and hydraulics at Vertol. He had overseen all the controls and hydraulics being designed for each helicopter. As a project engineer, he was responsible for engineers doing transmission design, electrical work, and power plant work. Mr. Dancik had intimate (or "tacit") knowledge of the variety of disciplines which he would later oversee as a project engineer. He was not simply a generalist, but someone whose career path had rotated him through the various specialties (Dancik, 1995).

As a result of such job rotation, Mr. Dancik was sufficiently familiar with each discipline "to know that when I asked a question, I got an answer and I wasn't snowed under." In contrast, the LRV project was headed by an engineer who understood specialized knowledge like dynamics and vibration, but had "no design experience." These constraints on the LRV manager's knowledge limited his ability to meet guidelines from outside the firm: "How can he choose the right box out of six different designs? Which one is the right one? Which one best meets the specification?" (Dancik, 1995). Thus, the "interpretive abilities" or diverse skills of Mr. Dancik as manager affected whether the CTA specification could be successfully applied. The value of customer intervention, the specification, and long-term knowledge embodied therein, was contingent on the social organization of work within Boeing-Vertol.

Integration through Direct Managerial Oversight

As noted earlier, Boeing senior management proposed the theory that engineering managers can not oversee work to make sure defense engineers conform to cost and design guidelines. Mr. Dancik engaged in such direct supervision and thereby extended the communication link between the civilian specification and implementation by defense designers:

> I spent ninety percent of my time over drawing boards with the designers. LRV didn't have people do that. I looked at every drawing that went out. Before I signed a drawing, I made sure that I went over it with the group engineer after it was signed by the checker. A group engineer is the supervisor of a specific design group, e.g., the transmission, circuits, truck or electrical equipment group. I knew what I was signing because by spending that much time over the drawing boards, I saw the designs being developed.

When the subway car developed a problem with its weights, Mr. Dancik assembled the weights man, the group engineer, the immediate designer and stress engineer and "stayed at the drawing board until a decision was reached as how to resolve the problem, even if it was for two hours or longer." Decisions were not delayed "because a delayed decision is almost as bad as no decision at all" (Dancik, 1995).

Mr. Dancik explains that the LRV project "had the wrong people for the jobs." After Mr. Dancik became a project manager, he worked with specific disciplines. By working with engineers attached to different disciplines such as electrical engineers and transmission engineers, Mr. Dancik gained intimate knowledge of the best and brightest in the reservoir of engineering talent in the firm: "I got to know the people who worked in their groups . . . I got to know all the designers including the top lead men and all the key men in the technologies. So when I got my own project, I got to know the cream of the crop." But the key project engineer of the LRV, "didn't personally know who the good designers were." He might have only gained knowledge of who the best person for each job was by word of mouth. In contrast, Mr. Dancik had the technical skills to understand each component and who was the best engineer to develop or design that component. When you select someone who is not the best engineer for a discipline, you risk delays and picking people who do not know how to come up with solutions to design problems. The better engineers would have more experience so they could explain why a given design might not work (Dancik, 1995). Mr. Dancik also relied on his associations with powerful managers to assemble the team he wanted and link that team to the CTA without interference.

"Geography" on a small scale, but more significantly, the organization of power relations, helped extend integration. On the geographic front, all the engineers charged with weights, stress, and other specialists "worked in the same area" and were not segregated in "cubby holes." Like the "skunk works" system, the specialists were physically grouped together (Dancik, 1995). Power extended integration as follows. There is an element of power in being able to: (a) gain one's choices for the best and brightest engineers, and (b) being able to effectively communicate with civilian parties outside the firm without internal interference or "buffers." That is, integration here, as elsewhere, depends on power. Originally, the CTA project at Boeing-Vertol had a program manager who knew nothing about design issues. His job was to provide coordination among the customer, the engineering staff and production. This manager could have created an internal buffer between the customer and supplier. But, Mr. Dancik worked out an arrangement with his boss who was also the boss of the program manager. Both Mr. Dancik and his boss did not allow this manager (who was eliminated from the program) to interfere with Mr. Dancik's direct relationship with Walter Keevil of the Chicago Transit Authority and the program being run at Boeing. The manager who was removed had been a "marketing type" i.e., a salesman (on how financiers and marketing managers can complicate production, see: Veblen, 1965; Melman, 1983).

Mr. Dancik's close associate was the director of engineering for transit business. He backed Mr. Dancik up in getting most of the personnel he wanted for the CTA project. His technology manager, David Bevan, was "outstanding." This director, Mr. Dancik's boss, had a superior whom he had known for twenty years. The superior was Carl Wiland who was the head of the transit business. Mr. Wiland was very close to the President of Boeing-Vertol. This chain of contacts helped Mr. Dancik get who he wanted for his team based on his recommendations. On the other hand, the project engineer for the LRV "would accept just about anybody he was given." When Mr. Dancik didn't get the top person he wanted, he picked the second best person, but refused "inadequate personnel" (Dancik, 1995).

Acquired Knowledge and Testing

Even with the benefits of the established specification, in some key areas Boeing-Vertol established novel technologies and applications (see below). As a result, the CTA established a rigorous testing protocol. The LRV combined novelty with poor testing procedures. In contrast, the CTA car faced much more rigorous testing protocols. Mr. Cihak, the former CTA manager, explains Chicago's practice. After the first four cars were delivered, they underwent "an operational test which required them to run 600 hours without failure before we would approve the manufacture of the remaining 196 cars." These four prototype cars were tested over a three month period. If problems developed on cars, the manufacturer was required to "investigate, evaluate and recommend corrective action for every failure." If a problem developed, the manufacturer was required to re-test the prototype before the rest of the cars were manufactured. But the cars went through the test period without any major problems. As a result, Mr. Cihak suggests that Boeing's internal testing procedures were sufficient to work out major problems. Mr. Cihak believes that CTA's testing requirements provided an incentive that such tests would be carried out. Mr. Cihak repeatedly told the supplier that until a car got through his test program, he would not accept a car and that the risk had to be theirs: "they could be holding up fifty or so cars back at the plant, until the first one got through the test" (Cihak, 1994).

Where the LRV got into trouble by using novel parts without extensive tests, the CTA insisted on a rigorous testing procedure and tried and true components. For example, the propulsion system of the CTA car was assembled at GE's Erie, Pennsylvania facility. This system was tested long before the cars were constructed. The new propulsion systems were an enhanced upgrade of an earlier GE model which had already been used successfully on 300 rail cars in Chicago's system (Cihak, 1994; Keevil, 1997a and b).

The major components like the air conditioning and doors were each individually tested and inspected in what was termed "first article inspection." These are the first articles which come off the production line. These inspections are generally performed at the factory of the manufacturer of the article. All the components are assembled in a mock-up "commensurate with how each will fit on the car" and the parts are tested in this configuration. In the first article inspection, Boeing and CTA witnessed the tests in the beginning and end of the process which brought together components. Many of the major subsystems can be tested before they are brought together in an entire car: if such a test is conducted and you have confidence that each component is tested properly, then the risk is low that components will not work when put in the rail car (Cihak, 1994). Yet, even first article inspections were not always helpful on components when they were novel and not tested in use. The first article inspection is generally performed after the design reviews and the hardware is manufactured. They are useful to make sure that the design was translated properly into the hardware. One area where the first article inspection did not help CTA was in its evaluation of air conditioning. The old system ran on DC current and the CTA wanted to run it off AC which it believed would be less expensive and more reliable. The CTA specifications called for a new air conditioning system. The "philosophy" of trying to get a new air conditioner that was more reliable and cheaper to operate made sense. But, the "execution" of this requirement led to problems.

At a test in the supplier's plant, the first article inspection showed that the unit worked well. Yet, neither the CTA, nor Boeing appreciated that the actual performance on the vehicle would be substantially different. A required engineering test discovered that the air conditioning did not work well in an actual vehicle. During the climate room test, the problem was discovered. This required test had to take place after a fully made vehicle was available. This test took place on about the fifth car (which already had its air conditioning unit installed). (This was early in the program.) The completed vehicle was placed in an environmental chamber and its air conditioning system was tested by varying both its internal and external conditions. Then, air ducting, ventilation and insulation were changed to meet specifications for cooling capability. This environmental test was only a static test, not a test of a car in service. The full extent of the problem was only discovered "after running the cars for several years." It turned out that the compressor motor assembly was too big and the condensing capability of the system was too small. These two factors led the motor compressor assembly to burn out very quickly in service. This led CTA to buy an entirely new air conditioning system to respond to what was a design flaw. Interestingly, this poor design can be traced to another aerospace firm which supplied the air conditioner, Lear Siegler. Boeing-Vertol had picked one of its commercial jet suppliers in Lear Siegler. Their claim to fame was that they had supplied the wayside air conditioning systems to keep planes cool on runways (Keevil, 1995). In sum, the CTA could not act as the enforcer of the tried and true when even it lacked direct experience with a component, like air conditioners.

The CTA Car and Novelty in Design

The CTA car did not succeed simply because it was not based on novel technologies. There were many other differences between it and the LRV car (see Table 18.1). Moreover, in a host of areas, CTA engineers introduced a series of novel technologies. The Boeing-CTA rail car case raises the important question as to what degree learning by diversifying defense firms can be "parthenogenic" i.e., learning without directly interacting with a customer who has acquired design knowledge of a given product line. It is also important to ask whether this case was simply one in which the CTA relied on tried and true designs and forced them upon Boeing-Vertol. In some areas, Boeing-Vertol introduced innovations into the rail car which were truly novel and broke with precedent. In other areas, the lines between what Boeing did successfully on its own and what it could not do by itself were more fuzzy. In still other areas, it is clear that when Boeing ventured into new ground without precedents and controls it failed miserably. With air conditioners, even the CTA lacked the benefits of long-term learning, allowing a subsystem to fail.

Mr. Keevil explains that the CTA subway cars did not simply mirror the specifications which the transit authority provided. Boeing-Vertol had an opportunity to propose innovations that went beyond the proven specification. Alternatives were conceivable especially since some time had passed between the first specifications and the winning of the bid. When the CTA "could not come up with good hard reasons to prevent them from doing something, we let them do it" (Keevil, 1995).

The specification required that the "riding qualities" of the 2400 series cars "had to be equal or better than the best Budd cars riding on the CTA property." Mr. Dancik says that when Boeing-Vertol "signed up for that specification, we signed up for a blank check." Even CTA didn't know the riding characteristics or vibration qualities of the best Budd cars. As a

Table 18.1. Key differences between the LRV and CTA projects

	LRV	CTA
Specifications	Performance	Detailed
Testing: Prototypes	No production prototypes	Production prototypes based on prior models
Testing: Protocols	Major modifications made after trolley care in service use	More rigorous acceptance policy
Project manager	Theoretical engineer	Generalist
Customer learning	Limited to ad hoc repairs in Boston, some evidence in San Francisco	Extensive design reviews with in house staff and learning from the specification
Consultants	Act as a buffer to producer	No consultants as intermediary between producer and user
UMTA role	Favored performance specifications and innovative designs to modernize trolley car technology	Helped on some components testing made use of by Boeing-Vertol, but no direct influence on CTA's more detailed specifications
Suppliers	Introduced novel, poorly tested systems	CTA's long experience with suppliers and rigorous inspection systems reduced chances for problem based on faculty components

result, Boeing-Vertol was venturing into unknown territory. Yet, CTA did not have data about how Boeing-Vertol could meet these riding requirements. Boeing-Vertol instrumented what CTA considered its best Budd car. By applying instrumentation to the car, Boeing-Vertol defined its own specifications for use of the product. These specifications helped them with their design (Dancik, 1995).

Boeing-Vertol also introduced the idea of a "tension field" used in airplane fuselage bodies to design a light weight car shell. This is a way to redistribute the tension of weight on the wall of a rail car. It provided a way "to optimize the thickness in the structure for the least weight" (Dancik, 1995). The tension field principle provided a superior way of calculating the relationship between required thickness and the loads applied to a structure. Cars could be designed with thinner skins rather than the heavier gauges that had been used before. Weight was reduced by the design of sections under a load in the wall of the car shell. This became the primary way in which the weight of the car was reduced (Dancik, 1995). This innovation was not a direct response to the specification and grew out of the application of aerospace-based technology. Boeing-Vertol was able to use this innovation to produce "a fairly light car body" (Keevil, 1995).

The CTA has always required a test of the wiring on its new rail cars so that they can insure that this wiring is done correctly. Boeing-Vertol introduced a new diagnostic technique to CTA which used its aerospace-based technology. As part of Boeing-Vertol's standard technique, they have a computerized system to check all the wiring on a helicopter. Vertol used this aerospace-based technique to check all the wiring on the CTA cars and successfully prevented any wiring errors. This system had an advantage in that it could make an assessment of wiring problems even before power was applied. Since no power had to be applied, this technology eliminated the risk of damaging the circuitry if there were problems. The diagnostic technology saved Boeing-Vertol time in examining each car. Companies which followed Boeing-Vertol in supplying CTA rail cars, and had wiring problems before, benefited by learning about this technology (Keevil, 1995).

There were some areas where Boeing Vertol got into trouble after it tried to go beyond the boundaries of acquired knowledge as formalized in the specification. For example, the specification for the test was clear, but not the procedures which governed how the tested element would be handled up to the point at which it was tested. The specification required that the car body shell be tested for water tightness before Boeing installed the interior trim so they could find any leaks that might create problems. Boeing-Vertol bought the shells in Portugal and shipped them across the ocean on ships. Shells were lifted on and off the boat, twisting the shells and stressing joints. The shells were then stored outside its plant by the Delaware River. The shells had been treated with an anti-corrosive substance and surface protectant in Portugal. In order to ready the cars for production, the inside and outside of the cars had to be steam cleaned to remove this protective coating.

The combination of the protective coating, the time between its application and removal, and the cleaning process resulted in a great number of leaks in the car shells at the time of testing. CTA believes that Boeing-Vertol tried to "avoid responsibility" on the water test "and not make the shell completely water-tight per our specification." Boeing-Vertol later made adjustments to correct the problem (Keevil, 1995).

Aftermath of the CTA Program

Boeing-Vertol lost a renewal contract for CTA rail cars in 1978. Boeing lost this renewal contract to Budd, but "Budd underbid the competition by over $100,000 a car" (Keevil, 1995). On the second rail car order, which

Boeing lost, the Budd Company's bid for 300 cars was $133.3 million, Boeing's bid was $174 million and Pullman's bid was $248 million. In fact, Budd lost about $40 million on their contract with the CTA. Initially, Boeing may have low-balled the first bid to get into the rail car business. They may have bid less than their estimated cost to get into the market (Cihak, 1994; Dancik, 1995, Keevil, 1997a and b). Yet, Boeing's defense overheads likely did not make the company unable to compete based on costs because their second bid actually reflected the true costs of the program (Keevil, 1995). In fact, Budd's "internal house estimate for the cost of the car was very, very close to Boeing's, but when they came to enter the bid, they put down a different number. . . . They thought they could make up the difference on the option cars" but Budd ended up losing money because their costs ran up high, in part because of severe inflation in the mid-1980s (Keevil, 1995). This suggests that Boeing learned to reduce costs by the time of the renewal contract. The second order of 300 cars was virtually identical to the first order that Boeing had won. Boeing should have had an improved opportunity with its second bid because they had carried out their tooling and had gained experience in rail car production (Cihak, 1994). Boeing didn't bid for any other prime contract transit work after they lost CTA's follow-on contract (Keevil, 1995; Confidential interview, 1997b). Boeing continued as a subcontractor on some transit work. In the CTA contract they originally won, Boeing-Vertol's costs were largely covered by their contract, and if they did not make a profit, the contract was close to covering their costs (Keevil, 1995).

Boeing-Vertol's costs were covered or largely covered, but the market was not profitable enough for its management. Mr. Cihak says that Boeing concluded that because they were underbid for a contract that would have produced a profit for them, they could not compete in the mass transit business. He argues that Boeing as a defense contractor was accustomed to operating in an environment where "cost was a relatively small consideration" because "technological performance" was emphasized. Most of the defense firms left the transportation business "because their cost structure was not comparable with this market. . . . That's the reason why Boeing left" (Cihak, 1995). His view is confirmed by a manager at Boeing-Vertol at the time who argued that the firm didn't "know how to play in a market that says the lowest price wins" because customers, particularly foreign transit makers backed by government subsidies, can always "come into the market and underbid the price." In sum, Boeing argues that it dropped out of the transit market because they "could not get a reasonable return" on their investment. For its part, Boeing-Vertol believed that it should have won the CTA follow-on contract because it offered the best vehicle Chicago had seen at a reasonable price (Confidential interview, 1997b).

While Boeing-Vertol may have judged the subway car market as less than profitable, the engineers working on the program showed that they accumulated valuable civilian production skills and enhanced their ability to develop civilian products during the life of the program. In fact, after Boeing-Vertol got out of the business, Mr. Dancik worked as a consultant for Morrison Knudsen to help them with their proposals. Morrison Knudsen won their bid and built about 257 CTA cars (Dancik, 1995; Keevil, 1997a and b).

CONCLUSIONS

Requirements for Production Success

Like the LRV program, the CTA project was physically separated from high defense overheads. Each program was part of a mass transit market which, like the Pentagon, represented a centralized customer base. Each factor proved insufficient for conversion success as one program was considered a huge technical success, the other a dismal failure.

The most important difference between LRV and CTA projects was that the former, under UMTA guidance and oversight, involved poor testing, novel designs and limited direct user input by customers who had formalized tried and true methods through specifications. In the CTA project, the successful conversion of skills required joint decision-making or teaming by civilian contract consumer and military-engineering-based producer. The CTA project was also successful because of Dancik's ability to know and claim the best engineers and integrate a variety of engineering disciplines, but the LRV project was led in part by a theoretical engineer who was not a generalist. Even if Mr. Dancik was one of Boeing-Vertol's most competent engineers, he was one of many highly qualified project managers with generalist knowledge (Bevan, 1997). If Dancik's management routines of monitoring and generalist knowledge embodied best practices found in Japan, then this says less about his being a "fluke," than about the need for retraining of defense engineers and managers, a point long underscored by conversion scholars (Melman, 1970b) and ignored by those who see conversion as a relatively easy process.

The CTA case shows that an independent design authority, having influence vis-á-vis defense engineers, was often critical to make sure the firm properly applied acquired knowledge. Such power and knowledge are part of the larger managerial equation that makes diversification successful. Nevertheless, in some subsystems, where it could tap analogous learning from aerospace practices, Boeing-Vertol was able to contribute novel design innovations without direct guidance, although such guidance was often more critical than not. Moreover, the consultants' role in the LRV case suggests that customer power without acquired knowledge prevents technology transfer, although San Francisco's consultants encouraged

MUNI to push Boeing-Vertol to make needed changes (Lewis T. Klauder and Associates, 1985a; Crugnola, 1979: 40).

Boeing-Vertol's Exit from the Mass Transit Industry and Military Specialization

Boeing-Vertol's CTA project succeeded technically, but appeared to fail as a going business concern. Yet, this failure says less about the limits of conversion caused by military specialization than more important, non-microeconomic factors. Boeing's decision to drop out of the subway car market suggests that the company was unable to generate a profitable business out of this venture in the short-run. Yet, it is not clear that they would not have been able to produce at a profit in the long-run. Several reasons suggest that the decision to drop out of the market was not a byproduct of the specialized skills of production and engineering staff.

First, the lost contract did not reflect weaknesses in the firm's technical competence. The second rail car order that Boeing lost was virtually identical to the first order it had won. The company clearly learned by its experience with the first set of cars and moved down its cost curve. After having established the necessary tooling and production routines to develop the cars, later orders should have been easier.

Second, Boeing's lost bid reflected advantages over other commercial competitors. Boeing's bid of $174,000 per car was considerably less than civilian firm Pullman's bid of $248,000 per car. Even though civilian firm Budd won its bid, they seriously underbid to win their contract and ended up losing millions.

Third, the decision to abandon this mass transit product line should be evaluated at a scale of economic analysis that is much broader than the firm, i.e., market success also depends on the larger system of innovation in which individual firms are a part (Edquist, 1997). The American innovation and procurement system has created incentives to keep military firms engaged in military markets and promoted technology transfer in closely related industries like commercial aircraft and satellites, but has been less successful in developing alternative markets in mass transportation, alternative energy and a variety of infrastructure needs (Mowery and Rosenberg, 1982; Rosenberg, 1987; Feldman, 1989; Markusen and Yuden, 1992).

The incentives for aerospace and military production have been greater than for mass transit production. Over a four year period, Boeing-Vertol expended $140 million worth of resources on the LRV and CTA programs. But, during this same period Boeing was selling about three 747s every two weeks for $50 million each. Boeing could make as much from selling planes in two weeks as Boeing-Vertol could make in four years from selling transit cars. Both Boeing-Vertol and Seattle headquarters decided to exit the transit business. Helicopters then cost on the order of $20 million. A single transit car costs about $300,000 each. The helicopter business could produce more revenue faster than the transit businesses. Mr. Dancik believes that "with the money made in the aerospace industry compared to what was made in commercial business," the transit projects could be considered "a waste of resources and management time" (Dancik, 1995). During the 1980s, "the top defense contractors performed substantially better than the market" in terms of return on investment (ROI) and return on assets (ROA) (Trevino and Higgs, 1992: 214, 217). Boeing's specialization may have led it to overestimate the size of the LRV market (cf. Young, 1993: 30, 35), but this market is small in any event with only about 600 rail cars produced annually (Cihak, 1995). Ultimately, the extent of mass transit procurements is partly a political problem depending on new federal priorities and coordination among divergent interest groups (Feldman, 1989, 1997b).

While defense industrial policies have been strong, competing industrial policies to support a *mass transportation industry* and a larger circle of competent producers have been weak and poorly designed. The U.S. government has underinvested in its mass transit industry compared with its counterparts overseas and even civilian firms like Budd Company have abandoned the production of rail passenger cars (Ullmann, 1993; cf. Edquist, 1996). UMTA wanted to modernize rail car technology and aerospace firms were equipped to do so, yet the government failed to create a program that continued that process. UMTA did nothing to provide a system of checks on the problematic designs of Boeing-Vertol. On the front-end, it did nothing to encourage that designs be informed by knowledgeable experts in the rail car field. On the back end, "it did nothing to enforce a requirement for prototype testing, and indeed approved a schedule so short as to preclude the possibility of testing in any serious way" (Melman, 1983: 258).

Appendix 1: Chronology of Boeing-Vertol LRV and CTA Programs

Mid-June 1971	Boeing-Vertol selected as systems manager for the U.S. Government's Urban Rail Vehicle and Systems Program
July 1971	MBTA requested that UMTA fund a demonstration program involving two articulated cars by Düwag Co. of Germany but program canceled in Fall of 1971 after President Nixon's new "buy American" policy implemented
1972	Early development of specification for LRV
May 1, 1973	Boeing-Vertol awarded a contract to design and manufacture 150 LRV cars for MBTA and 80 cars for MUNI
December 1973	Boeing-Vertol wins CTA contract

August 30, 1974	First prototype LRV
1975	First new LRV car shipped to Boston and vehicle tested for 11 weeks
Fall 1975	Three vehicles sent to the U.S. Department of Transportation test track in Pueblo Colorado
1976	Boeing-Vertol completes the design, development and delivery of 200 subway cars
January 6, 1976	MBTA reports that there are still many uncertainties in the overall test program for the LRV
April 1976	MBTA receives two LRV "pilot" cars for testing purposes
October 1, 1976	A "Quarterly Progress Report" by MBTA submitted to UMTA states that LRV design is 97% complete and production is 23% complete
December 30, 1976	First four Boston LRVs began public operation
April 6, 1977	Boeing-Vertol is performing substantial modifications on LRVs in Philadelphia, prior to delivering them to Boston
December 5, 1978	MBTA announced that it was returning 35 LRV cars to Boeing-Vertol because the trolleys didn't work
December 1978	Boeing-Vertol loses new contract bid for CTA cars
Mid-1979	LRV first introduced into revenue service in San Francisco
Summer 1979	Final test results of LRV published
February 10, 1982	Memo from Richard Sklar, General Manager, Public Utilities Commission, City and County of San Francisco, to Boeing-Vertol *et al.*, expressing concern about an increasing number of in-service delays with LRV cars in San Francisco
1988	MUNI announced that its LRV cars are beginning to fall apart

Sources: Bei, 1978; Crugnola, 1979; Fischler, 1979; Radin, 1980; Sklar, 1982; Lewis T. Klauder and Associates, 1985a; Young, 1993; Keevil, 1995.

Appendix 2: Problems with the San Francisco Trolley Car

The Boeing car from a maintainer's standpoint is a nightmare for several reasons, mostly based on the car's "design." The trolley car was not designed to facilitate maintenance in the fashion usually found in urban mass transit. The trolley cars are instead designed for the kind of maintenance job usually found in the aircraft industry. There are several problems with the San Francisco cars:

- Many of the components are not easily accessible and take a lot of time to reach or change.

- The trolley cars were assembled with many foreign components which come from manufacturers which either went out of business or no longer supply the required parts.
- The Boeing-Vertol design failed to anticipate the frequency with which cars would be coupled and uncoupled which is hundreds and hundreds of times a day. The couplers fall apart because they are not sturdy enough but overly complex in their design.
- The traction motor was a terrible design and required major modifications.

One major modification was to change the way the motor was wound. Another major modification was to change the operation of the cars' propulsion system. When the trolley cars were held still using the traction motor system, the armatures would overheat and the commutator bars would rise and this would "chew up brushes." The armature is the rotating part of the motor that turns the gear box. The commutator, attached to the armature, transfers electricity. The brushes rub against the commutator and armature as they rotate to help transfer electricity. If the commutator is not perfectly round and smooth, it will chip, crack and break brushes and eventually this will lead the carbon on the brush to be used up and create sparks that reduce the life of the motor. In sum, these motor problems were greater than MUNI could deal with, given its resources. Even with substantial changes, the motor's life is less than it should be. At one point, the LRV would travel for 38,000 miles before its traction motor would have to be replaced with new armatures and commutators. In contrast, trolley coach traction motors would operate 150,000 to 250,000 miles before such replacements would be needed (Miller, 1996).

REFERENCES

Adelman, K. L. and Augustine, N. R. "Defense Conversion: Bulldozing the Management." *Foreign Affairs*, Spring, Vol. 71(2), pp. 26–47, 1992.

Bei, R. "San Francisco's Muni Metro, A Light-Rail Transit System." In *Light-Rail Transit: Planning and Technology*, Proceedings of a conference by the Urban Mass Transportation Administration and conducted by the Transportation Research Board in cooperation with the American Public Transit Association, Special Report 182, Transportation Research Board, Commission on Sociotechnical Systems, National Research Council, National Academy of Sciences, Washington, D.C.:, pp. 18–23, 1978.

Bevan, D. Former technology manager, Boeing-Vertol. Phone interview with author, Media, Pennsylvania, August 11, 1997.

Bowen, H. K., Clark, K. B., Holloway, C. A. and Wheelwright, S. C. *The Perpetual Enterprise Machine: Seven Keys to Corporate Renewal Through Successful Product and Process Development*. New York: Oxford University Press, 1994.

Child, J. *Organization: A Guide to Problems and Practice*. London: Harper & Row Ltd., 1984.

Cihak, F. Chief Engineer and Deputy Executive Vice President-Technical Services, American Public Transit Association. Interview with author, Washington, D.C., October 21, 1994.

Cihak, F. Phone interview with author, Washington, D.C., May 22, 1995.

Confidential interview. Interview by Ann Markusen, Michael Oden, Elyse Golub and author of Boeing officials, Seattle, Washington, July 20, 1993a.

Confidential interview. Interview by author of General Dynamics managers, 1993b.

Confidential interview. Former employee of San Francisco MUNI. Phone interview with author, 1997a.

Confidential interview. Manager present at Boeing-Vertol during trolley and subway car projects. Phone interview with author, December 23, 1997b.

Crugnola, C. M. "History of the Light Rail Vehicle (LRV) Program at the MBTA (1967–1979)." Systems and Methods Office, Operations Directorate, Massachusetts Bay Transportation Authority, Boston, January 1979.

Dancik, P. Former Design Manager, Boeing-Vertol. Interview with author, Broomall, Pennsylvania, June 28, 1995. Phone interview with author, Broomall, Pennsylvania, July 27 and July 29, 1995.

DeGrasse, R. W. "Corporate Diversification and Conversion Experience". In Lynch, J. ed., *Economic Adjustment and Conversion of Defense Industries*. Boulder, Colorado: Westview Press: pp. 91–120, 1987.

Dodgson, M. "Technological Collaboration and Innovation." In Dodgson, M. and Rothwell, R., eds., *The Handbook of Industrial Innovation*, Aldershot: Edward Elgar, pp. 285–292, 1994.

Dumas, L. J. *The Overburdened Economy*. Berkeley and Los Angeles: University of California Press, 1986.

Edquist, C. "Government Technology Procurement as an Instrument for Policy Coordination through Task Forces." Tema-T Working Paper No. 163, Department of Technology and Social Change, Linköping University, Linköping, Sweden, June 1996.

Edquist, C. "Systems of Innovation Approaches—Their Emergence and Characteristics." In Edquist, C., ed., *Systems of Innovation: Technologies, Institutions and Organizations*. London: Pinter, Chapter 1, pp. 1–35, 1997.

Feldman, J. "Constituencies and New Markets for Economic Conversion: Reconstructing the United States' Physical, Environmental and Social Infrastructure." In Bischak, G. A., ed., *Towards a Peace Economy in the United States: Essays on Military Industry, Disarmament and Economic Conversion*. New York: St. Martin's Press, pp. 202–241, 1989.

Feldman, J. "Diversification Success and Failure After the Cold War: Results of the National Defense Economy Survey." Working Paper No. 112. Center for Urban Policy Research, Rutgers University, New Brunswick, New Jersey, 1997a.

Feldman, J. "A New Transportation Infrastructure Agenda for the New York Metropolitan Region." New York: Business, Labor and Community Coalition of New York, September 1997b.

Fischler, S. I. *Moving Millions: An Inside Look at Mass Transit*. New York: Harper & Row, Publishers, 1979.

Fiske, J. *Introduction to Communication Studies*. London: Methuen & Co. Ltd., 1982.

Freeman, C. "Critical Survey: The Economics of Technological Change." *Cambridge Journal of Economics*, Vol. 18(5), pp. 463–514, October 1994.

Gorgol, J. F. *The Military-Industrial Firm: A Practical Theory and Model*. New York: Praeger Publishers, 1972.

Hedrick, I. "Grant". Former Chief Technical Engineer, Grumman Aircraft Corporation. Phone Interviews with author. Great River, NY. May 24, June 9, and July 13, 1995.

Imai, K., Nonaka, K. and Takeuchi, H. "Managing the New Product Development Process: How Japanese Companies Learn and Unlearn". In Clark, K. B., Hayes, R. H. and Lorenz, C., ed., *The Uneasy Alliance: Managing the Productivity-Technology Dilemma*, Boston: Harvard Business School Press, pp. 337–375, 1985.

Keevil, W. Chief Rail Equipment Engineer, Chicago Transit Authority. Chicago, IL. Phone interview with author, Skokie, Illinois, July 7, 1995.

Keevil, W. Phone interview with author, Skokie, Illinois, September 10, 1997a.

Keevil, W. Communication with author, August 20, 1997b.

Kelley, M. R. and Watkins, T. A. "The Myth of the Specialized Military Contractor." *Technology Review*, Vol. 98(3), pp. 52–58, April 1995a.

Kelley, M. R. and Watkins, T.A. "In from the Cold: Prospects for Conversion of the Defense Industrial Base." *Science*, Vol. 268(5), pp. 52–58, April 28, 1995b.

Lazonick, W. "Organizational Integration in Three Industrial Revolutions". In Heertje, A. and Perlman, M., eds., *Evolving Technology and Market Structure*. Ann Arbor: The University of Michigan Press, pp. 77–98, 1990.

Lazonick, W. "Strategy, Structure, and Management Development in the United States and Britain." In *Organization and Technology in Capitalist Development*. Brookfield, VT: Edward Elgar, pp. 219–257, 1992.

Lewis T. Klauder and Associates. "San Francisco Municipal Railway: Light Rail Vehicle Improvement Program." Final Report, Philadelphia, May 1985a.

Lewis T. Klauder and Associates. "San Francisco Municipal Railway: Light Rail Vehicle Improvement Program." Appendix, Philadelphia, May 1985b.

Lundquist, J. "Shrinking Fast and Smart in the Defense Industry." *Harvard Business Review*, Vol. 70(6), pp. 74–85, November–December, 1992.

Mackay, R. Chief Mechanical Officer, Massachusetts Bay Transportation Authority. Interview with author, Brookline, Massachusetts, May 12, 1987.

Malerba, F. "Learning by Firms and Incremental Technical Change." *The Economic Journal*, Vol. 102(413), pp. 845–859, 1992.

Markusen, A. and Yudken. J. *Dismantling the Cold War Economy*. New York: Basic Books, 1992.

Markusen, A. "The Post-Cold War American Defense Industry: Options, Policies and Probable Outcomes". Working Paper No. 108. Center for Urban Policy Research, Rutgers University, New Brunswick, New Jersey, 1996.

Markusen, A. "Understanding American Defense Industry Mergers". ms. New York: Council on Foreign Relations, 1997.

Massachusetts Bay Transportation Authority and the San Francisco Municipal Railway in Cooperation with other Transportation Agencies and Consulting Engineers. "Standard Light Rail Vehicle Specification: Technical Section,

Revised to include Agenda 1 through 4". Sponsored by the United States Department of Transportation, Urban Mass Transportation Authority, October 1972.

Melman, S. *Decision Making and Productivity*. Oxford: Blackwell, 1958.

Melman, S. *Pentagon Capitalism: The Political Economy of War*. New York: McGraw Hill, 1970a.

Melman, S. "Characteristics of the Industrial Conversion Problem." In Melman, S. ed., *The Defense Economy: Conversion of Industries and Occupations to Civilian Needs*. New York: Praeger Publishers, 3–17, 1970b.

Melman, S. *The Permanent War Economy*. New York: Simon and Schuster, 1974.

Melman, S. "Industrial Efficiency under Managerial versus Cooperative Decision-Making." In Horvat, B., Markovic, M. and Supek, R., eds., *Self-governing Socialism*. White Plains, New York: International Arts and Sciences Press, Inc., Vol. 2, pp. 203–220, 1975.

Melman, S. *Profits without Production*. New York: Alfred A. Knopf, 1983.

Merton, R. K. "The Machine, the Worker, and the Engineer." *Science*, Vol. 105(2717), pp. 79–84, January 24, 1947.

Miller, J. General Superintendent of LRV and Trolley Coach Maintenance, San Francisco MUNI. Interview with author, San Francisco, November 19, 1996.

Mowery, D. C. and Rosenberg, N. "Technical Change in the Commercial Aircraft Industry, 1925–1975." In Rosenberg, N., principal author, *Inside the Black Box: Technology and Economics*. New York: Cambridge University Press, pp. 163–177, 1982.

Neat, G. Former Manager of the Urban Rail Supporting Technology Program, Transportation Systems Center. Phone interview with author, Cambridge, Massachusetts, August 28, 1997.

Nelson, R. R. and Winter, S. G. *An Evolutionary Theory of Economic Change*. Cambridge, MA: The Belknap Press, 1982.

Oden, M. "Cashing-in, Cashing-out and Converting: Restructuring of the Defense Industrial Base in the 1990s." In Markusen, A. and Costigan, S., eds., *Arming the Future: A Defense Industry for the 21st Century*. New York: Council on Foreign Relations, 1997.

Pages, E. R. "Weathering the Defense Transition: A Business Based Approach to Conversion." Washington, DC: Business Executives for National Security, November 1992.

Pages, E. R. "The Future of U.S. Defense Industry: Smaller Markets, Bigger Companies and Closed Doors." *SAIS Review*, Vol. 15(1), pp. 135, 151, Winter-Spring 1995.

Radin, C. A. "MBTA: Not Out of the Woods Yet." *Mass Transit*, June: 10–13, 1980.

Rogers, R. Former General Superintendent, Maintenance, San Francisco, MUNI. Phone interview with author, Gresham, Oregon, August 21, 1997a.

Rogers, R. Former General Superintendent, Maintenance, San Francisco, MUNI. Communication with author, December 28, 1997b.

Rosenberg, N. "Learning by using." In Rosenberg, N., principal author. *Inside the Black Box: Technology and Economics*. New York: Cambridge University Press, Chapter 6, pp. 120–140, 1982.

Rosenberg, N. "Civilian 'Spillovers' from Military R&D Spending: The U.S. Experience Since World War II." In Lakoff, S. and Willoughby, R., eds., *Strategic*

Defense and the Western Alliance. Lexington, MA: Lexington Books, pp. 165–188, 1987.

Rumelt, R. P. "Diversification Strategy and Profitability." *Strategic Management Journal*, Vol. 3(4), pp. 359–369, October–December, 1982.

Sabel, C. F. "Learning by Monitoring: The Institutions of Economic Development." In Smelser, N. J. and Swedberg, R., eds., *The Handbook of Economic Sociology*, Princeton, NJ: Princeton University Press, pp. 137–165, 1994.

Sayer, A. and Walker, R. *The New Social Economy.* Cambridge, MA: Blackwell, 1992.

Sklar, R. General Manager, Public Utilities Commission, City and County of San Francisco. "Memorandum to: John Hayden, Boeing-Vertol Company; Pete Venuti, Boeing-Vertol Company; Rich Rogers, MUNI; *et al.*," February 10, 1982.

Trevino, R. and Higgs. R. "Profits of U.S. Defense Contractors." *Defense Economics*, Vol. (3), pp. 211–218, 1992.

Ullmann, J. E. "Tasks for Engineers—Resuscitating the U.S. Rail System." *IEEE Technology and Society Magazine*, pp. 13–20, Winter 1993.

Ullmann, J. "Conversion Problems in Changing Economies." In Dumas, L. J. ed., *The Socio-Economics of Conversion from War to Peace.* Armonk, NY: M. E. Sharpe, pp. 303–322, 1995.

Veblen, T. *The Engineers and the Price System.* New York, A. M. Kelley, booksellers, 1965.

Vincenti, W. G. *What Engineers Know and How They Know It: Analytical Studies from Aeronautical History.* Baltimore: Johns Hopkins University Press, 1990.

von Hippel, E. *The Sources of Innovation.* New York: Oxford University Press, 1988.

Vuchic, V. Communication with author, 1996.

Weidenbaum, M. *Small Wars, Big Defense.* New York: Oxford University Press, 1992.

Wynn, G. "Metro Carborne Maintenance Manager, San Francisco Municipal Railway." Memorandum to Rich Rogers, Electric Transit Equipment Superintendent, Re: Boeing-Vertol Letter, March 15, 1982.

Young, A. "The Boeing-Vertol Light Rail Vehicle." *Locomotive & Railway Preservation*, pp. 28–39, January–February, 1993.

ACKNOWLEDGMENT

This chapter builds on early research by the author while a Corliss Lamont Fellow in Economic Conversion at Columbia University and collaborative research with Dr. Ann Markusen and Michael Oden at the Project on Regional Industrial Economics, Rutgers University. The author would like to thank Dominic Bertelli, Leif Hommen, Seymour Melman, Gerald Susman, John Ullmann, Vucan Vuchic, Todd A. Watkins, Britt Östlund and participants in the Second Klein Symposium on the Management of Technology for many helpful comments. This research was funded in part by the Joyce Mertz Gilmore Foundation, the MacArthur Foundation, the National Science Foundation and residency as a Postdoctoral Fellow at the Systems of Innovation Research Program at the Department of Technology and Social Change, Linköping University. The author also extends a special thanks to the defense managers and transit industry personnel who spent many hours helping him on this project.

PART FOUR:
GOVERNMENT PERSPECTIVE

19

POST-COLD WAR DIRECTIONS FOR DEFENSE AND INDUSTRY: A VIEW FROM THE PENTAGON[1]

PAUL G. KAMINSKI

As I look back over the last couple of years, I often run into people who ask me what it was like to have had this sub-cabinet position in the Department of Defense, having the responsibility to oversee $125 to $150 billion of annual expenditures and to answer to the Congress and to the public for those issues that develop. I have to say that that experience reminds a little bit of the drunk who was arrested for smoking in bed and having started a fire. It seems that the drunk, as he was appearing before the judge, said "Your Honor, I admit to being drunk—I was—but I swear that the bed was on fire before I jumped into it!" I would say that it is very nice now to be out of the burning bed. I have a little more control of my own life, and I now have the opportunity to be with you all to reflect on some of those important issues. The theme of this symposium is to examine the effect on defense-related companies of the end of the Cold War: to attempt to gain some assessment of the impact of the business and technology strategies that were used, and to look at some lessons learned. I'd like to break from the precedence of keynote speakers and actually talk on the symposium theme today.

Let me first start with my own reflections about the change in the defense industry. I chose as a reference window the time between 1989 and 1995, because I think that window really captures the principal first phase of the post-Cold War transition. If we look first at the investment dollars associated with the defense industry, procurement expenditures by the Department between 1989 and 1995 went down about two-thirds from about $130 billion to about $40 billion. That was a dramatic decrease—what I call the "planned pause" in procurement. The RDT&E budget went down less—about 15 percent or so—it is now sitting at a number in the high $30s. Putting those two components together, there was a reduction of about 50 percent in the Department's investment budget, a very dramatic change.

Let's go now to the impact of this change on the people in the industry. In 1989, the entire aerospace industry employed slightly less than 1.3 million people. In 1995, that number went down to about 780,000—a reduction of about 550,000 people, or about 40 percent. There are two reasons why the number of people did not go down by 50 percent, in direct proportion to the Department's investment. One is that the commercial portion of the aerospace industry did not go down to the same degree as defense, and the other is that there was some DoD investment made in support systems and logistics that did not go down in the same proportions as other investment. One of the exacerbating elements in this post-Cold War transition was that the commercial cyclic downturn was in phase with defense. The recession in the airline industry—the airlines lost about $15 billion in this period—caused a substantial reduction in commercial airline orders.

Most companies in their diversification strategies would hope that the defense and the commercial business cycles would be out of phase, and they have been in most situations. Unfortunately during this period, they were nearly in phase, and that's what led to this huge downsizing. This was the biggest downsizing in people we'd seen in more than forty years. So let's examine how this reduction was spread among various disciplines. The overall reduction in people was about 40 percent. According to the AIAA, which maintains some records, administrative staffs were cut a little more than the average (about 44–45 percent), assembly workers were cut 43 percent. It is interesting to note, that scientists and engineers were cut 37 percent and technicians were cut only 32 percent. This is an indication that the companies tried to hold on to the family jewels—tried to keep critical skills in the face of downsizing.

Let's go now and look at the companies associated with this base. Here I chose the statistics that are kept by the Defense Contract Management Command—they looked at the companies that make up the top two-thirds of the supply base in investment resources in the Department. In 1989 there were seventeen firms that took up the top two-thirds of the business; in 1995 there were eight. That was reduced by a little more than a factor of two—not out of proportion to the reduction of people and to the size of the investment. We have now seen the completion of the first phase of consolidation in the defense industry.

We are still today in the midst of the second phase, which began around 1991. In my own opinion, consolidation at the prime contractor level has just about run its course, with the Lockheed Martin and Northrop Grumman merger still being considered by the Department of Justice. But I believe there is still some more consolidation ahead in the supporting tiers of the industry. There is one relevant issue that I have raised, but I don't have time to talk about it in any detail today. What I think bears watching in the second wave is the potential for excessive vertical integration by the platform suppliers. In essence, reaching down into the supplier base, which concerns me because it has the potential to stifle competition at the supporting tiers of the business—where we still have the opportunity today for vigorous competition. This becomes more problematical as the number of primes is reduced.

Let's go from this background now to the current situation. I see the investment accounts in the Department of Defense now stabilizing and in fact, if anything, I see the investments increasing due to this "pause in procurement" that I spoke about. This strategy was laid out by Secretary Perry whose view was, since we were downsizing our force structure in response to a greatly reduced threat, we should take out the old pieces of equipment—the one-third showing the most age—and take a pause in procurement for a few years as we were taking down the force, and that the average age of the equipment would not go up. And that, in fact, did happen.

But this pause in procurement works only in the transient phase of the downsizing. Once you've taken out a third of the force and reached a steady-state, if you're not buying new equipment, any equipment you have is going to go up one year per year in age. Procurement investment does now have to increase. But one of the challenges the Department has had has been the difficulty in bringing to fruition its plans to increase procurement investment. These plans have been in the offing for three or four years now, without much success in seeing them implemented. I've described this challenge as analogous to Charlie Brown and Lucy with the football. Lucy promises this time she's going to hold the football steady as Charlie comes running up to kick it, but it never happens, she always moves the ball. For the last two or three years, that's what has happened in the Department of Defense. We have projected an increase in our investment accounts for procurement next year, and, as the budget year came around, the out-year projection was not realized.

A major reason for the problem is that we have not yet fully dealt with the excess infrastructure that remains in Defense. We have taken the force down by one-third, taken our overall budget down by one-third and our investment accounts down by more than one-third; but we've only taken our infrastructure down by about 20 percent. We have excess bases, we have excess laboratory facilities, and unfortunately the portion of the

Quadrennial Defense Review dealing with infrastructure, which Secretary Cohen just released to Congress, in my opinion, was dead on arrival. Congress has not been willing to face up to another round or two of base closures because it is politically very expensive to deal with such issues. But my opinion is that until such time as we do deal with these issues as a country, we are going to be shorting our forces of the equipment they need. This is a zero-sum game. We must deal with these issues. I do predict that we will begin to have some success in dealing with the infrastructure issues and our investment accounts will increase, but it will take some time to build the political will to do this.

There is a better news story on the commercial side of the aerospace industry. Looking at the forecast, the AIAA predicts that employment will be going up from about 786,000 to 830,000 in 1997, driven mostly by the growth in commercial business. If you look at commercial aerospace, it is interesting to have a private company like Boeing to look at. They are targeting by early 1998 that they will be producing commercial jets at the rate of about 43 per month. They are having difficulty in meeting that target. To put it into perspective, in 1996 they were making 19 per month. In fact, they announced that they are having difficulty and have been late on delivery of twelve aircraft due to parts shortages from suppliers. This bring us back to think some more about the supplier tiers and delays in training people. I think, in looking back on this situation, the consolidation of the defense industry has been orderly thus far. It has also been very beneficial in taking out unnecessary overhead. As a result, we will be, over the next five years, effectively buying three billion dollars more of equipment than we would had we kept the overhead in place and not consolidated the industry. As a result, we have an industry that is more right-sized to our needs.

From the macro now to some of the micro changes in the industry and its infrastructure, I devoted a lot of my attention during my past three years of government service along with Secretary Perry to this program of acquisition reform. I don't have the time today to go through the whole litany of things that were done and I don't believe that for this audience I need to go through the why of acquisition reform—there has been very broad agreement on the why. The key issue is how to bring this kind of reform across to both government and the industry. I put a lot of personal attention into the "how."

There are two or three aspects to this that I would like to address today. One has to do with the whole program of mil-specs. We need to move to a more commercial way of doing business. Buying things that meet commercial standards. I will tell you that Bill Perry and I were serving on a Defense Science Board Task Force—this must have been about 1990–91. This Task Force was looking at buying commercial semiconductors. In briefing after briefing the case was made that it would be impossible to buy commercial semiconductors for most of our DoD applications: plastic packaging was absolutely out; we needed to have ceramic packages. It just so happened that Bill Perry and I were doing some consulting work for General Motors, and about two weeks after those briefings, we were at Delco Electronics in Kokomo, Indiana and looking at some production lines. We saw a line that was producing engine spark controller chips for automotive applications in plastic packages. So we asked where were these packages actually going to be mounted. They were going to be mounted on the engine block—in the presence of fuel, electromagnetic emissions from spark plug wires, high temperature conditions and engine vibrations. But these plastic packages actually worked. So the two of us both got the bit in our teeth and decided that we ought to look at this hard.

We undertook at the beginning of our acquisition reform program to do a wholesale review of our policies on mil-specs. I think I can summarize this best by saying that what we tried to do is put the shoe on the other foot. The program managers in the past used to have to justify using a commercial spec. The shoe was on the foot leading to a mil-spec unless the program manager could justify otherwise. We put the shoe on the other foot—we would have a commercial spec unless there was a good reason why a mil-spec should be required.

There are cases where we do need a mil-spec, but we wanted the onus to be on the person recommending the mil-spec. The impact of this has been very significant. I would draw your attention to one or two programs that evolved over that period of time. One is called JDAM (Joint Direct Attack Munition). The idea behind this program was to make an electronics kit that would be applied to a large number of our dumb unguided bombs and turn a dumb bomb into a smart bomb, that is, one that was precisely guided to the target. The concept is to insert the set of coordinates where you want the bomb to hit. The kit then uses a receiver for the global positioning satellite so the bomb would know where it is. It then uses a guidance program in a small computer that will steer guide vanes to direct it to the location that you wanted. The idea here was to turn ordinary dumb bombs into PGMs with 13-meter accuracy in all weather conditions. We actually started to procure this the old way and the reason I cite this example is because I have a before and after acquisition reform comparison.

The old way, the winning bid on this award, which will ultimately produce 18,000 of these kits, was $42,000 per kit. We were happy with that; if you look at the cost to change to smart bomb accuracy from dumb bombs, that was a good investment. But we made JDAM a pilot acquisition reform program and we did that procurement again. We put in place all the acquisition reform initiatives and the commercial spec initiatives. We went from an RFP that had a 100 page work statement to an RFP with a two-page performance spec—not how to build this, but what we wanted it to do with no mil-specs required. The winning bid on that round was

$18,000 a kit. Less than one-half of the original cost. When we go through this whole production program this will save the Department $2.9 billion. This program is now in test and it is beating its CEP estimates—it's guiding more accurately than 13 meters in all weather conditions.

Just to prove this wasn't a fluke, we did another one just like it on something called Wind Corrected Munitions Dispenser. In that case, our estimated cost going in was $25,000 per kit, and the winning bid using the reformed commercial approach was $9,000 per kit. So I think this is going to have a very significant impact.

The next key issue I wanted to highlight is the Single Process Initiative. One of my frustrations in the Department was that while all the new acquisition reforms were being applied in the new contracts—what about the existing contracts? How could we go back and impact the existing base? So we launched the Single Process Initiative. This was brought home to me by a visit I made to a Raytheon missile production facility. I learned that Raytheon was operating using eight different soldering processes. That means documenting each one of these, using statistical controls, training people for each of the processes, etc. Three of these processes were commercial and somehow we, in the Department of Defense, decided that we needed five more. This was probably decided by program managers who never talked to one another. Our objective was to see how few soldering processes we could get down to, and could we use the commercial ones—was there any reason for us to do otherwise? As we looked into this, we got the typical bureaucratic answers you get when you look into anything at the Pentagon. The answers were "well, this is too hard—you can't change this one contract at a time; you have to go back and change multiple contracts with multiple program managers and that's just too hard to do." So we went off and worked on this one and we started the Single Process Initiative program off with something I call the "Mother of all Block Changes." We signed one change agreement with Raytheon that modified 880 contracts at sixteen production facilities—giving us all a common process base.

When that block change was signed, our industry started to pay attention. People began to realize that this wasn't just another smoke and mirrors initiative—we were serious about trying to move towards a commercial base. Some CEOs needed a little nudging. I will admit that I called about eight or ten in the slower moving companies to tell them that their competitors were moving on the Single Process Initiative and I wondered why they weren't. I asked them if there was something that we were doing in the Department to discourage them—if so, I would work to change it. That helped to get some energy behind this program.

Where we stand today is that we now have 800 of these block changes approved by 170 different companies. The issue was not just manufacturing processes, but business processes—proposal writing processes. This is having a big impact on the industry and I will share with you just one story about this. One of the CEOs that I called had a division that makes helicopters. I told him that I thought he was moving a little slowly on this initiative. His response was that they were already now producing these helicopters on a commercial line—same line for DoD and the commercial users. What they were unable to do was support the DoD helicopters using their commercial logistics procedures—that the DoD had imposed so many onerous provisions in logistics and support that there was no way that they could do that work. I will come back to that in a minute because there is a big piece of unfinished business in reforming our logistics systems.

As I look at other aspects associated with trying to drive our defense industry away from a separate commercial base to a common national industrial base, there are three or four other initiatives being pushed by the Department. I won't talk in detail about these, but I will just highlight them.

One is the work we must do in dual-use technology. The TRP program was a key element of this. Unfortunately that program got tangled in all kinds of political issues and didn't get the support that it deserved. I will illustrate for you an example of the kind of work we were trying to pull off. One involved a technology called multi-chip modules. The idea was to be able to package bare chips rather than put each semiconductor into its own package. We would then mount these chips one on top of the other in a very dense package. This gave us excellent packaging for very high performance applications for the Department, and we had the lead use for the application ahead of the commercial needs for the process. What we did was partner with commercial industry for some of the front-end investments. The initial use was 90 percent defense and the cost per unit was something like $10x$. Now partnering with commercial suppliers over a period of five years led to the development of commercial markets for the multi-chip modules and the DoD share is now only about 10 percent. Our cost is now down from $10x$ to x and the commercial line is enabling the process. And we now have one national industry doing this work for high performance modules and that is the model for what I think this dual-use technology program should be doing.

Dual-use technology is not enough. We also have to be looking at dual-use products for the Department of Defense produced by a common industry. I will share with you another short story. Secretary Perry and I were having a discussion just before we deployed our U.S. forces to the IFOR support task force in Bosnia in 1995. We had sent the Defense Science Board Task Force to the field in Bosnia to look at the kind of intelligence support that our forces were going to need in the field. That task force came back with the good news that the intelligence we had in our major command centers was exquisite—it was everything we needed. The bad news was that when it came time to supply intelligence to our forces in the field, we were in very bad shape.

Our forces in the field were depending on 9.6 Kbytes per second modems. You can go to your local computer store and buy a 28 Kbyte per second modem. So our troops were waiting 30 minutes to see a map or an image. I authorized the expenditure of between $80 and $90 million to address the problem. What we did was sign a lease to use a transponder on the Orion direct broadcast satellite that was already operating in Europe. And we only had to make two small modifications. We had to move a beam a little bit so that the energy from the beam came over the area our forces were operating. We then supplied our forces with 20 inch satellite dishes that you can buy commercially and we added encryption hardware and software so that this could be a secure link.

This capability was fielded in a period of seven months. Now the capacity was not 9.6 Kbytes per second but 30 Megabytes per second—over a 3000 fold improvement, obtained in a period of seven months by buying and using a commercial product—actually in this case a commercial service. I will tell you honestly if we had gone about this in our usual way of doing business, in seven months we would have still been arguing what the requirements were. We would have gone off and built our own satellite system and it would have taken probably twelve years. What we did was leverage commercial development to supply what our forces needed in the shortest possible period of time. This needs to become a common business practice.

Now the area I promised to go back to: logistics and support. Of the life cycle cost of our major weapons in the Department of Defense, 60 percent of that life cycle cost is incurred after the system is fielded. We have an exquisite system to oversee and control the cost of the development and procurement of our systems—Congress watches that—so we have wholesale management controls in place. But we have no systematic disciplined oversight of the 60–70 percent of the cost of these systems after they are fielded. A very decentralized system is in place to oversee and control the cost of logistics and support. The DoD logistics system operates much like the way our commercial industries were operating in the 1950s using inventory at various points and we are operating what I call a "just-in-case" system rather than a "just-in-time" system. I don't want to be flip. We do have to have a stable base to rely on to support our forces in the field; going to a pure just-in-time system would not be the right choice to provide that support. But what we have today is a system that is too far over in the area of just-in-case and needs to move closer to that of just-in-time.

We started with one of our advanced concept technology demonstration programs here to combine fast transportation and information systems to be able to address this problem. We developed a demonstrator to use in Bosnia called Joint Total Asset Visibility to avoid the problem we had in Desert Storm. In Desert Storm we actually shipped twice the number of shipping containers that we needed to support our forces in the field. Why did we do that? We had no idea what was where. A commander would call and ask "where is x" and we would reply "we shipped x two weeks ago," and he would say "I don't have it" so guess what we would do—we would ship x all over again. Meanwhile we built a mountain of containers in the theater—18,000 of them with no idea what was inside without opening them. No commercial operation would operate that way today. The technology to deal with this need is already developed and on the shelf, it is just a matter of integrating it to field what we need.

We started an initiative this year analogous to the TRP program to look at using commercial products to reduce the support costs of our systems. Unfortunately that was not well supported in the Congress. Something called COSSI (Commercial Operational Support Savings Initiative) that in essence showed that taking an existing part that was developed commercially, and with the industry qualifying that part for DoD use, it could be used as a replacement part and reduce life cycle cost. For example, we have a lot of trucks today that are twenty-five years old that still have universal joints. This results in tires that are replaced at a very high rate. If those joints were replaced by constant velocity joints, the return on investment would pay back what you spent in a year and half on tires. All you would need to do is take those constant velocity joints and qualify them for use in our equipment to achieve that kind of payback. COSSI really needs a lot more push with the Congress.

My overall observation is that the initiatives that are working very well are the ones we developed at the acquisition front-end, but our initiatives to reform our support and logistics systems are where we were about three or four years ago in acquisition reform. There is still a whole lot of work to be done there. Some of this goes back to reducing the infrastructure still in place that is not needed.

The last piece is some lessons and some directions for the future, and there is a big long list. But I thought I would highlight the three that are most important to me. The first one is dealing with what I call "open systems," which are open in the sense of being able to use commercial off-the-shelf components, software and sub-systems able to allow for future innovation and improvement so that software packages can be updated and rehosted. They must not be specific to a particular computer host. Today, we invest a huge amount of money in software that is not portable. As hardware improvements are realized, we need to be able to transport it to new hosts. This is a field of growing importance.

When you think about platforms like our B-52 which will be in our inventory for an estimated 90 years and information technology turning over every two or three years, if we want the equipment to be leading-edge, we have to take advantage of the technology to do those upgrades. The commercial technology is every bit as available to our adversaries as it is to us. To stay ahead of adversaries we need the best cycle-time to pick up the leading-edge technology and get it fielded. If we do not take an open-systems approach we will lag behind forever. What we are really looking for here is system engineering skills and the development of applications integration expertise (I call it the applications "glue") to turn developed commercial products, either software or hardware, into useful applications for commercial or defense use. It is a premium skill base that we should be developing in this country. It is a very high leverage for us. It is one of the most critical capabilities to maintaining our combat edge.

In order to deal with those systems applications I also think it is helpful to consider this in the context of the diversification discussions we have been having. If you go back to the old classic Marketing 101 courses, the idea is to take existing products and apply them to new customers or new markets, or alternatively to take new products and apply them to existing markets or customers. The thing to avoid is to take new products and apply them to new customers. As I look at this combination of diversification activities I think we will see more of taking new products to existing customers. I point you to the satellite example, where we take commercial product or service that we can then apply to defense. I think that will be seen more often than taking existing products to new customers—not that that won't remain important.

Another area of very high importance is the issue of the life-cycle costs—fundamentally a logistics and support issue. In many commercial businesses that is where companies make their money—in the after-market. Something like a commercial jet engine for an aircraft is a losing proposition if you look at the cost and return of the development and production phase—most companies don't make money in that phase. Where they make their money is in the spare parts and the support. We need to start moving to a cradle to grave arrangement where we look at the whole life-cycle cost and manage that in the most efficient way. There is growth in that piece of the business. There is often too much emphasis on the front-end and a lack of emphasis on the back-end.

The last area I would like to highlight is probably the most important. It's the whole issue of partnerships. Developing skills and partnering and being able to operate together. One of the first things I did when I got to my job in the Pentagon was to put in place this integrated product team concept, and not just in words, but seriously operating the Department just that way. From personal experience when I was directing the Stealth program back in the 1980s—we had a small team of people in the F-117 program who covered all the disciplines. There were less than ten people. There was the Lockheed program manager, the GE engine program manager, the government program manager, the Air Force command that was going to use the airplane, the head of our test force, the head of logistics, the personnel specialist, and maybe one or two others as subjects came up.

That group met once a month and decisions were made at the meeting. People who couldn't decide at the meeting ended up being fluff and soon fell out. People who came to the meetings made decisions at the meetings—and crisp decisions were made once a month. I will tell you that not every decision was right, but very seldom did a bad decision stand for over two months.

This program moved on a very tight aggressive schedule and it moved very well. My whole thesis in acquisition reform was to take this team experience and put it in place for all of our programs. That is basically what has been done on most of the large programs and has not been fully implemented yet on some of the smaller programs. It takes a while to get through the system. The team approach is a wonderful substitute for layering. I think Norm Augustine said it, "if you add enough layers to a bureaucracy in the supervision of a program, you're not going to leave failure to chance."

Our acquisition system was clearly headed in that direction. So the idea of streamlining this—having the team include the customer, the user and the developer—was very key. There are programs in the Department that I had little to do with starting, but I am very proud of them because the services have now picked up this lesson and now starting to develop ownership. There is a program in the Navy called Smart Ship in which the late Mike Boorda, Chief of Naval Operations, devoted a ship to this concept. In eighteen months the Navy has fielded forty-seven different prototype systems on this cruiser in which the crew was intricately involved. The result is that they will be able to reduce the size of the crew by about 20 percent and improve the fighting effectiveness of the ship. The crew had to be involved in all of this technology and not all of it has worked—so it is critical to be working in that kind of team. The Army is doing something similar today in Force XXI.

There are other partnerships that I would like to highlight. We started something in the Department called the Government/Industry/University initiative. I think of this as a three-corner partnership—government, industry and a university. I proposed this in a Semiconductor Technology Council that I co-chaired with Craig Barrett, who is the COO of Intel. I made the proposal to Craig that if the Semiconductor Industry Association were willing to put up a matching investment, we in the DoD would invest to set up one or two semiconductor technology centers of excellence in a university to be awarded competitively. What I wanted to do here was recognize the fact that the commercial research and development funds are vastly exceeding DoD funds, it's about a two to one margin now. What's happening with these research funds is that the DoD has had a good horizon, we have looked out five years or more. Our commercial companies are looking at shorter term horizons. In this partnership, I was going to let our commercial brethren have a big say about the areas of investment, but we in DoD would have our say about the time-frame to push out this horizon as a compromise. The Semiconductor Industry Association came back to me. I was thinking about an investment of about $20 million. They said fine, we're prepared to invest a $100 million. I said we might have trouble in DoD to match that, so they said we'll match you two to one. So this is one initiative and we should have some in several other areas. Space is such an area.

There is one other aspect of this government/university/industry partnership that I think is critical. We often think of our research work in universities as a commodity. We develop this intellectual property, tie a bow around it, and put it on the shelf until it's ready to be used. I think that is missing the picture. Half of what we are getting from the university environment is not intellectual property tied up in a bow, it is people who are trained and could be productive in our operations. We sometimes don't consider our most important asset—the people we are developing. And that is something we need to address as part of the return on investment.

This idea of partnerships—the combinations of large and small. Large companies sometimes lack incentives for innovation. They have plenty of mass but not much velocity. Yet there are a number of small companies which have great velocity but do not have mass. We need to apply our physics—momentum = mass × velocity. Partnerships between small and large entities are needed and we also need to look at creating international partnerships. That is a whole area that needs attention. If you look at commercial business today, it has a global, multi-national character. The defense industry has much less of that character and it needs to be moving in that direction.

I would close by commending this particular symposium for providing an outstanding forum for the discussion of so many of the issues that I have highlighted today. It's been an honor to have been a part of a great team, and to share with you some of my experiences as a member of that team.

NOTES

1. Keynote address at Second Klein Symposium on the Management of Technology, Penn State University, September 17, 1997. Dr. Kaminski made similar remarks in an address at the Bantle Symposium on Industrial Diversification at Syracuse University, October 27, 1997.

20

THE PERENNIAL PROBLEMS OF DEFENSE ACQUISITION[1]

JOHN A. ALIC

INTRODUCTION

No other nation can claim military power remotely comparable to that of the United States. Other countries have large armies; only the United States has a military in which offensive power and defensive capabilities reside in a large and diversified inventory of high-technology weapons systems. The United States aims to hold on to this superiority, while spending less money. That will require attention to an acquisition system plagued by a complex of problems that has changed relatively little, despite repeated efforts at reform, since the 1950s. Indeed, many acquisition problems can be traced back to World War I, when, for the first time, fast-paced technological change saw new generations of weapons, notably aircraft, introduced during the course of a conflict.

World War II brought more rapid and radical technological change, illustrated by radar, the proximity fuze, digital computers, and the atomic bomb. Long-range ballistic missiles and the jet engine also appeared during the war (though they had little impact on its course). U.S. experience during the first half of the 1940s laid the basis for a new set of R&D policies and organizations that rapidly and totally transformed the country's "national system of innovation." From the beginning, the new system spanned military and civilian institutions. In 1946, for example, Congress established the Office of Naval Research (ONR), in many respects a prototype; by the next year, ONR funds were supporting 2500 science and engineering graduate students in American universities (Sapolsky, 1990: 44).

If university research and training were seen from the start as a critical part of the foundation for a high-technology military, the decisive change in acquisition policy came, not in the immediate postwar years, but in the aftermath of the Korean War and the Soviet Union's 1953 hydrogen bomb test. In Korea, U.S. forces fighting with obsolescent equipment were badly mauled. Much more than the first Soviet fission tests in 1949, which had been anticipated (if not quite so soon), news of a Soviet H-bomb, preceding the first U.S. thermonuclear explosion, created consternation in Washington. Among the results was a rapid increase in R&D on technology-intensive weapons systems of all kinds.

Military R&D rose from about $2 billion in 1955 (4.8 percent of defense expenditures) to $5.7 billion in 1960 (12.4 percent of defense expenditures). In this early period, DoD and its contractors explored a wide variety of technologically-based systems. Some, such as nuclear-powered submarines, highly accurate intercontinental missiles, and new generations of supersonic aircraft, became staples in the nation's military arsenal. Other programs are now largely forgotten, e.g., Dyna-Soar, a space plane intended to skip along the earth's outer atmosphere like a stone across a pond. No doubt there were "black" projects, shrouded in secrecy, that even the historians combing DoD's archives have yet to learn much about.

Since the 1950s, R&D (more properly RDT&E, for research, development, test, and evaluation) has never accounted for less than 10 percent of the defense budget. Although the Cold War arms race ended nearly a decade ago, and the procurement portion of the acquisition budget, which pays for production of weapons systems, is down, RDT&E has remained roughly constant (in nominal dollars). Overall U.S. acquisition policy has changed relatively little, with modest regulatory reforms joining the push for "dual-use."[2]

ACQUISITION FUNDAMENTALS

As Table 20.1 shows, the DoD budget approached $300 billion per year in the late 1980s, with acquisition (RDT&E plus procurement) around 40 percent of the total. The remainder goes chiefly for personnel and for operations and maintenance (O&M). The O&M portion of the budget includes substantial spending for at least one major technology category—computer software. According to informed observers, DoD pays for something like three-quarters of its software expenditures from O&M categories. In part this is because a great deal of DoD software spending, as in the civilian economy, goes for maintenance and upgrades. RDT&E, in turn, averaged about 40 percent of procurement during those years.

Finally, with a swelling federal budget deficit putting a brake even on military spending and the Soviet threat collapsing, procurement funds—but not RDT&E—began to fall. In 1998, RDT&E will reach an astounding 80 percent of procurement (after which current projections show procurement

Table 20.1. Defense department outlays.[a]

Fiscal Year	RDT&E (Billions of current dollars)	Procurement	DoD total
1980	$ 13.1	$ 29.0	$ 130.9
1981	15.3	35.2	153.9
1982	17.7	43.3	180.7
1983	20.6	53.6	204.4
1984	23.1	61.9	220.9
1985	27.1	70.4	245.2
1986	32.3	76.5	265.5
1987	33.6	80.7	274.0
1988	34.8	77.2	281.9
1989	37.0	81.6	294.9
1990	37.5	81.0	289.8
1991	34.6	82.0	262.4
1992	34.6	74.9	286.9
1993	37.0	69.9	278.6
1994	34.7	61.8	268.6
1995	34.6	55.0	259.4
1996	36.5	48.9	253.2
1997[b]	36.0	45.6	254.3
1998[b]	34.6	43.1	247.5
1999[b]	35.2	44.6	249.3
2000[b]	34.0	47.6	255.2
2001[b]	33.2	51.6	256.2
2002[b]	33.6	55.4	261.4

[a]Department of Defense only. The Energy Department's defense programs spend another $10–12 billion annually on nuclear weapons. Also note that published figures omit black programs, which evidently added something like $10 billion to DoD's annual RDT&E totals (alone) during the mid-1990s. See Bill Sweetman, "The Budget You Can't See," *Washington Post*, July 7, 1996, p. C3.
[b]Estimated.
Source: Historical Tables, Budget of the United States Government, Fiscal Year 1998, Table 3-2.

growing once again and the ratio of RDT&E to procurement slowly declining). To set these figures in context, note that R&D seldom rises above 10 percent of revenues even in high-technology civilian sectors like microelectronics or biotechnology. Admittedly, such comparisons are not *apropos*, given that defense RDT&E is so unlike most civilian R&D. But that is the point: defense, where the ratio of RDT&E to procurement has not fallen below 40 percent since 1985, is *sui generis* vastly more R&D-intensive than other sectors of the economy. Comparisons with civil R&D spending can also be misleading because DoD's RDT&E categories do not correspond to those commonly used in industry. In particular, DoD designations 6.4–6.6 (category

6.4 covers the engineering of particular weapons systems, while 6.5 and 6.6 are labeled "management support" and "operational systems development"), amounting to some two-thirds of the RDT&E budget, include some items that commercial firms might not consider R&D at all.

It is in fact impossible to directly compare military and civilian R&D spending. Only the nation's space program, which began as an appendage of the military, bears comparison with defense. Compared to other countries, the United States directs a larger share of government R&D funding to defense and a smaller share to other R&D categories (with the exception of health-related research). These patterns are a direct consequence of the Cold War. As further context, note that the government share of all U.S. R&D has been steadily declining, and is now down to 30–31 percent. Three decades ago, federal agencies accounted for fully two-thirds of national R&D. In recent years, real R&D spending by private firms has grown strongly, while federal spending has drifted downward (Payson, 1997).

There is a further message behind the numbers in Table 20.1. Two-thirds of the RDT&E money goes to industry, primarily for the engineering design and development of weapons systems. Procurement outlays go primarily for weapons (e.g., guided missiles), electronics (radars, electronic countermeasures, or ECM, command, control, communications, and intelligence, or C^3I, electronic warfare suites), and platforms (tanks, aircraft, satellites, ships). Platforms typically carry both weapons and electronics. Together, these "weapons systems" (platform plus weapons plus electronics) account for the great majority of DoD development and production contracts. What does it mean if RDT&E remains high even though procurement has slowed? It means that the United States is continuing to design and develop future systems almost as if the Cold War was still underway. In the years ahead, there will be heavy pressures to purchase the weapons systems being developed with today's RDT&E funds. Congress and the administration will face choices that have always proven extraordinarily difficult: to kill systems that have already been developed; to cut production volumes to low levels to save on procurement dollars; or to go ahead with weapons that appear to have little justification given likely future threats to U.S. security.

Most of the acquisition budget goes for what DoD refers to as "major" programs—those valued at $1 billion or more in 1980 dollars or $200 million or more in RDT&E funds. These get most of the attention from policymakers and account for most of the technical and managerial problems in acquisition. At any one time during the Cold War, DoD had around a hundred major programs underway, either in the development stage or in production. These have been, and remain, the centerpiece of the nation's technology-intensive national security policy. As defense spending falls, effective management of the major programs that remain—e.g., the F-22—will become, if anything, more important. Begun in 1981 with the objective of staying ahead of the Soviets in fighter aircraft, the first F-22 prototype flew in 1990. The Air Force originally hoped to buy 600–700 F-22s at a total cost of nearly $100 billion. Planned production has since dropped to 339, reducing total spending while doubling the price per plane (largely because the fixed RDT&E are spread over fewer units). Meanwhile, critics have claimed the F-22 is no longer needed, given that the 900 F-15s in the U.S. inventory far overmatch the capabilities of any imaginable adversary (see *Transforming Defense: National Security in the 21st Century*, December 1997: 49).

Major systems normally push technology hard, not only in search of the last available increment of performance (whatever that might mean in actual combat), but because acquisition cycles are lengthy. DoD has not only driven the state of the art but has often refused to trade off cost—both first cost and life-cycle cost—against function, even though analysts have long held that the last few percent of performance may increase system costs by 30 to 50 percent (see, for instance, *A Quest for Excellence*, 1986). If 10 or 15 years elapse between the start of RDT&E and the start of production, the initial goals must be very ambitious or the system risks being outdated before reaching the field. Once operational, moreover, systems remain in service for periods ranging from perhaps 20 years for tactical missiles to 40–50 years for some ships and planes. Very long life-cycles generate pressure to incorporate the newest technology as late as possible during RDT&E. Nonetheless, at some point it is necessary to stop iterating on the design and start producing it.

Once equipment has entered service, the questions turn to modifications and upgrades, as well as redesigns during lengthy production runs. The B-52H, which remains the Air Force's primary heavy bomber, looks much like a B-52 rolled out in 1955, the first year of production, while incorporating better materials, more powerful, more efficient engines, and vastly superior electronics. Of course, the overall design—the "system architecture"—eventually becomes enough of a constraint on upgrades that the entire system comes to be seen as obsolescent.

Major programs have historically suffered developmental problems ranging from cost and schedule overruns to disappointing performance (for example, 41 of 83 systems evaluated by GAO showed "system effectiveness problems" classed as "operational/performance limitations" (*Major Acquisitions: Summary of Recurring Problems and Systematic Issues: 1960–1987*, 1988: 18). On the other hand, performance as measured against cost and schedule targets has improved since the 1960s. Moreover, DoD's clouded reputation has never been entirely deserved. The Pentagon's track record in managing large projects seems to be somewhat better than that of other federal agencies (or the defense agencies of most other countries), and large-scale civilian projects have frequently experienced greater cost and schedule overruns (*A Quest for Excellence*: 61ff; also see Rich, Drews, and Batten, 1986; Biery, 1992).

If schedule overruns have been reduced, schedules have lengthened. Acquisition cycles—the time it takes to get new equipment into the field—have been increasing since the 1950s (Gansler, 1989: 173). The reason is ever higher levels of technical complexity: most of the extra time comes during the early stages of RDT&E, between program initiation and the beginning of engineering development. *Transforming Defense*, like many earlier DoD documents, urges the necessity for shortening acquisition cycles. The report calls for emulation of civilian practices that enhance flexibility, especially through information technology, but offers no specifics on how this might be accomplished. (Nor does the report ask to what extent these practices have actually proven effective in the private sector.)

Unit costs, moreover, have been rising for decades at two to three times the rate of inflation, and even faster for the most technology-intensive procurements (Adelman and Augustine, 1990: 90–93). Even antitank missiles now have price tags too high for more than the occasional practice firing (so that training must rely heavily on simulators and dummies). If the Air Force eventually takes delivery of 21 B-2s, the cost per plane will come to about $2.1 billion, three-quarters of this representing distributed RDT&E expenditures. (The Air Force originally planned to buy 132 B-2s; the number was cut to 75 in 1990 and reduced further in 1992.) Indeed, the B-2 is so expensive that it seems hardly worth the risk of flying, given that accidents claim 50–75 military aircraft each year.

High technology also contributes to high maintenance costs and low reliability. Almost a third of the military's enlisted personnel now work in maintenance and repair. The F-15 is a well-proven design, but its ratio of maintenance hours to flight hours is 20 or 30 to 1, and even then the mean flight time between *un*scheduled maintenance is only two hours.[3] And the "stealthy" skin on a B-2 must be repaired after nearly every flight.

It was probably the very large and seemingly ever-escalating price tag on the B-2 that began to drive home the point in Washington: the country was paying more and more for fewer and fewer planes, ships, and tanks. Such a trend could not continue indefinitely. And if the costs for complex systems kept rising, performance often seemed to lag behind the promises of the Pentagon and its contractors. Although B-2 production continues, in 1991 then-Secretary of Defense Richard Cheney canceled the Navy's A-12 attack plane—at $50-plus billion, the most expensive weapons system ever halted in midstream. At the time, the A-12 program had been running for three years, had fallen at least a year behind schedule, and was 50 percent over budget (Bond, 1990). This was hardly the first program to encounter these kinds of difficulties. The difference, as Cheney and other decision-makers saw it, was that the nation could no longer afford to bail out programs in this much early trouble. The larger question, of course, is how to avoid such situations in the first place.

The quandaries outlined above are easy to describe and hard to resolve.

Ever more complex technologies require ever greater expenditures on RDT&E. High front-end costs leave less money for production. Congress and the administration stretch out programs to limit annual outlays. With money spread thinly, development takes longer and production rates are inefficiently low. Even before the recent declines in procurement outlays, something like half of all major systems were produced at less than the "minimum economic rate" (*Effects of Weapons Procurement Stretch-Outs on Costs and Schedules*, 1987). Trying to keep to schedules and show progress, too many systems are pushed into production before development is complete, leading to costly redesigns and modifications. DoD and its contractors sometimes skimp on prototype testing and evaluation. They may fail to specify realistic and demanding tests: since programs compete with one another for scarce funds, failing a test can put the program in a position of fatal weakness; the incentives are to cover up problems rather than fix them. With DoD depending ever more heavily on software-intensive information systems, testing promises to become still more of a problem. Because computer programs are themselves abstract symbolic constructs, many of the familiar tools of engineering analysis cannot be applied. It is difficult to visualize, model, or simulate software behavior—or to test and verify computer systems functionally.

Some of the difficulties outlined above follow from the lack of a market for military equipment. Because government cannot simply purchase the systems it wants from those on offer, DoD must manage the process. Before World War I, governments everywhere internalized a good deal of design, development, and production, typically in their own shipyards and arsenals. But rapid technical change, notably in aviation, forced a turn to industry, since only private firms had the necessary technical expertise. The difficulties of the U.S. Army in buying aircraft during World War I not only prefigure many of the issues of acquisition planning and management with which DoD continues to struggle, they led directly to the regulatory structures that continue to govern the "military–industrial complex."

THE HERITAGE OF TWO WORLD WARS

Aviation was an infant technology when World War I began. Reconnaissance flights led to aerial combat and then to experiments in bombing, all in a literal hot house. A U.S. industry that had produced a cumulative total of 200 or so aircraft since the Wright brothers flew at Kitty Hawk turned out nearly 14,000 during the 19 months prior to the armistice (and some 42,000 engines) (Lee, 1984). All told, World War I combatants built some 200,000 airplanes. Rapid technical advances rendered the latest designs obsolete in six months or less. The most successful aircraft were built by firms that learned to quickly incorporate the lessons of combat. As I.B. Holley (1964, p. 512) put it in his classic account:

> Continuing superiority requires continual change. Every innovation introduced by the enemy must be outmatched. Superior performance in aircraft is the sum total of many components—range, speed, climb, maneuverability, fire power, and the like—each conditioned by thousands of features of design: here a change in engine cowling to improve cooling and increase horsepower, there a better gun mount to enlarge the field of fire, and so on in an endless procession suggested by experience in the field and innovations on the drawing board.

Military aircraft could not be put out for bid like uniforms, or produced to standard designs in government facilities; efforts to do so saw the Army buying aircraft that proved outdated before the first deliveries. If flown in combat, these planes would have been literal death traps.[4] Governments everywhere had to turn to the handful of private firms with established capabilities in aircraft design and manufacture.

World War I thus marks the beginnings of the modern-day defense industry (i.e., an industry populated by private firms developing and manufacturing high-technology weapons as opposed to small arms): Lockheed Martin, for example, descends along one branch from the Glenn L. Martin Company and from Wright-Martin, both of which grew rapidly during these years. World War I also marks the beginnings of the modern-day acquisition system, as Washington struggled with issues ranging from accounting standards to policies for reimbursing firms for the costs of R&D.

Today, as in the past, defense firms normally make their profits on production contracts, with RDT&E a lead-in to procurement. Usually, the firm that does the R&D gets the contract to build the system. But not always. During World War I, Congress held, in essence, that airplanes should be purchased like rifles or machine guns. These were put out for competitive bids. Why not planes? The congressional position was simple. The Army had been buying and evaluating prototypes from aircraft companies. The government owned these prototypes and by extension the rights to their designs. Once the Army had selected a design, based on the prototype, any and all firms should be able to bid on the production contract and the low bidder should get the order. This was "free enterprise."

But not all the early aircraft firms had engineering capabilities. Some sought only production work. Those able to design planes argued that they properly retained rights to their technical work notwithstanding the sale of a prototype to the Army. In any case, they argued, another company could not do a satisfactory job of reproducing the design—prototypes were typically delivered without drawings, much less the underlying engineering calculations. Far too much would be left to chance: the prototype embodied company-specific practices and tacit know-how. Changes were often made on the shop floor. Another firm could not possibly understand the multitude of design decisions that had gone into the completed prototype, even after tearing it apart and making a new set of drawings. And how could a low-bidding firm—which might have production skills but no substantive technical expertise—be expected to improve the design, as would surely be required?

Where cost-based competitive bidding was implemented, delays proved the norm as low bidders struggled to replicate prototypes designed by others. In a number of cases, the low bidder ended up redesigning the entire airplane, in which case the manufacturer lost money and the government received what was in essence a different plane than it had ordered. In one notorious case, in 1919, the Army took delivery of 50 Orenco-D pursuits—designed by Ordnance Engineering Company but built by Curtiss—and then had to destroy them as unsafe to fly. Meanwhile, Ordnance Engineering, having lost the production contract, went out of business (Vander Muelen, 1991: 59; Holley, 1964: 85–86). Nonetheless, Congress periodically renewed its push for "competition," arguing that by decoupling R&D from procurement and giving the production contract to the low bidder, government could save money. Defense firms, on the other hand, continued to hold that their designs were proprietary, that RDT&E contracts do not and cannot be expected to fully and fairly reimburse them for the true costs of system design and development—particularly for complex systems that draw on deep reservoirs of experience in the contractor's engineering organization—and that follow-on production is essential if government expects firms to devote their best efforts and most talented employees to RDT&E. The military services, for their part, have generally sided with industry, in the belief that negotiated cost-plus contracts and close working relationships with contractors are most likely to yield the systems they want.

Most of the issues faced by the Army in World War I resurfaced after 1935, although World War II lasted long enough for the United States to put its enormous productive capacity to work. Weapons systems had become much more complicated. Acquisition managers and administrators in defense firms had to deal with huge flows of engineering change orders as technological improvements worked their way into aircraft, radars and other electronic systems, submarines, even landing craft. Armies of clerks and bookkeepers, accountants and auditors were kept busy recording modifications (essential for field repairs), monitoring inventories and purchase orders (millions outstanding at any one time), tracking costs, and determining allowable expenses.

It proved impossible to estimate in advance what it would cost to produce a new aircraft design. Sometimes companies underestimated, losing large sums after bidding on fixed-price contracts. Other times, production costs proved lower than anticipated, or declined rapidly with volume because of learning-curve effects, leading to profits viewed by Congress and the public as exorbitant. The alternative was cost-reimbursable contracts, but the War Department found it difficult to structure these so that firms would have incentives to minimize expenditures. (With profits calculated as a percentage of costs, a company that saved money could find its profit reduced.) Cost-reimbursable contracts also posed a host of accounting and allocation problems. Should government pay the full costs of tooling that might be used after the war for commercial production? What should be done about companies that gave their executives huge raises once war work began? Holley (1964: 384) reports a case in which the president of an aircraft firm boosted his annual salary from $20,000 in 1941 to $50,000 the next year. All these problems continue to plague acquisition, from allegations of waste, fraud, and abuse to the lack of incentives in cost-based contracts.

THE POLITICS OF PROCUREMENT: "WASTE, FRAUD, AND ABUSE"

Competitive bidding in federal procurements bears little resemblance to contracting in the private sector. The purchasing department in a private firm weighs the past performance and reputation of bidders before awarding a contract; the low bid does not automatically win. Federal contract law, based on an idealized, legalistic model of "full and open competition," leaves government managers little latitude for weighing past performance.[5] Congress, often aligned with the Office of Management and Budget (OMB) and other executive agencies, has held that any and all "qualified" bidders should be allowed to seek any and all government contracts, notwithstanding the risk that a firm may underbid and then fail to deliver. In part, congressional insistence on competitions open to all comers is a straightforward response to constituent pressures—there are always firms angling for a piece of the government's business. Not a few members of Congress have also shared the impression, widespread among the public at large, of an unholy alliance between contractors and the services—a military–industrial complex in which billion-dollar decisions get little scrutiny by outside parties, with sole-source and cost reimbursable contracts an open invitation to waste, fraud, and abuse. Competition has then seemed the necessary prophylactic.

Certainly it take a strong political and organizational coalition to initiate and sustain a major defense program.[6] One of the services must want the system—and want it more than others competing for a limited pool of funds. The Office of the Secretary of Defense (OSD) and OMB must sign on. Boards, panels, advisory bodies, and committees must approve the technical requirements. Congress must appropriate money and will no doubt ask for multiple reports both from the administration and from its own General Accounting Office. There is in fact plenty of sunshine on acquisition.

On the other hand, large-scale programs will inevitably be opaque to outside observers, especially those without deep understanding of the technologies involved. Add the many layers of bureaucracy, conflicts within government (intra-service as well as inter-service rivalries, friction between OSD and the military) and within the defense industry, and it is easy to understand why some observers see conspiracy where it might be more accurate to see confusion, bureaucratic maneuver, and as-yet unresolved technical questions. (For a range of perspectives, see Rosen, 1973; for a thorough examination of congressional influence over procurement decisions, see Mayer, 1991, who finds that members generally vote in accord with their view of national security imperatives, rather than, for example, trying simply to steer production contracts to their districts. This could be changing. With the Cold War over and no serious threats to U.S. military security in sight, members of Congress may be less likely to obey their sense of what is best for national security and more likely to obey their sense of what will reward their constituents.)

In any case, Congress has never fully trusted the executive branch and the services to manage acquisition efficiently and effectively. Long-running conflicts between Congress and the military over procurement law and policy have created a thicket of precedents and legal interpretations.[7] In recent years, procurement practices have been covered by nearly 900 different laws and more than 4500 pages of regulations, constraining decision-makers and driving up overhead costs. DoD's complicated paperwork requirements (bid and proposal packages, accounting and auditing, specifications and standards) cost the treasury billions of dollars each year (see, for example, Fox and Field, 1988: 322; Gansler, 1989: 338–345; 1995: 119–146 and 162–168). Some estimates put the costs of paperwork at 40 percent of the procurement total (Borrus, 1993). Viewed differently, DoD-specific practices and procedures raise costs by about 18 percent compared to the private sector (*Directions for Defense: Report of the Commission on Roles and Missions of the Armed Forces*, 1995: 3–23). Making procurement information available electronically will be of only limited help.

During the Cold War, the United States could live with, if uncomfortably, the conflicting views of Congress and the military on acquisition policy and management. Money was little object. Despite continuing if sometimes *ad hoc* involvement by Congress in major acquisition decisions—and continuing complaints from DoD and the defense industry of micromanagement and meddling—most of the time the services got what they wanted. Whether this was what the nation needed was another question. Too often it seemed that weapons cost too much or didn't perform very well. As discussed below, this was often the result of technical requirements that were overambitious, ambiguous, or in conflict with one another.

In fact, the many studies of defense acquisition over the past several decades have generally found waste, fraud, and abuse to be relatively minor concerns, frequently overstated.[8] These studies also suggest that simulating competition in the absence of a functioning market is ineffective.

The alternative to competition is more-or-less direct management and oversight by the government. On the other hand, many studies have also criticized the ability of defense agencies to effectively manage negotiated relationships with defense firms. With a shrinking acquisition budget, the dilemmas of cost and accountability become still more serious.

REQUIREMENTS

Dual-use policies can help DoD control costs and, in some cases, improve performance through purchase of materials, components, and subsystems originating in the civilian economy. Dual-use technologies, latent if not realized, can be found nearly everywhere. Defense systems should incorporate them wherever possible. Dual-use exhibits a pronounced asymmetry: there is much more for defense to gain, at least in an immediate way, from commercial technology than for commercial industry to gain from military technology. At the same time, defense support for the underlying knowledge base through R&D, and especially for the tools and techniques of engineering and science that firms everywhere employ in their day-to-day technical activities, has had enormous impacts (for further discussion, see Alic, 1994). But making use of commercially available integrated circuits or ball bearings should be viewed as a minor, though not insignificant, reform. The big problems in acquisition have to do with the overall design of complex systems, their "architecture," not with the pieces from which such systems are built.

When major programs encounter major technical problems, it is safe to say the reasons will be found in the requirements laid down by DoD. These requirements govern conceptual design decisions and most of what follows. The B-52H, for example, remains at the core of the nation's bomber fleet because it has been an excellent airplane from the very beginning. The decades-newer B-1B, though crammed with the latest "technology," will never be a very good airplane because of compromises and design trade-offs made to satisfy incompatible technical requirements prescribed by DoD.

The initial requirements for what became the B-1B, dating from the 1960s, called for high-altitude flight at more than twice the speed of sound coupled with the ability to fly a terrain-following course at treetop levels beneath Soviet radar (see Kotz, 1988). And in fact, the genesis of the B-1B predates this set of requirements, going back to the B-70, first proposed in the 1950s as a replacement for the B-52. The B-70 was to reach Mach 3, fly at 70,000 feet, and have a range of 7000 nautical miles. President Eisenhower tried to kill the B-70 and President Kennedy succeeded, although two prototypes were built. The idea of a high-speed, high-altitude penetrating bomber then re-emerged during the Nixon administration labeled as the B-1. President Carter tried to kill the B-1, and nearly succeeded. But 100 B-1Bs were finally produced as part of the Reagan defense buildup, with the rationale that the B-1B could be placed in service before the stealthy B-2, supplementing elderly B-52s and heading off a U.S. "bomber gap." In the event, it was B-52s and not B-1Bs that pummeled Saddam Hussein's forces in 1991.

For the B-1B, the critical requirements began with mission profile— high-altitude supersonic flight coupled with low-altitude bombing. The aborted B-70 had been designed to fly above the assumed ceiling for Soviet anti-aircraft missiles and faster than Soviet fighters. Perhaps because the Soviets had managed to bring down Francis Gary Powers' U-2 in 1960, after many unsuccessful attempts to reach high-flying U.S. spy planes, the B-1's requirements called for it to fly *beneath* enemy radar, i.e., at altitudes of a few hundred feet. Nonetheless, the Air Force continued to insist on high-altitude supersonic flight, although never presenting a persuasive rationale for this requirement, which was in fundamental conflict with the dynamic responsiveness needed for low-altitude terrain-following. To achieve both high-altitude supersonic capability and low-altitude maneuverability, the B1-B needed a moveable wing, one that could be swept back at high speeds. The bulky, heavy, and costly wing structure compromised the plane's overall design and also its low-altitude bombing mission: a B-1B cannot carry a full load of bombs and a full load of fuel and still maneuver as needed during low-attitude penetration *(The B-1B Bomber and Options for Enhancements*, 1988). More than likely, no B-1B will ever fly supersonically on an operational mission.

As the B-1B example suggests—and many other cases could be adduced—the services and the Pentagon determine design requirements and contractors must live with them. Overambitious requirements have not been limited to aerospace. The Army once asked for a "universal tractor" that could be dropped by parachute and would have a mean-time between failures (MTBF) of 2000 hours (Sammet and Green, 1990: 299). This requirement is absurd on its face—the equivalent of a working year of 8-hour days without a mechanical problem. MTBFs for Caterpillar's D-7 bulldozer, a workhorse used all over the world, fall in the range of 60 to 80 hours, suggesting that an Army requirement of even 200 hours would have been difficult to achieve. Because the services have sought performance at almost any cost, the imperatives facing defense contractors are quite different than those for firms working on commercial products: in essence, military contractors must go to almost any lengths to satisfy DoD, bending if not breaking the laws of physics. In contrast, when a company like Boeing designs a new commercial transport, it works closely with its customers, the airlines, each of which has somewhat different needs and desires (depending on current routes, plans for the future, and so on), trying to satisfy as many potential purchasers as possible. The design process, in other words, is in part a search for evaluative criteria: What do customers want or need? Product planners pose such questions as: What premium will customers pay for a particular feature (the folding wing tips optional on Boeing's 777)? How will they weigh first cost against operating costs? Boeing makes the final decisions on range, takeoff weight and payload, cruising speed and rate of climb. It pays the bills for design and development, betting that customers will purchase the final product. Airlines that place orders tell Boeing how many seats to install, whether they want the folding wing tips, where to locate the galleys and lavatories. Over time, customers for this or any other class of products discover which of those available meet their needs the best, function well or poorly, need frequent and expensive repairs, wear out prematurely or last indefinitely. Companies learn to design good products (and to control their manufacturing costs) or else they disappear.

These considerations do not apply in the same way for military systems. There is no market test. The Pentagon pays the bills for RDT&E. Requirements are the subject of ongoing negotiations. Contractors try both to shape and respond to the military's desires. But the contractor does not make decisions on major technical parameters; those remain with DoD.

Good initial design decisions largely determine downstream outcomes (Alic, 1993) This is just as true for defense systems as commercial products. If requirements are never pinned down, or if they include unresolved conflicts, a defense program may never converge on a satisfactory technical solution. That was the case with the B-1B.

In theory, major weapons programs emerge from assessments of future threats and U.S. capabilities to respond that have been analyzed, discussed, and debated, then translated into requirements that reflect the best available judgment concerning the actual needs of combat units. This is no easy task. Although relatively small differences in system performance can have life-or-death consequences—a compelling concern which, unfortunately, sometimes becomes a rationalization for arbitrary and unrealistic requirements—it is not always clear, even to putative experts, just how the performance of weapons systems should be measured and compared, or even what "performance" means under battlefield conditions (Spinney, 1980: esp. 96–121). These uncertainties invite ongoing disputes over requirements, making it difficult to reach closure. In practice, threat assessments have too often been open-ended and vague, with decisions on weapons systems the outcomes of bureaucratic infighting and politics.[9]

In these debates, military officers typically argue that they and only they can understand what is needed for war-fighting, a valid claim but also one that deflects outside scrutiny and can obscure the inter-service rivalries that have been endemic in the United States as elsewhere. The Air Force, for example, has been reluctant to buy planes and train pilots to provide ground support for Army troops. Among other consequences, this has led the Army, barred from operating fixed-wing planes, to acquire heavily armed and armored helicopters as a partial substitute. It is by no means clear that the end result approaches anything like an optimal mix of fixed-wing ground support planes and helicopters; indeed, some might argue that helicopters are fundamentally a second-best weapon system for this mission. Within the Air Force, moreover, advocates of fighter-interceptors have long opposed advocates of strategic bombers, both groups seeking more and newer planes—a conflict alive and well in the middle 1990s as the Air Force tried to balance the claims of the B-2 and F-22 on scarce dollars (while arguing for both against the Navy's assertion that the nation should instead buy three new aircraft carriers to gain flexibility in mounting air strikes (Fulgam, January 16, 1995 and January 23, 1995). Such debates can go on for years. They may involve not only OSD, the services, OMB, and Congress, but defense contractors and any number of nominally disinterested experts from inside and outside government (intelligence agencies, advisory boards, consulting firms, think tanks). OSD, for example, has repeatedly sought Air Force and Navy agreement on common purchases of aircraft engines and subsystems, if not airframes. Rarely has the Secretary of Defense prevailed. When he has, the consequences have sometimes been unfortunate, as when Robert McNamara forced the Air Force and Navy to share the TFX/F-111 with minimal modifications for their differing missions (fleet defense, air superiority) and operating environments (carriers, land basing). The missions proved too different, the F-111 too much of a compromise to serve either effectively (Coulam, 1977). At the end of 1994, to take a more recent example, OSD killed the Tri-Service Standoff Attack Missile program because, after ten years of development, there seemed little likelihood of meeting all the requirements posed by the three services. These called for a missile that could be carried by eight different planes and could itself carry five different types of warheads (Graham, 1995).

Examples like those above can be multiplied almost endlessly. On the other hand, some programs have been notably realistic, setting objectives but not "requirements" and putting the priority on budget and schedule. The Navy's several generations of Polaris missiles (and the submarines to carry them) proved successful in part because Polaris was intended as a deterrent, threatening Soviet cities but not "hard targets" that called for high accuracy. Given this, along with the Navy's desire to head off an Air Force monopoly over strategic weaponry, technical performance became less important than getting a functioning system into the fleet as quickly as possible (Spinardi, 1994).

There is a final point concerning requirements. Doctrine and training stand alongside technological capabilities as guarantees of national security. (Doctrine refers to a military's codified prescriptions of how to fight; doctrine is discussed and debated, formalized in written manuals, and inculcated during training.) Given that ambiguity and self-interest invariably color accounts of both victory and defeat, it can be difficult to sort out the contributions of technology relative to other factors. But the common perception of the 1991 Gulf War as a victory won by technology should be set against the views of experienced officers who argue that, more than anything else, doctrine and training accounted for the ease with which U.S. troops and their allies drove Iraqi forces backwards. "As Desert Storm proved (at least to the military professional), realistically trained troops were more important than having the most advanced technology" (Dunnigan and Macedonia, 1993: 248).

ACQUISITION MANAGEMENT

For decades, proposals for acquisition reform have circled around the twin poles of fostering competition among defense firms and improving management by DoD. It is certainly possible to do a better job of managing major programs, but the idea that competition could cut through the dilemmas of acquisition, although attractive in principle, seems misguided (for major systems, not necessarily for smaller procurements). Full-blown competition would require, first, design and development of prototypes by two or more firms (or groups of firms) so that tests, trials, and fly-offs could be used to select the better system. This is the course being taken with the Joint Strike Fighter. To save on production costs, however, production runs must be large enough to justify multiple sourcing. Splitting a procurement among two or more contractors makes sense only when each can achieve scale economies—a condition rarely met even during peak periods of the Cold War, and then for relatively simple systems produced in large volumes such as tactical missiles.[10] The more the defense budget shrinks, and production volumes with it, the less practical competitive procurements become.

Thus there seems no getting away from the ultimate dilemma in acquisition policy: lacking the discipline of a marketplace, how to manage a process involving such huge sums and huge stakes? The first step would seem to be straightforward: DoD should focus on active management of major programs (as opposed to auditing and oversight), recognizing the unique character of the contractor relationship and seeking to develop the kinds of in-house skills and expertise needed to achieve better outcomes.

Once again, there are a host of practical obstacles. Throughout government, agencies operate at a consistent disadvantage in the day-by-day give and take with contractors.[11] DoD operates with further handicaps (Fox and Field, 1988). Military officers run almost all major programs. Many have little experience or training. (While the Pentagon would not put a battalion commander into the field who had not proven himself at lower levels, there appear to be few such compunctions when it comes to acquisition.) For an upwardly mobile officer, acquisition holds few rewards: the road to the top passes through operational commands. With acquisition a detour, tenures are short. During a five-and-a-half year period, five managers rotated through the Stinger missile program; during the 18 years following the first attempts to define parameters for what became the Army's current M1 tank, six generals headed the program office.

Civilian managers stay with programs longer than military personnel. But pay ceilings make it hard to retain civil service personnel in the higher grades, while at lower levels civilian employees may not have the education, training, and experience needed to oversee multi-billion dollar weapons systems. Furthermore, the services, probably rightly, have been reluctant to trust civilian managers, given the life-and-death consequences of developmental decisions.

Whether military or civilian, DoD managers often find it easier to skirt or even hide problems than to confront and solve them. Asking hard questions risks bureaucratic and political trouble. Given inter- and intra-service rivalries, both DoD managers and their counterparts on the industry side must constantly go to bat for their programs against all others. With more systems in development than the defense budget can support, at least at levels commensurate with efficiency, programs compete continually for their share of funds. No one can admit to serious technical trouble without jeopardizing their entire program. Both program office and contractor, then, have incentives to over-promise on performance, cost, and schedule. Given the political and bureaucratic realities, a better trained, more professional corps of acquisition personnel would probably not make much difference when it comes to major programs. Better management would have its primary impacts at the level of routine, day-to-day decisions.[12]

CONCLUSION: PROSPECTS FOR REFORM

If there were a way out of the predicaments described above, it would have been found during the past eight decades. Still, a declining defense budget makes a renewed attack necessary. The United States simply must do a better job of designing and developing its major weapons systems. Perhaps the only path that has not been extensively explored would entail substantial steps toward deregulation of the defense industry.

At the heart of the acquisition problem lies the stranglehold the Army, Navy, and Air Force maintain over requirements, the unwillingness of high officials to delegate authority to program managers, and the unwillingness of program offices to delegate responsibility to contractors. Reform must first get at the processes through which the services, OSD, and Congress settle on mission needs and system requirements, forcing the services to end their long-standing practice of specifying system requirements that are all but impossible to meet. This, not greater competition, is what deregulation would mean in defense.

These decision processes, the difficulties created by rigidity in requirements, the tendency of the services to insist on unrealistic performance goals, and their frequent unwillingness to consider trade-offs—especially of performance, real or imagined, against cost or to permit systems to be designed to meet the need of more than one of the services—have been discussed and debated since the 1950s. Rather than giving contractors latitude to solve a broadly defined problem, making the best use of available technology and the creativity of their best engineers and scientists, DoD forces them to meet artificial challenges posed by a bureaucracy that may not fully grasp either technical complexities or military realities. Rather than permitting program managers to make decisions on technical grounds, U.S. law and a political setting in which members of Congress continually seek reassurances concerning "fairness" and "equity" conspire to force managers into self-protective and sometimes self-defeating choices.

The only way out is to give program managers discretion in resolving technical matters with contractors—accepting or rejecting their choices (e.g., to use commercial components), negotiating compromises and trade-offs as design and development proceeds—without the risks of politically or bureaucratically motivated reversals and second-guessing.[13] This, in turn, means creating an organizational structure with less dispersion of authority, one with shorter and more direct channels between OSD and the program offices (and more insulation from Congress). It also means developing a corps of military and civil service personnel with technical competence on a par with their industry counterparts—program staffs with the ability to make tough decisions and stick to them.

The defense industry is so heavily regulated because Congress, trusting neither the services nor their contractors, and mindful of the responsibility for maintaining civilian control over defense, sees no other route to account-ability. Each case like the A-12 or the DIVAD (Divisional Air Defense rapid-fire antiaircraft gun) reinforces perceptions of mismanagement, leading to calls for more oversight, new rules to avoid conflicts of interest, greater distance between DoD and its contractors. Reversing directions would not be easy. The services have lived through many past shifts in policy. No other part of the public sector can match the Army, Navy, and Air Force in the ability to look out for institutional interests.[14] It will take a concerted effort by Congress and the administration, acting together and pursuing consistent policies over a decade or more, to persuade the services that times have truly changed.

The first step in this process must be a compelling redefinition of the nation's future needs for military systems. This can only come as a result of discussion, debate, and some level of consensus on what national security means and how it can be ensured in a post-Cold War world. That debate has hardly begun. The 1997 Quadrennial Defense Review did not resolve the issues, any more than the earlier "Bottom-Up Review." Indeed, both these reviews ducked the major questions. *Transforming Defense: National Security in the 21st Century*, released at the end of 1997, puts great stress on information systems and "information warfare," but at a very high level of generality.

The consequences of failure to conduct a fundamental reevaluation of security needs include (1) a gross mismatch between the likely levels of future procurement budgets and the force structures that the services seek to maintain, and (2) an equally obvious mismatch between U.S. military capabilities and likely future missions in a world with one rather than two superpowers. The simple fact is that the United States is going along much as if the Cold War was still on. As Table 20.1 showed, defense RDT&E remains at a high level. If the nation were to actually build the systems being designed with this RDT&E money, the procurement budget would have to double, roughly speaking, over the figures listed in the table. There is almost no prospect of this kind of procurement spending. Nor has anything approaching a compelling case been made that the United States needs whole new generations of high-technology military systems of the sort that were built during the Cold War. On the other hand, there are almost certainly new needs, suited to a world of small-scale conflicts, peacekeeping, and the prospect of non-conventional warfare (info-warfare, bio-warfare, terrorism carried out with weapons of mass destruction). These needs are getting lip service and limited funds, while the bulk of acquisition dollars go for weapons that continue to reflect missions inherited from the past.

NOTES

1. This chapter is based on a book manuscript in progress under the working title *Deadly Skills/Useful Knowledge: Military Technology and the Civilian Economy*. The author especially thanks the participants at the Klein Symposium for helpful comments.
2. *The Defense Reform Initiative Report* issued by Secretary of Defense William S. Cohen in November 1997 promises substantial changes in routine procurements—e.g., through electronic commerce—but has nothing significant to say about the major systems acquisition programs that are the focus of this chapter.
3. *IEEE Spectrum*, November 1988: 64, gives a 20:1 ratio for F-15 flight hours to maintenance hours. Martino (1993) puts the ratio at 33:1 (p. 149). Neither figure stands out as unusual in the group of 19 fighter aircraft for which Martino presents data.
4. Indeed, the DeHaviland DH-4, built by the thousands, came to be reviled as a flying coffin. Vander Muelen, 1991: 38–39, writes:
   ... production of the DeHaviland observation plane ... increased from approximately 1,800 in September 1918 to the incredible figure of 4,000 in November. However, this cumbersome craft only underscored the flawed approach taken by the Americans to providing themselves with air power. The DH-4, already obsolete in summer 1917, was built by Fisher Body and Dayton-Wright Corporation, which had been organized to mass-produce the aircraft for the war. The great resources poured into it were clearly driven more by the desires to get something—anything—to

the front and so salvage as much of the program and various reputations as possible. So strong were these pressures that Dayton-Wright shipped DH-4s to the front after randomly testing one in six.

5. On the frequently bizarre workings of federal procurements, see Kelman, 1990. Although Kelman treats non-defense procurement practices, most of the problems he describes can also be found in defense, sometimes in heightened form. Also see McNaugher, 1989.

6. "... to get a proposed C^{4}I system through the evaluation phase, it is necessary to orchestrate the inputs of at least 20 different organizations. Simply managing the logistics of such a process requires organization and entrepreneurship". Crecine, 1986: 102.

7. Holley writes (p. 147) that, by the late 1930s, "Every statute upon the books was encrusted with an intricate overlay of judicial decisions, Judge Advocate General and Attorney General opinions, and Comptroller General rulings as vital to the procurement process as the statute itself." Some disputed claims from World War I were still before the courts during World War II.

8. For a review of 15 major studies, see *Assessing the Potential for Civil–Military Integration: Technologies, Processes, and Practices*, 1994: 171–175. For more detail, see Kovacic, 1990.

9. According to a retired Air Force officer with extensive acquisition experience:
   There were offices with the words plans over their doors and many published documents with this word in their titles, but I was to learn that none of these had anything to do with the actual decisions . . . The planning world and the budget-decision world were separate . . . Budget decisions were based on pragmatic, near-term considerations, such as the degree of political support for a particular program.
   Burton, 1993: 29. *Transforming Defense*: p. 82, puts it succinctly: "Since its creation in the early 1960s, critics have pointed out that the first 'P' [in Planning, Programming, and Budgeting] is silent."

10. The Sidewinder is perhaps the best-known example of successful competition in procurement. For a brief summary, see Morrison, 1990.

11. Kelman (1990: 34) relates the following, from a confidential interview with a "top government marketing manager for one major computer vendor":
    Our attitude is "Bid what they ask for, not what they want." We look at a government specification and see it has loopholes and errors. We don't tell the government that the specification won't do the job. You win the contract, and then you go back to them afterwards, and say, "By the way, the thing you specified won't work. We need a change order."

12. This is the focus of Secretary Cohen's recent *Defense Reform Initiative Report*.

13. Kelman (1990) based on his study of non-defense procurements, concludes (p. 1) that "The problem with the current system is that public officials cannot use common sense and good judgment in ways that would promote better vendor performance. I believe the system should be significantly deregulated to allow public officials greater discretion." Gansler (1995: 109–118) advocates contracting out many of the functions DoD now provides internally.

14. Congress gave DoD more money in fiscal 1997 than the administration had requested. According to Graham (1996):
    Pentagon officials had asked that if funds were added to the defense

budget, they go toward accelerating the purchase of items already in the administration's five-year plan. Congressional members said they tried to honor that request, basing their choice largely on wish lists submitted by the Army, Navy, Air Force and Marine Corps. Only belatedly have senior administration officials discovered that about 40 percent or more of the things included on the Army, Air Force and Marine Corps wish lists were not, in fact, in the Pentagon's long-term plan. Defense officials are now trying to find out why . . .

REFERENCES

A *Quest for Excellence*. Final Report by the President's Blue Ribbon Commission on Defense Management. Washington, D.C.: Office of the President, June 1986.

Adelman, K. L. and Augustine, N. R. *The Defense Revolution: Strategy for the Brave New World*. San Francisco: ICS Press, 1990.

Alic, J. A. "Computer Assisted Everything? Tools and Techniques for Design and Production." *Technological Forecasting and Social Change*, Vol. 44, pp. 359–374, 1993.

Alic, J. A. "The Dual Use of Technology: Concepts and Policies." *Technology In Society*, Vol. 16, pp. 155–172, 1994.

Assessing the Potential for Civil-Military Integration: Technologies, Processes, and Practices. Washington, D.C.: Office of Technology Assessment, September 1994.

Biery, F. P. "The Effectiveness of Weapon System Acquisition Reform." *Journal of Policy Analysis and Management*, Vol. 11, pp. 637–664, 1992.

Bond, D. F. "A-12 Cost Overruns Misjudged; Navy Removes Top Program Officials." *Aviation Week & Space Technology*, pp. 26–27, December 10, 1990.

Borrus, A. "The Godfather of Stealth Won't Slip This One By." *Business Week*, pp. 60–61, September 6, 1993.

Burton, J. G. *The Pentagon Wars: Reformers Challenge the Old Guard*. Annapolis, MD: Naval Institute Press, 1993.

Coulam, R. F. *Illusions of Choice: The F-111 and the Problem of Weapons Acquisition Reform*. Princeton, NJ: Princeton University Press, 1977.

Crecine, J. P. "Defense Resource Allocation: Garbage Can Analysis of C^3 Procurement." In March, J.G. and Weissinger-Baylon, R. eds., *Ambiguity and Command: Organizational perspectives on Military Decision Making*. Marshfield, MA: Pitman, pp. 72–119, 1986.

Directions for Defense: Report of the Commission on Roles and Missions of the Armed Forces, Washington, DC: Government Printing Office, 1995.

Dunnigan, J. F. and Macedonia, R. M. *Getting It Right: American Military Reforms After Vietnam to the Persian Gulf and Beyond. New York*: William Morrow, 1993.

Effects of Weapons Procurement Stretch-Outs on Costs and Schedules. Washington, D.C.: Congressional Budget Office, November 1987.

Fox, J. R. and Field, J. L. *The Defense Management Challenge: Weapons Acquisition*. Boston, MA: Harvard Business School Press, 1988.

Fulgam, D.A. "B-2 Buy Tangled in Mission Rivalries." *Aviation Week & Space Technology*, pp. 61–62, January 16, 1995.

Fulgam, D. A. "Navy Rebuts USAF on B-2s Versus Carriers." *Aviation Week & Space Technology*, p. 65, January 23, 1995.

Gansler, J. *Affording Defense*. Cambridge, MA: MIT Press, 1989.

Gansler, J. *Defense Conversion: Transforming the Arsenal of Democracy*. Cambridge, MA: MIT Press, 1995.

Graham, B. "Missile Project Became a $3.9 Billion Misfire." *Washington Post*, pp. A1, A8, April 3, 1995.

Graham, B. "Panels Challenge Clinton on Defense: Republicans on Hill Seek to Accelerate Spending in Coming Years." *Washington Post*, p. A6, May 3, 1996.

Holley, I. B., Jr. *United States Army in World War II, Special Studies – Buying Aircraft: Materiel Procurement for the Army Air Forces*. Washington, D.C.: Office of the Chief of Military History, Department of the Army, 1964.

IEEE Spectrum. "Special Report, Military Systems Procurement: The Price for Might," November 1988.

Kelman, S. *Procurement and Public Management: The Fear of Discretion and the Quality of Government Performance*. Washington, D.C.: AEI Press, 1990.

Kotz, N. *Wild Blue Yonder: Money, Politics, and the B-1 Bomber*. New York: Pantheon, 1988.

Kovacic, W. E. "Blue Ribbon Defense Commissions: The Acquisition of Major Weapons Systems" and "The Sorcerer's Apprentice: Public Regulation of the Weapons Acquisition Process." In Higgs, R., ed., *Arms, Politics, and the Economy: Historical and Contemporary Perspectives*. New York: Holmes & Meier, pp. 61–103 and 104–131, 1990.

Lee, D. D. "Herbert Hoover and the Development of Commercial Aviation, 1921–1926." *Business History Review*, Vol. 58, pp. 78–102, 1984.

Major Acquisitions: Summary of Recurring Problems and Systemic Issues: 1960–1987, GAO/NSIAD-88-135BR. Washington, D.C.: U.S. General Accounting Office, September 1988.

Martino, J. P. "A Comparison of Two Composite Measures of Technology." *Technological Forecasting and Social Change*, Vol. 44, pp. 147–159, 1993.

Mayer, K. R. *The Political Economy of Defense Contracting*. New Haven, CT: Yale University Press, 1991.

McNaugher, T. L. *New Weapons, Old Politics*. Washington, D.C.: Brookings, 1989.

Morrison, D. C. "Two for the Money." *National Journal*, pp. 1343–1346, June 2, 1990.

Payson, S. *R&D Exceeds Expectations Again, Growing Faster than U.S. Economy During the Last Three Years*. Data Brief NSF 97-328, Arlington, VA: National Science Foundation, November 5, 1997.

Rich, M., Drews, E. and Batten, C. L., Jr., *Improving the Military Acquisition Process: Lessons from Rand Research*, R-3373-AF/RC, Santa Monica, CA: Rand, February 1986.

Rosen, S., ed., *Testing the Theory of the Military–Industrial Complex*. Lexington, MA: Lexington Books, 1973.

Sammet, G., Jr. and Green, D. E. *Defense Acquisition Management*. Boca Raton, FL: Florida Atlantic University Press, 1990.

Sapolsky, H. M. *Science and the Navy: The History of the Office of Naval Research*. Princeton, NJ: Princeton University Press, 1990.

Spinardi, G. *From Polaris to Trident: The Development of US Fleet Ballistic Missile Technology*. Cambridge, UK: Cambridge University Press, 1994.

Spinney, F. C. "Defense Facts of Live". Staff paper, U.S. Department of Defense, December 5, 1980.

The B-1B Bomber and Options for Enhancements. Washington, D.C.: Congressional Budget Office, August 1988.

Transforming Defense: National Security in the 21st Century. Report of the National Defense Panel, December 1997.

Vander Muelen, J. A. *The Politics of Aircraft: Building an American Military Industry*. Lawrence, KS: University Press of Kansas, 1991.

21

ISSUES IN ACQUISITION REFORM

GENE PORTER

INTRODUCTION

The first problem in discussing the current state of acquisition reform within the Department of Defense (DoD) is one of definition. Almost any five defense acquisition officials will have five very different interpretations.

Over the past several months there has been a concerted effort to identify, collect and collate all of the initiatives that form this body of official activity that constitutes "acquisition reform." There are well over thirty distinct initiatives that have, at one time or another, been characterized under the umbrella of "acquisition reform." The most recent DoD compilation is shown in Table 21.1 (see glossary of acronyms on page 373).

As David Berteau indicates in chapter 22, all proposals for acquisition reform did not originate suddenly with the start of the Clinton administration. Many go back to the Packard Commission; indeed some of them go back to the Revolutionary War.

GOALS OF ACQUISITION REFORM

The current Department of Defense acquisition reform initiatives generally pursue two goals: cost savings and improved access to leading technologies.

Table 21.1. Change elements that define Department of Defense acquisition reform
(Source: Coopers and Lybrand / SRC briefing to SAEs dated 10/97)

Discipline	Change element	Description
Contracting	Improved pre-solicitation phase communication	Increased communication to provide potential suppliers greater understanding of Government's needs and Government greater understanding of supplier capability (include conferences, bulletin boards, requests for information, Comm Advocates Forum, draft RFPs).
	RFP streamlining	Reduction in the size and complexity of RFPs due to elimination of unnecessary SOW complexity and contract clauses.
	Elimination of military specs and standards/use of performance-based requirements	Changing the way DoD states its requirements in solicitations and contracts by: establishing a performance-based solicitation process; implementing standardization document improvements; creating irreversible cultural change.
	Government encouragement of contractor-proposed cost/performance tradeoffs	RFPs shall include a strict minimum number of critical performance criteria that will allow industry maximum flexibility to meet overall program objectives.
	Use of past performance/best value evaluation criteria	FASA and subsequent memoranda require use of past performance evaluation criteria in source selection decisions. The criteria use past performance information to select the best sources, and motivate contractors to perform better on their contracts.
	Streamlined pre-award process	Use of tools and methods to decrease time and effort required by both Government and industry from solicitation to contract award, including: IPT type activities (alpha contracting), oral presentations.
	Use of EDI to streamline procurement process	Initiate, conduct, and maintain business related transactions between the government and its suppliers without requiring the use of hard copy media, including electronic source selection.

Table 21.1. Continued

Discipline	Change element	Description
Contracting	Performance based service contracting	SOW for services—"what" not "how"; minimize reliance on intrusive process-oriented inspections and oversight.
	Improved communications related to potential disputes during contract execution	More thorough, timely communications during contract execution, including use of ADR, avoiding unnecessary litigation.
	Use of commercial warranties and other product liability issues (risk management)	FASA requires contracting officers to take advantage of commercial warranties
	Rights in tech data and computer software	DoD acquires only tech data and software rights necessary to satisfy needs; contractor retains rights if data developed at private expense.
Engineering	Use of open systems approach	Integrated business/engineering strategy to choose specs and standards adopted by industry standards bodies or de facto standards for selected system interfaces.
	Use of quick (rapid) prototyping in software development	The creation of a working model of a software module to demonstrate the feasibility of the function. The prototype is later refined for inclusion in a final product.
	Contractor maintains configuration of the design solution	Use of performance-based acquisition reduces oversight of contractor configuration management practices; allows technology updates, other changes without extensive contract change.
	Streamlined procedures for review/approval of engineering change proposals (ECPs)	In performance-based acquisitions, ECPs are restricted to those affecting DoD's performance requirements with concurrent elimination of CL II ECPs.
	Simulation as a replacement for some engineering tests	Use of modeling techniques to test and evaluate design without building hardware prototypes.
	Survivability/lethality testing below end-item level	SECDEF may issue waiver allowing survivability/lethality testing of components, systems and subsystems.

Table 21.1. Continued

Discipline	Change eElement	Description
Engineering	Concurrent developmental testing (DT)/operational testing (OT)	T&E programs structured to integrate all DT&E, OT&E, live fire, and modeling and simulation activities conducted by different agencies.
	Use of commercial engineering drawing practices	MIL-STD-100 being revised to eventually convert to ASME Y 14.100; also, reduction in level of detail required in drawings due to revision of MIL-T-31000 to conform with MIL-STD-961D; also, use of CALS CITIS will help resolve issue of data detail required.
	Use of EDI to streamline engineering design and testing (e.g., JEDMICS, CMIS)	Use of automated tools enable government–contractor interface in standardized manner and operate in integrated database environment. Eliminate lost aperture cards; contractor receives/ delivers drawings in digital format.
Finance	Use of risk-based approach to DCAA financial oversight	Tailoring scope of DCAA audits based upon risk assessment methodology; provided and discussed with contractor executives annually. Objective—work with contractor to correct deficiencies.
	Use of tailored negotiation of forward pricing rates	Establish tailored FPRAs for smaller contracts when facility-wide agreement not possible; renegotiate elements of FPRA versus total agreement.
	Direct submission of cost vouchers to DFAS	Contractors with adequate billing systems authorized by DCAA to submit direct costs (other than first and last).
	Use of commercial and other exemptions for cost or pricing data	Created exemptions to requirement for cost or pricing data for services and modifications to commercial items: also, for noncompetitive buys for commercial items.
	New order of priority for information/adjustment of TINA threshold	FASA recognized reliance on unnecessary cost or pricing data increases proposal preparation costs, extends acquisition lead times and wastes resources.
	Use of parametric cost estimating	Use of parametrics on firm proposals submitted to Government.

Table 21.1. Continued

Discipline	Change element	Description
Finance	Reduced number of TINA sweeps	Use of agreed cut-off date to eliminate endless TINA sweeps prior to contract signing.
	Use of performance-based progress payments	Contract financing based on output/outcome versus input (labor, materials and overhead costs)—applicable only on contracts for non-commercial items awarded non-competitively.
	Use of EDI to facilitate contractor payment	Use of EDI for business transaction information in accounting and vendor pay systems reducing data errors and transaction costs; use of DFAS Major Contract Payment System for progress payments and commercial invoices; DFAS major contract payments by EFT.
Manufacturing	Use of commercial soldering/other commercial manufacturing practices	MIL-STD 2000A was canceled 6/95—no longer required on new contracts. SPI is being utilized to remove off existing contracts. The use of existing manufacturing processes shall be capitalized upon whenever possible.
	Commercial standards/practices for calibration	DSIC cancellation of MIL-STD-45662A. Contractors given choice of ANSI/NISC 2 540-1, ISO 10012-1 or any comparable standard.
Plant-wide	Single Process Initiative new requirements/reprocurements and prime/subcontracts	SPI supports MIL-SPEC and STD reform in DoD by providing a process to do block change removal of government unique requirements off all contracts in a facility; later memos addressed new requirements, subcontractor issues impeding full implementation of SPI.
	Program stability	Use of recent statutory and other means to provide increased stability to DoD programs (increased use of multi-year contracting)—increased stability will reduce program restructuring and associated changes in quantities and/or schedules.

Table 21.1. Continued

Discipline	Change element	Description
Plant-wide	Streamlining procedures/controls related to administration of Defense Industrial Security Program	Efforts to put in place a more simplified, uniform, and cost-effective industrial security program, while ensuring the security of sensitive information and technologies.
	Use of "Other Transaction Authority"	Prototype projects conducted using "cooperative agreements and other actions" versus contracts using FAR/DFARS; PL 104-201 expanded authority to military services, requires competitive procedures to the maximum extent practicable.
	More thorough post-award debriefings	More thorough, timely communications, including debriefings to losing competitors, to reduce reliance on other means of getting info, such as protests.
	Streamlined Government Property Management	Modifying requirements in FAR Part 45 to account for and maintain government furnished property. FAR deviation allowing contractors to refrain from tracking gov. property valued below $1,500 issued 31 Mar 95. Total rewrite of FAR Part 45 is ongoing.
	Reduction/elimination of contractor purchasing system reviews	Reviews based solely on risk assessments; no time requirements; conducted only when necessary; limited in scope to those areas where sufficient data is not already available; maximum use of existing contractor data; summary report generated.
	Streamlined contract close-out	Various PAT recommendations affecting both internal government operations and contractor operations. These include changes to interim final billing rates and an increase to the quick close-out threshold.
	Elimination on non-value added packaging requirements	Ease packaging specifications to allow use of more commercial-type packaging where appropriate.
	Use of commercial procedures and EDI related to; shipping documentation, GBLs. etc.	Use of commercial practices and modern technology (e.g., TRAMS, CFMS) related to shipping documents; enhanced vendor delivery—use of third party traffic management on FOB origin contracts and use of commercial GBLs.

Table 21.1. Continued

Discipline	Change element	Description
Plant-wide	Commercial sourcing—reduction in applicability of certain laws Reduction of multiple SCEs	Reduction in restrictive laws and domestic source restrictions that limited contractors from using commercial sources. DCMC lead in effort to coordinate software capability evaluations—provide feedback to contractors
Program management	Use of joint Government–industry IPTs	IPPD concept includes joint government–industry IPTs, focusing on program execution and identification/implementation of AR. Initiative would resolve program issues in a more timely manner through increased communications.
	Elimination of redundant oversight (Program Office, Services, DCMC)	Reduction of redundant oversight by DCMC, service buying activities and program offices. Citations provided guidance for roles played by various government activities and use of a risk management approach to contract administration activities.
	Alignment of oversight with program risk	Tailoring contract administration based on risk assessment methodology. Transition of government unique requirements on existing contracts to commercial/contractor specs and standards (DCMC)
	Tailoring cost/schedule reporting standards to industry guidelines/ reduction of contractor mgt system reviews	Modification of C/SCSC to accept industry's earned value management criteria. USD (A&T) memo cited stated the industry guidelines (drafted by NSIA, AIA, EIA, SCA and ABA) as acceptable substitutes. DoD PM can tailor K data to specific program needs.
	Use of EDI to facilitate information between Government and contractor	Beginning FY97, all new contracts require on-line access to, or delivery of, their programmic and technical data in digital form. Preference is on-line access to contractor developed data through contractor information system.
	Elimination of non-value added reporting requirements/CDRLs	Review and cancellation of obsolete/unnecessary DIDs by services, DLA and OSD; management data items limited to those essential for effective control.

Table 21.1. Continued

Discipline	Change element	Description
Program management	Cost as an independent variable	Meeting aggressive cost targets through use of cost/performance trade-offs and making process changes to eliminate non-value added activities.
QA	Use of commercially accepted quality program standards (e.g., ISO 9000 series)	Recognition of commercially accepted quality program standards (e.g., ISO 9000 series) in place of MIL-Q-9858 A, MIL-I-45208, etc. This would reduce unnecessary paperwork and eliminate redundant quality assurance systems (both government and commercial).
	Elimination of non-value added receiving/in-process/final inspection and testing	Elimination/conversion/revision of multiple MILSPECs and STDs—883D; 454; O-38535; I-45208; 781; 415; 2165; 810E; most government unique requirements eliminated from RFPs; SPI being utilized to change existing contracts.
	Streamlined documentation/ resolution of non-conforming material issues	Cancellation of MIL-STD-1520A allows contractors in initiate less costly but effective procedures to identify and correct non-conforming parts and materials. This eliminates unnecessary paperwork related to MIL-STD-1520A and reduces cycle times.

Cost Savings

It is widely acknowledged that the Department of Defense will be unable to keep its current military force structure—its Army and Marine divisions, Air Force squadrons and Naval battle groups—supplied with modern equipment at the levels of defense spending that the body politic seems willing to support, unless more money can be freed up for acquisitions by streamlining support activities. The cost of the Department of Defense acquisition process itself has long been identified as excessive, making it a prime target for streamlining. This applies to both the direct cost of the goods and services the Department buys, and to the "overhead" costs incurred within the government.

The expected fruits of acquisition reform constitute a significant fraction of the savings wedge that this administration is counting on as a source of funds for recapitalizing the combat forces. Several issues that complicate the Department's efforts to reduce acquisition costs are discussed in subsequent sections of this chapter.

Improving Access

The other goal of several segments of the acquisition reform community is to improve DoD's access to "cutting edge" technologies—technologies that are developed primarily in the civil sector. Historically, many high tech companies outside the mainstream of the traditional defense industrial base have eschewed DoD business, particularly during the early phases of development. Even when some new technologies have been converted to useful non-military products, DoD's rules and business practices have not only seriously delayed their introduction into the combat forces, but more than doubled their cost. But early integration of key new civilian technologies into future generations of combat systems may even constitute **the** defining military advantage for the United States in an era of global availability of such technology. Of course the efficient introduction of commercial products into military systems also tends to lower their acquisition and operating costs.

The difference between access to products and access to engineering work isn't widely discussed or understood. Many of the acquisition reforms that have been put in place in the 1990s have dealt with access to commercial products, not engineering. It is much easier now for the government to buy commercial products or for government prime contractors to embed commercial components in our ships and tanks and airplanes as a result of the changes that have been made in the federal acquisition regulations.

What is not easier is to contract for the engineering talent of the leading-edge commercial product developers to be applied to specific DoD design problems. This is primarily because, under current rules and practices, DoD must contract for such work using cost-type contracts that are foreign to the commercial marketplace. The private sector overwhelmingly contracts with other companies on a fixed-price basis.

The Defense Department is currently experimenting with some new approaches to gaining access to cutting edge commercial engineering work without imposing the intrusive federal cost reporting rules that otherwise inhibit such access. Congress has recently extended to the Military Departments the authority to enter into fixed-price funded "agreements" for new technology work, while avoiding the formal contract rules of the Federal Acquisition Regulations and Cost Accounting Standards (FAR). This "Other Transaction Authority" had previously been restricted to use by DARPA.

Additionally, recent relaxation of the FAR definition of "commercial items and services" has permitted limited fixed-price contracting for R&D work—most in the form of fixed-price payments to industry to provide certain capabilities needed by DoD in their next generation commercial products. But movement in this area is very cautious—if only because of the failure of several major fixed-price R&D contracts during the 1980s. Indeed, one such contract (for the Navy's A-12 aircraft) continues to be disputed in federal court, with several billion dollars of taxpayer and stockholder money at stake.

IMPEDIMENTS TO ACQUISITION REFORM

Missing Incentives

One major set of impediments is the lack of specific personal reform incentives in both the industry and government segments of the acquisition community. Symposia speakers frequently point out that getting the incentives right is a better way to make acquisition reform work than by punishing slow progress. But this is a general proposition about motivating change that is much more easily advocated at symposia than accomplished in the field, particularly when the most obvious and productive steps entail major disruptions to close-knit working communities and personal relationships. And once selected for a long-term contract, particularly a "cost-plus" contract, industry will typically work hard to cut costs only if rewarded by a large share of the savings—an incentive largely missing from most DoD contracts.

The Lessons of Past Wars

In addition to the general problem of missing personal incentives, there are some very legitimate military concerns about the ongoing course of acquisition reform. What does it mean for the war fighters if the current enthusiasm for "commercial off the shelf" (COTS) equipment is taken too far? Many of the U.S. successes in the Gulf War resulted from having equipment that was built on the lessons of past conflicts. Some of these lessons involved bitter wartime losses, and many responsible officials are loath to trade too much of the demonstrated reliability and robustness of current battle tested equipment for the promise of less costly commercial approaches. For example, there is reason to doubt that the USS *Princeton*, which was heavily damaged by a mine in the Gulf War, would be afloat today if she had been built to commercial standards rather than to the combat shipbuilding standards that evolved from the lessons of World War II.

The "Requirements" Process

A major driver of the acquisition reform discussion—particularly the attempt to get some cost out of the process—is the traditional military "requirements" process. The conventional approach to establishing "requirements" involves estimating the likely future threat in some detail; envisioning something that could be invented to "better" deal with that threat than can the current equipment; formalizing this as a Required Operational Capability; writing a detailed equipment specification and statement of work; giving it to industry, getting it priced; and then going to Congress to get the money. Some believe that this historical approach should be, and is, changing in a fundamental way.

In this view, the "requirements" process (at least for most major military equipment) is shifting from an explicitly threat-driven approach towards age-driven modernization planning. For example, we clearly will need to do something about our warships as they approach, say, fifty years old. In the absence of some overriding change in the perceived nature of maritime conflict or in the nation's need for "forward presence," the logical action will be to buy replacement ships that are "affordable" and that embody proven technologies. Perhaps this has parallels in the way one replaces the family car when it becomes uneconomical to operate. The new car almost always has "better" features, which are not usually driven by any definitive new "requirement" of the owner, unless he has switched his line of work in a way that impacts his transportation needs.

This shift from threat-oriented acquisition to time-phased modernization has significant implications for some of the acquisition reform initiatives, particularly those that are intended to reduce the time it takes to field new equipment. Threat-driven modernization planning has historically imparted a strong sense of urgency. As a result of Soviet secrecy, during the Cold War it was normal for new threats to be discovered by U.S. intelligence quite late in their development cycles—sometimes not until new equipment was already in the field. Responding to such discoveries with urgent new U.S. equipment developments would still typically take years. To the extent that such threat-driven urgency is waning in importance, perhaps we can plan to accomplish our modernization over longer periods by doing things in a more orderly fashion.

In particular, we could plan and execute the engineering work even more carefully and thoroughly with the goal of more often getting it right the first time instead of rushing equipment to the field and having to fix it later, because we are desperately afraid that an enemy is going to get ahead of us. Such a more deliberate approach could also allow more attention to limiting the operating costs of the equipment when fielded.

In general, there is considerable tension between the orderly planning, programming, and management of equipment development programs and the enthusiasm for getting new technology fielded quickly. The weakness in the latter approach is evident in the Department's failure to field a significant number of operationally useful unmanned aircraft, despite decades of attempts. One explanation is that DoD has failed to invest in anything like the serious engineering development activities that have proven to be needed for "real" aircraft, and has rushed to field unproven, immature equipment, to the long-term detriment of real progress.

Cost as an Independent Variable (CAIV)

Entwined with the formal "requirements" process is the increased interest in treating cost as an independent variable. There is more to this approach than just the notion that budgets are tight so we can no longer afford to spend what is *really* needed to meet *the* "requirement." As a practical matter, there have always been tradeoffs made between what is technically possible or operationally desirable and the intended cost of new equipment.

There is some confusion about what cost is being varied in CAIV deliberations. The acquisition community has historically focused on development costs and unit production cost, in large part because those costs are embodied in the instant contracts. But CAIV is intended to be used early in the planning of an acquisition program to establish intended **ownership** cost. If acquisition managers (and Congressional Appropriators) were to focus more sharply on ownership cost and recognize more fully that 70 percent of such cost is not the equipment, but in the pay and training of the operators and in the fuel and spare parts and likely repairs, there would likely be much greater willingness to robustly fund the development and testing phase of such programs, wherein the subsequent operating costs are largely ordained.

CAIV concepts are just now starting to get infused into the formal DoD acquisition system. Very few weapons system contracts have contractually enforceable ownership cost targets. The Navy's Arsenal Ship contract was a notable attempt—it used crew size as a rough proxy for ownership cost. To formulate a contract to build a ship operable by a crew of fifty instead of 400 constitutes real acquisition reform. Although the Arsenal Ship itself has fallen victim to the budget process, the Navy seems determined to employ this same approach for its next class of major combatants.

A looming impediment to implementation of CAIV principles is the growing practice of legislatively capping R&D and/or unit production costs of major weapons systems. This is a very perverse management practice that is increasingly being imposed by the Congress, albeit with some acquiescence by the defense acquisition community. The result in the case of the B-1 was fielding a bomber that couldn't be effectively employed due to the lack of self-protection equipment. In the case of the admittedly expensive Seawolf submarine, cost caps are preventing the insertion of new technology whose future savings would much more than pay for the initial costs.

Oversight Costs

An important issue in reducing the overall cost of acquisition involves the reduction in the direct and indirect cost of government oversight.

Several years ago DoD funded a systematic field investigation of the differential cost of doing business in the commercial world and with the government for similar products and services. Experienced analysts went out to industry to find those who were building similar equipment for the commercial and the government market and found that the difference in cost is about 18 percent. Although difficult to extrapolate to dissimilar products, and politically impossible to carry to the limit, it is clear that significant costs could be saved if the government could act more like a commercial buyer. Many of the results of that earlier study are reflected in the initiatives summarized in Table 21.1.

A major contributor to the 18 percent differential is the legislated requirement that the government obtain certified cost data from its sole source suppliers, of whom there are many. Full disclosure to his customer of a supplier's actual detailed costs is a practice that is foreign to the commercial marketplace, so setting up accounting systems to meet this unique government demand entails considerable expense. One approach to this problem has been a push (not very successful to date) to permit the government to accept the commercial company's internal cost accounting as adequate to protect the taxpayer, rather than requiring a separate cost reporting system.

More broadly, there have been many efforts started to "get the government off the backs of industry." Several of the acquisition reform initiatives shown in Table 21.1 are moving in that direction. But the Nation still has a way to go to reach a working consensus on "how much is enough" in the field of government oversight.

Program Stability

What else drives up the cost of weapons acquisition programs without adding to combat effectiveness? Program instability is a primary culprit. Almost any perturbation to a previously planned program tends to increase its total cost. Therefore, every program manager wants a funding profile that he or she can count on over the next five to ten years. The fact that they almost never get such stability turns out to be due not so much to technical problems or the competition of other acquisition programs for money. Instead, instability in ongoing acquisition programs results primarily from the systematic under-budgeting of outyear costs in non-acquisition accounts. These shortages become "must pay" bills as the budget years dawn, at the inevitable expense of ongoing acquisition programs, which can in fact be more easily stretched than can the paychecks of the troops.

Program instability is a big management issue within the Department of Defense. There are several palliatives that are being or that have been tried. One is multi-year funding. Once the Congress commits to a multi-year contract, with its bigger front-end costs that promise future savings, it becomes sort of a contract between the executive branch and Capitol Hill. This tends to dampen enthusiasm for making changes to the program plan, absent any technical problems. Considerable savings have already been realized as a result of multi-year contracts.

Another approach is to do a better job of budgeting realistically for risk. But there is a long history of skepticism concerning the appropriation and distribution of unallocated management reserve funds to and within DoD despite its proven importance in all other complex undertakings—and some that are not so complex, such as having a house built to order. This is a difficult area in which communication and education may have some positive impact.

The costs of typical defense acquisition programs end up growing only about 20 percent over the development and production phases of the program. This is a lower percentage growth than for any comparable major industrial activity for which records are available. Projects like the North Slope oil line or the space shuttle, or major federal buildings have all experienced greater cost growth than most DoD weapons programs. What we don't know in any collective way is the cost growth incident to major internal development programs in major high tech companies. Some major failures have been reported anecdotally, but there is little systematic data available.

Another contributor to program instability is the inefficient management practices engendered by the artificial "color of money" rules. It is certainly not congruent with good management practices to manage research and development money totally separate from procurement money. Lectures and seminars routinely emphasize the importance of integrating development and manufacturing. Yet currently legislated rules punish the managers of major weapons systems severely for co-mingling the last ten dollars of R&D money with the first ten dollars of procurement money.

In addition to the foregoing incremental initiatives, the Department recently declared war on the root cause of instability identified above—the systematic under-budgeting of non-acquisition programs. Hopefully this initiative will be carried through this time with resulting improved stability in the acquisition program budgets.

Profit Structures

Another problem for acquisition reform is the existing government rules and practices concerning profitability. As a practical matter, most of DoD's major acquisition spending is sole-source in nature. This spending is dominated by long-running follow-on production and modification contracts with suppliers who may well have won a "winner take all" competitive contract years (or even decades) earlier. F-16s, aircraft carriers, and many other systems fit this mold. Contractors have a fiduciary responsibility to their stockholders to drive up stockholder value. In cases of long running sole-source contracts on which profits are set (and limited) as a fraction of costs, this responsibility would seem to conflict directly with pressure from the government to drive down reimbursable costs.

Overhead Costs

Many of the acquisition reform efforts displayed in Table 21.1 deal with costs that show up in contractor's overhead accounts. And yet most DoD acquisition management and oversight efforts focus only on direct costs— costs that account for only half of the total cost. DoD does not contract directly for "overhead." Any such costs that are reasonable, allowable, and allocable, can be billed to the government. When the Department sought to reduce overhead costs at one major contractor by encouraging adoption of a single soldering process, about eight hundred separate contracts had to be changed.

The "schoolroom" approach to overhead management is to encourage and rely on competitive pressures to motivate contractors to reduce their overhead costs without much direct government involvement. But concerns are growing, as competition appears to be lessening, that insufficient progress is being made in reducing overhead costs in industry. The new Single Process Initiative (SPI) that more directly involves the government in the approval of industry overhead activities is making some progress in this area.

Another government approach to reducing industry overhead rates has been to encourage diversification and growth in ways that would help spread the overhead costs over a greater sales base. Such efforts continue in the form of foreign sales for some classes of military equipment, but defense industry specialization has limited the effectiveness of this approach with regard to commercial diversification.

Governance

There are also some real issues about governance. What is the proper role of the United States government and the Department of Defense in running the acquisition process? How much governance is enough?

Because there are many federal jobs at stake in this issue, it is one of the least tractable elements of acquisition reform. But, federal jobs aside, there is also a lack of accepted methodology for even roughly sizing the "proper" levels of oversight employment. The degree to which government employees should perform such "oversight" introduces yet another facet of this complex issue.

And always lurking in the background of the "governance" aspects of acquisition reform is the belief that there will be another $400 hammer story that will undo all the recent progress. This threat could be reduced, if not eliminated, through more open discussions and analysis of the marginal cost of oversight versus the marginal cost of fraud. Such trades are relatively common in the commercial world.

Looking Ahead

Much of the talk about defense acquisition reform has been about fighter planes, ships, and similar major weapons, and some real progress has been made in this arena—particularly in the shift away from mil-specs and in the growing inclusion of ownership cost considerations in the development and selection process. And yet much of the real acquisition reform "walk" is taking place in the less visible areas of small purchases and process improvement. One such improvement is the deployment of standardized procurement software throughout all elements of the DoD procurement community. Such mundane changes exert high leverage on the whole system, but it will be several years before the real impact can be seen.

There seem to be three major thrusts emerging for the next phase of acquisition reform. First, it is worth noting that DoD's contracting officers spend more time buying services than they do buying hardware. Reform of services contracting has yet to be seriously addressed. Competing software support on a man-day rate basis is a poor way to drive down *total* software support costs. Second, the Defense Science Board continues to promote a shift away from cost-based R&D contracting as a way to get earlier access to leading-edge technology. Dr. Gansler, the new Under Secretary of Defense for Acquisition and Technology, has been an active supporter of this initiative in the past, and it will be interesting to see how far this approach can proceed. The third thrust is the movement toward outsourcing much of the acquisition support work now performed by government employees.

The route to outsourcing leads through the political minefield of public/private competition—a field that continues to claim victims. But one conceivable outcome could be the establishment of a sufficiently strong entrepreneurial spirit and competence within the competing government support organizations to lead to *voluntary* privatization. That would set a new standard of success for acquisition reform.

GLOSSARY OF ACRONYMS

ABA: American Bar Association
ADR: Alternative Dispute Resolution
AIA: Aerospace Industries Association
AR: Acquisition Reform
ASME: American Society of Mechanical Engineers
C/SCSC: Cost/Schedule Control System Criteria
CALS: Computer Aided Acquisition and Logistics Support
CITIS: Contractor Integrated Technical Information Service
CDRL: Contract Data Requirements List
CFMS: Communications Facilities Management System
CL II ECP: Class II (Does not affect perform)
DCAA: Defense Contract Audit Agency
DCMC: Defense Contract Management Command
DFAR/FAR: Defense/Federal Acquisition Regs
DFAS: Defense Finance and Accounting Service
DID: Data Item Deliverable
DLA: Defense Logistics Agency
DT: Developmental Testing
ECP: Engineering Change Proposal
EDI: Electronic Data Interchange
EFT: Electronic Funds Transfer
EIA: Electronic Industries Association
FASA: Federal Acquisition Streamlining Act
FPRA: Forward Pricing Rate Agreements
GBL: Government Bill of Lading
IPPD: Integrated Product and Process Development
IPT: Integrated Product/Process Team
ISO: International Standards Organization
MIL-STD: Military Standard
MIL-T: Military Test Standard
NSIA: National Security Industrial Association
OT Operational Testing
QA: Quality Assurance
RFP: Request for Proposal
SCA: Subsidiary Communications Authorization
SCE: Software Capacity Evaluation
SOW: Statement of Work
SPI: Single Process Initiative
T&E: Test and Evaluation
TINA: Truth in Negotiations Act
TRAMS Texas Research Administrators Group

22

DEFENSE CONVERSION AND ACQUISITION REFORM

DAVID J. BERTEAU

INTRODUCTION

On December 31, 1992, the congressionally-mandated Defense Conversion Commission issued its report to the Secretary of Defense. Because the commission completed its work after the 1992 elections, some observers believed that its recommendations would have little effect, having been begun under a Republican administration and completed before the new Democratic administration was sworn in. Notwithstanding this ambivalent posture, a review of the commission's recommendations five years later shows that most of them were implemented to a large extent, and the results of America's "conversion" from a Cold War defense posture can lead to the conclusion that the efforts were successful.

The reasons for that success are harder to discern, but an assessment based on the Commission's recommendations concludes that most of the success at defense conversion was because of the underlying viability of the national economy rather than any specific government programs. While the Clinton Administration's defense reinvestment and conversion program, announced in March of 1993, touted $20 billion over five years, few of these programs contributed significantly to the basic growth pattern that has dominated the U.S. economy over the last five years.

One notable place where deliberate government actions may have produced real defense conversion benefits is in the area of acquisition reform. This chapter looks at the commission's recommendations for acquisition reform, reevaluates how they would contribute to successful defense conversion, and examines the actual acquisition reform initiatives of the Clinton Administration for their impact on defense industry and their attempts to diversify.

BACKGROUND

The Bush Administration initially resisted attempts by Congress to study defense conversion needs and to develop programs to assist industry and workers in diversifying away from dependency on defense funding. In 1991, as part of the FY 1992 Defense Appropriations Act, Congress set aside $5 million for the purpose of a commission on defense conversion. Following a series of exchanges between the Pentagon and the Appropriations committee, Secretary of Defense Dick Cheney agreed to establish a commission, but using executive branch officials rather than outside members appointed in consultation with Congress.

The basis of this resistance and compromise was rooted in the Bush Administration's opposition to any active "industrial policy." The Bush Administration felt strongly, as a matter of basic policy, that the federal government should not intervene in the economy or in the business of individual defense contractors any more than necessary to conduct defense procurement activities. In a celebrated case, the Pentagon declined to take any position on the proposed merger between the last two U.S. ordnance manufacturers, Alliant and Olin. The lack of a Defense Department position on the merger led the Justice Department to conclude that antitrust concerns were considerable, and the merger never took place, despite the Army's expressed support.

As a result, the Defense Conversion Commission's challenge was to undertake a fair assessment of the needs of companies, communities, and workers under the ongoing defense spending reductions, while responding both to the Congress (which felt that something more needed to be done) and the Bush Administration (which felt that enough was already being done). Its recommendations addressed this challenge by finding or developing a solid analytical basis for each position the commission took. In order to do this, the commission drew on dozens of analyses and performed additional studies of its own.

The Commission's analyses were based on assumptions that defense spending would level at approximately $240 billion in FY1997 (in 1997 constant dollars). Actual defense spending has now reached a plateau at a level nearly 10 percent higher, and the 1997 balanced budget agreement projects no substantial reductions from current levels (though inflation may produce reductions in real value).

Based on these defense spending figures, one can conclude that the post-Cold War drawdown is complete and that future defense spending reductions will be driven by factors other than the collapse of the Soviet Union and the demise of the Warsaw Pact. As a result, this is a good time to examine the results of federal government actions that were taken in support of defense conversion. One framework for that examination is the forty-three recommendations of the Defense Conversion Commission, because most of them have been implemented. In order to focus that framework, this chapter evaluates the subset of the commission's recommendations that impact acquisition reform.

COMMISSION RECOMMENDATIONS

The commission's report was issued at the height of the post-Cold War drawdown in defense spending. Its recommendations addressed four goals: (1) facilitating the economy's transition to lower levels of defense spending, (2) preserving and enhancing defense industrial capability, (3) easing the impact on defense-dependent workers, communities, and companies, and (4) improving the integration and operation of federal government programs.

One of the mainstays of the commission's approach was to focus recommendations toward achieving measurable results. Too often, the effectiveness of federal government programs is measured by the level of inputs. For example, the effectiveness of workforce training and assistance programs will be measured by the amount of budget funding provided over a given period, and few if any statistics are available that detail the number of workers who obtained new jobs as a result of this training assistance.

Overall, the commission made forty-three specific recommendations that addressed the four goals outlined above. Eighty percent of those recommendations dealt with the broad area of relations between the federal government and U.S. industry, both from a technology development and a contractual point of view. Those recommendations form the nucleus of the commission's proposals that have been subsequently addressed by the Clinton Administration's acquisition reform efforts. The sections below list each relevant recommendation, describe the actions taken by the government, and assess the results of those actions. In some cases, the assessment is subjective, but in many cases, there are adequate hard data to support conclusions.

A BRIEF HISTORY OF ACQUISITION REFORM

The history of problems in defense procurement and acquisition is older than the Republic itself. During the American Revolution, the Continental Army was plagued with merchants who sold them spoiled food and shoddy supplies and who failed to deliver promised goods. In the Civil War, contractors were guilty of delivering substandard weapons to the U.S. Army, reclaiming those weapons from the disposal yards, and then reselling the same weapons back to the Army again. In fact, the statutory basis for the *qui tam* lawsuits under which recent whistleblowers have collected large cash settlements date from Civil War-era legislation.

With the development of a permanent defense industrial base during and following World War II and into the Cold War, the calls for acquisition reform have developed a permanence that was not present in earlier years. Beginning with the Truman Commission in 1942, various bodies of experts have detailed recommendations to make the federal and defense acquisition processes more effective and efficient, while protecting the interests of the taxpayers.

Over time, the relative importance of effectiveness vs. protection against fraud and waste has varied. Celebrated cases of fraud and abuse force Congress and the executive branch to focus more on detection and punishment of mistakes and criminal mischief, sometimes at the expense of the larger question of the most effective value for the taxpayers' dollars. For example, while the implementation of acquisition reform initiatives in the early and mid 1980s may have achieved some beneficial results (based on the 1981 Carlucci Initiatives, the 1982 Grace Commission reports, and the 1986 Packard Commission recommendations), they were overshadowed by the revelations of illegal activities in Operation Ill-Wind, beginning in 1987.

By the beginning of the Clinton Administration, however, there were a number of forces that supported renewed emphasis on defense acquisition reform. First, both the defense industry and the Pentagon recognized that there were essential elements of technology in which commercial applications were outstripping defense developments. Among these were communications, electronics, and information systems, including computing capability. Second, the declining defense procurement and research budgets would not continue to support a defense-unique industrial base that was robust enough to provide for all of defense's needs. Third, the expanding global economy meant that the definition of competition was shifting, from a purely domestic issue to one that depended more on global competitiveness as the relevant measure of true competition.

Fourth, and perhaps most importantly, the Congress began to shift away from a posture of attacking defense spending levels through procurement rules and toward a posture that recognized that defense spending and the overall economy were more closely related and therefore that the defense-unique rules and regulations were more of a burden on the overall American economy than they were a protection of the taxpayers' interests.

Because of these reasons, the Clinton Administration and the Pentagon were able to initiate new acquisition reform efforts that held promise to meet the specific recommendations that the Defense Conversion Commission had proposed.

THE COMMISSION'S ACQUISITION REFORM RECOMMENDATIONS

The commission referred to this interrelationship between the defense and commercial economies as "commercial–military integration." Its recommendations were aimed at strengthening the basis of this integration. They included the following, with a brief summary of the implementing actions that were subsequently taken, the current status and results, and the prospects for further progress:

(1) Efforts to foster commercial–military integration should be strengthened, expanded, and accelerated. Under William Perry as Deputy Secretary and later Secretary of Defense, DoD adopted a broad approach to achieve greater commercial–military integration. Most of the actions undertaken are listed below, in the summaries of specific recommendations.

(2) DoD should work closely with the Congress and with other Federal agencies to reduce or eliminate the statutes and regulations that prevent greater commercial–military integration.

In 1994, Congress passed the Federal Acquisition Streamlining Act, or FASA, and in 1995 followed it with the Federal Acquisition Reform Act, or FARA. These acts accommodated a number of reforms to make it easier for the federal government to purchase commercial products. In subsequent years, the changes needed in the Federal Acquisition Regulation, or FAR, to implement those legislative changes have been developed and are now being implemented. Not all of the changes have been completed yet.

In the field, the results of these statutory changes have been less far-reaching. Contracting officers still require adherence to regulations that are on the way out, oversight of private sector providers still is too often based on the assumption that everyone is out to defraud the government, and DoD calls for cost and pricing data that no commercial vendor even keeps, much less would make available to its customers.

Notwithstanding these shortfalls in implementation, there is progress being made. What is still missing is any meaningful set of measures of success for the changes in statute and regulation. DoD needs to develop these measures, in concert with industry, and use them to track progress.

(3) The Secretary of Defense should require that DoD use commercial specifications, standards, and buying practices for all procurement actions except those for which he or she has approved military-unique practices on the basis of a demonstrated, compelling need.

In November of 1993, Deputy Secretary of Defense Perry issued a landmark memorandum which basically implemented this recommendation. In the following four years, progress toward implementing this policy change has been made in places. Many procurements are now being issued with fewer military specifications and standards called for. DoD is participating more effectively in the voluntary standards setting organizations, and there is greater cooperation between the public and private sectors. As with the regulatory changes outlined above, cultural barriers still exist, and it will take considerably more time before the policy is deeply rooted and on the way to success.

(4) DoD should issue a statement to all DoD prime contractors and DoD auditors that emphasizes current policy on allowable IR&D activities and highlights the fact that DoD permits companies to use IR&D funds to develop commercial products.

    The DoD policy which permits companies to use IR&D funds for developing products and processes with commercial applications was relatively unheard of in 1992. Its use is more widespread today, based on anecdotal evidence, but DoD has little data to support this conclusion.

(5) DoD should change overhead allocation policies to remove disincentives to the development of commercial products. This recommendation would permit contractors to use available capacity to develop commercial applications of defense technology without unwarranted reductions to their existing overhead allocations, provided that no performance degradation occurred on existing defense contracts.

    This is the one area that acquisition reform has not touched. In fact, the entire issue of cost controls has been largely omitted from acquisition reform, except for portions of the FAR that have been changed for small commercial purchases. It is an area crying for action.

CONCLUSION

The overall conversion required by the post-Cold War defense drawdown, to the extent that it was ever in doubt, has been completed. Government programs instituted in support of such conversion efforts are no longer needed for those purposes. In one area, acquisition reform, the impetus provided by the defense drawdown and the need for diversification by defense-dependent companies has contributed to a greater success than in many previous decades. Progress, while steady, is still incomplete, and continued work is needed.

23

REINVENTING THE PENTAGON

ERIK R. PAGES

My remarks will have a somewhat different focus than those of Gene Porter and David Berteau. They addressed many of the more traditional issues of acquisition reform: how we can buy things smarter, how do we buy commercial-off-the-shelf, things like that. I think that these are important questions, but I think that many of these problem areas have already been resolved, at least on paper. A lot of the statutory changes and regulatory reforms are in place.

What is needed now is for the message of acquisition reform to float from the top down. In other words, what we are looking for is a change in the corporate culture of the Defense Department. This is going to be hard and won't occur quickly, but I think we are on the right path. Given these trends, I think it's helpful to examine another issue. How does DoD buy services, not things?

Specifically I want to talk about the need for aggressive privatization and outsourcing at the Pentagon: what Defense Secretary William Cohen is calling a revolution in business affairs.

At Business Executives for National Security (BENS), we like to talk about this issue in terms of the ratio of tooth to tail. While our military leaders have continued to downsize and reorganize the fighting force (tooth), spending on support functions (tail) has remained steady and actually, in some cases, it is growing. We need to cut back on unnecessary support structures, bureaucracy, and overhead within the military departments and in the civilian side of the Pentagon.

Let me give you a bit more detail on the extent of the problem. We have all talked about the downsizing at DoD over the past decade. Spending has gone down about 40 percent since 1985. What is less noticeable is that the department's oversized support structure has been resistant to change. The traditional measure of support versus war fighting capability (tooth to tail) has become skewed.

Throughout the Cold War, our defense spending patterns were generally at 50 percent of spending on tooth (combat capability) and 50 percent on tail (support). We are now at almost 70 percent for support and administrative functions, while only slightly more than 30 percent goes to war fighting. Whether you want more or less defense spending, it's clear that we're spending too much on capabilities that do little for the national security bottom line: defending our nation.

We need to correct this imbalance if the military is going to afford the modern equipment that it claims it needs to carry out its mission. As you know, the consensus in Washington is that we need to spend roughly $60 billion per year on procurement (roughly an additional $15–20 billion per year over current procurement spending levels) to recapitalize the force. Whether you agree with that number or not, it is the consensus opinion across the political spectrum.

Where are we going to find $20 billion in a political environment where the prospects for new defense spending are limited? In the past, we could throw money at this problem. But I do not think that the political environment is amenable to that right now. And, it also seems unlikely that we'll significantly reduce our force structure, which would be the other way to save money.

The only way to proceed is by a third path and that is to aggressively attack bloated infrastructure in the Pentagon right now. It remains largely unchanged from Cold War levels. DoD can save billions of dollars by aggressive outsourcing of functions like military housing and information technology, conversion of military bases, and improving inventory management. If we are willing to turn to the private sector for some of these functions, we can not only improve operations but also meet this $20 billion figure that we ostensibly need for recapitalizing the force.

Let me turn now to some specific lessons from the private sector. I think the Pentagon would benefit greatly by looking at the experience of major American companies during the 1980s, companies like IBM, AT&T and Chrysler. Remember the competitiveness crisis of the 1980s when Japan, Inc. and a united Europe were going to take over and American companies were on a permanent path of decline? That is no longer the case.

We never hear about this threat anymore. Why is that? One of the primary reasons is that American companies transformed themselves. It was a very painful transition process, but the American corporate sector has reorganized and reformed itself to become efficient, productive and competitive. What happened? To survive, the private sector merged and consolidated, restructured and reengineered, and unfortunately, downsized jobs. They outsourced non-core functions in order to tap services from providers who were the best in the business.

If you want to find one message in this experience it is this: if someone else can do it cheaper and better, let them. DoD ought to do the same. What the Pentagon needs to do is focus on its core business, which is war fighting. It should at least consider outsourcing all activities that are tangential to that core business.

The candidates for outsourcing and privatization are huge. Nearly every business function from health care to finance and accounting, logistics and base operations should be considered potential candidates for outsourcing. When you look at candidates, I think you should operate according to two basic principles:

First, search for areas that offer the best prospects for savings and or improved operations. You really ought to do this for operational purposes to improve the efficiency of services provided to the troops and the military. Savings ought to be the secondary goal. In Washington, unfortunately savings becomes the primary goal because of budgeting concerns. Yet, we should really outsource to improve service and efficiency, not merely to save money.

Second, search for candidates that offer the best prospects for quick success. You want to search for the low hanging fruit. In other words, focus on areas that are far removed from combat operations. Try to build on small successes one step at a time. Don't touch health care; don't touch commissaries—things that are essential to quality of life, even if they are good candidates for outsourcing. Don't go to areas that are very closely related to combat or to weapon systems.

Today, much of the privatization debate in Washington focuses on Air Force depots. This dispute even held up the defense bill this year. While I support depot privatization, repair and maintenance depots are not good early candidates for outsourcing based on my earlier criteria. Privatizing these facilities is a very difficult political challenge. It has a big payoff monetarily, but in terms of the structure of the debate around the depots, it is a very, very hard political sell.

Where would I go? I think we should look at payroll processing, information technology, housing, travel, utilities—none of these is essential to war fighting. These are areas where we can find huge improvements in services and huge cost savings. The good news about going to these areas is that these are also the areas that the private sector has already "been there and done that." They are also areas where there is a competitive private sector to provide these services.

When DoD was first established after World War II and we faced the threat of a global war from the Soviet Bloc, it made sense to do a lot of these jobs in-house. In fact, in many cases, there was no other way to do it. Private firms to provide these kinds of support services didn't exist.

There weren't companies like Federal Express or Computer Sciences Corporation (CSC) or even SAIC—a lot of this had to be done in-house. Internal bureaucracy developed over time.

Forty years later, we have strong private firms in most of these sectors which are very competitive markets. Why not take advantage of this? There is no need to keep all of this in-house.

The other danger for DoD in this case, and it relates to what Dave Berteau was discussing: a lot of these firms are not going to want to do business with the Pentagon if it stays wedded to arcane business practices. This story may be too good to be true, but I've been told that some parts of the military still do their inventory control on 5×8 cards. Why would a company like Dell that can deliver a specially designed computer to your home in 36 hours want to work with a defense agency that does its inventory control on 5×8 cards? It's not worth the headaches.

Many benefits are to be realized for DoD to move away from these primitive business practices. And, as I said, I hate to make the argument based purely on savings, but the prospects for real savings are there when we begin to outsource a lot of these business functions.

Let me give you a couple of anecdotes. Let's take payroll processing at the Defense Department—just to process a paycheck in the defense department costs about $4.75, almost $5.00. Private sector firms do it for less than $2.00. If you are paying a million people every month, just do the calculations. Outsourcing can save hundreds of millions a year and allow all of you to get paid more quickly and be provided with much more than you are getting now.

Travel processing is another example. I was shocked to find that the Pentagon spends one billion dollars a year just to process travel vouchers. We trust people in the military to drive the most expensive equipment in the world, but we can't trust them not to overspend their per diem. It is crazy that we have this kind of extensive oversight of travel processing. The Pentagon's administrative costs on travel are 33 percent. The worst performing private sector firms do it at 6 percent while the industry leaders have cut administrative costs to 2 percent. So 33 percent versus 2 percent is pretty significant savings again on a billion dollars.

Overall the Defense Science Board thinks you can save about $30 billion a year through aggressive outsourcing and management reform; this includes things like another round of base closures as well. The key point in all of this, whether the government or private sector does this, (I happen to think the private sector should do a lot of it) is that you want to introduce competition. That is the key to it in every case, even if the work stays within the government.

Today, think of any organization whose name begins with a D—Defense Finance and Accounting Services, Defense Logistics Agency. The military services have to go to these defense agencies for certain functions—it's an internal monopoly. When you have a monopoly, costs are not going to be reduced and services are not going to be improved in the way they are when you have competition. The Center for Naval Analyses did a recent study of the impact of competition—every time you added competition (whether in the public or private sector) you saved 30 percent, regardless of who won a competition between the public and private sector. Every time you replace a monopoly with a thriving competitive marketplace, you will save money.

A couple of quick concluding comments. Of course, I think DoD needs to be run in a more business like manner. But DoD is not a business, I understand that. You cannot simply mandate downsizing; you can't mandate outsourcing like you might be able to do in the private sector. You have to understand the political reality and tailor the approach. So let me just make three observations about how the department might move forward in a manner that allows you to make these changes while recognizing the political realities.

First, I would caution that we have to be careful with the numbers. We can't simply trust bean counting or the number of A-76 studies (the process for doing public private competition—those of you who are familiar with it, it has probably given you many gray hairs), but you can't simply say we have done this many competitions or x number of people have been outsourced. You can't use that as the metric for success. What happens often is that a government agency will graft a new business practice onto itself and say we have changed—and check the requisite box in its list of performance measures.

We can't just list outsourced areas and say we have succeeded. We have to make this a continuous process. All of us have to be rethinking how to introduce competition, how to structure contracts, who can do the work better for you in-house or the best outsourced contractors.

Second, we need to change how we think about contracts. Almost all the rules for contracting right now are designed to buy things and not services. We have to train people to think about buying services not things. This is very, very hard.

One organization that I've taken an intense look at is part of the Defense Logistics Agency (DLA)—the Defense Reuse and Marketing Service (DRMS), which sells excess property. They keep claiming that they are going to outsource their business. They have put out dozens of Requests for Proposals (RFPs) to outsource some of their work. They can never get any companies to bid because companies can't make any money on the RFPs. Then the organization says "look the private sector can't provide this service to us." It is not that the private sector doesn't want to put in bids. The problem is that the contracts are written in such a way that no one is going to take that risk. The agencies want the private sector to take all the risk and make no money on the deal. This won't work—a real partnership is needed.

Finally, the last thing I would say is that you do need to take care of people. There is no denying that behind every current policy or process or function that is a good candidate for outsourcing is somebody's job. So whatever solutions we craft, we have to carefully design the job transition approach. I would commend you all to look at the experience of the Indianapolis Naval Air Warfare Center, where they have done a very good job in moving it from a government facility to a facility run by Hughes. They are going to have new jobs in the private sector with nearly all of the old government employees retaining their jobs with the new contractor. Almost everyone was taken care of in that transition.

The importance of transitional support is something we have learned from the base closure experience. If you come in and invest in the transition and develop special programs to help people in the immediate aftermath of the downsizing, most communities and people take advantage of them and prosper.

I'm not going to say it's easy. Its easy for me to spout off about why I believe outsourcing and privatization makes sense. But, with the budget situation that faces DoD, there really is no other choice but to pursue these options. The good news is that we are no longer talking about whether to outsource or privatize. That debate is over and I think the good guys won. The hard part is designing these programs in such a way that we effectively make it happen.

PART FIVE:
CASE PRESENTATIONS

CASE PRESENTATIONS

JOHN STUELPNAGEL, NORTHROP GRUMMAN CORPORATION

Northrop Grumman relied on its expertise in the production of airborne radar systems to enter civilian markets for radar technology, law enforcement, and parcel handling. It applied its engineering expertise to solve the windshear problem, but it did not take market conditions adequately into consideration. The problem was that there was no FAA requirement for predicting windshear. Thus, even though Northrop Grumman manufactured excellent radar, it had a problem getting FAA certification. Its competitors, however, knew a great deal more about getting the certification than it did. Also, since these competitors had numerous other related products, they were able to cut prices by bundling their products. Thus, Northrop Grumman was not successful in this venture.

The company also entered law enforcement, with a focus on multi-sensor airplanes, residential security, access card technology, law enforcement surveillance products, and integration of police mobile information systems. Again, it did not do well. The main reason for the failure was that the company did not understand the distribution channels in this market nor some of its financial aspects.

The company learned that it needed to understand more about markets when it succeeded with parcel handling. Despite the fact that it did not have a technology base in this area, through working together with an overnight delivery system company that wished to put in an automated parcel handling system, Northrop Grumman successfully integrated the operational system required by this kind of a business. Using its expertise in system integration, software and contracting, and working very hard at it, it was able to satisfy its customers and foresee excellent prospects for future business.

One of Northrop's most important strategic decisions was to acquire the Grumman Corporation. Northrop explored the alternatives it could follow in response to the changes in the marketplace. It considered three options. The first was to continue doing business the same way it had been doing, which was very risky. The second was to drastically change the nature of the company. The third was to abandon the defense business, which would have meant liquidation. It ended up going with the second option. It acquired electronic systems integration from Grumman and the Westinghouse Electronic Systems and sought to significantly reduce its dependence on military airframes by going into electronics. As a result, Northrop Grumman will be 50/50 electronics and airframe related activities within the next couple of years.

In terms of strategy, the company decided that (1) it needed to focus resources on an area and be committed to it, but first understand the total customer environment thoroughly; (2) it needed to understand how customers selected things, how they bought them and what their needs were; and (3) it needed to understand customers' needs better than it did, as well as how to apply technology to solve those needs. One important realization was that, in its business, technology did not have a significant value of its own since perceptions also come into play. Instead, the value was in meeting the customers' needs by applying the appropriate technology.

The company does not have a standard policy for intellectual property or technology to determine whether it will commercialize them or not, because over time it discovered that it is not good at technology transfer as a business operation. In addition, it is clear that it will remain focused on the defense business with some occasional and moderate initiatives into non-defense activities. It is a common belief within the company that it will be better off by staying with what it has known best and extrapolating from the culture in which it has done very well.

WILLIAM T. HANLEY, GALILEO CORPORATION

Galileo Corporation, located in Sturbridge, Massachusetts, was founded in 1959. It is the oldest independent fiberoptic company. It was founded as a fiberoptic component manufacturer for military night vision applications. In 1982, 85 percent of its business was still in military night vision components. The deviation from military applications started in 1966 when the company was acquired by Bendix Corporation and became the Electro-optics Division of Bendix. It started manufacturing single channel detectors. Those detectors were launched in the late 1960s and analyzed solar wind for about 20 years.

Today, although it is a small business on a worldwide scale, it is the

leader in fiberoptic and electro-optic technology. The company has a vertically integrated business structure all the way from applications engineering
to product development, manufacturing and marketing. Today, Galileo
Corporation is in four different businesses, two of which are mature and
profitable. The other two are still in the development stage with expected
future profits.

In the kind of business in which Galileo Corporation is involved, product
life-cycles are very long. For example, the company began to develop
Time-of-Flight detectors in 1977 and it took twenty years to commercialize
them. In 1997, the product is still a very important part of life sciences
technology and biotechnology development. However, Bendix management
wanted faster profits so five members of Bendix management bought out
the business through a leverage buyout in 1973. At that point in time, 95
percent of the company's business was in military night vision components.
The management decided to diversify the company but they were not able
to recognize the distinction between diversification and conversion, so what
they did in essence was to convert the company by going into telecommunications fiber optics. The only thing this technology had in common
with what the company had been doing was physics, which was not solely
enough to build a business on. As a result of the failure of that diversification
attempt, the company came to the verge of going out of business in the
early 1980s.

William Hanley's mission in joining the company as Vice President of
Manufacturing was to get its manufacturing capability up to the required
level and get back on track in making its remaining components and
products. He found out that the company did not have the capability and
know-how to do what it was doing and that competitors were in the same
situation also. It was an advantage for the company to recognize that fact
before the competitors did.

The first thing Mr. Hanley did was to focus on improving the quality of
fiberoptic and electro-optic technology for military night vision applications. That effort alone was enough to increase the company revenues from
$14 million to $27 million in just a couple of years. He also recognized the
fact that although the company had been involved in this area, the climate
of the business was inclined to change quickly so he insisted that the
company also set about converting itself into a structure that would be in
line with the changes in the industry.

Despite the fact that the company was doing very well in the night vision
market, Mr. Hanley decided to exit this business. When he looked at the
market conditions, he found out that what the company had been doing
had no further applications other than the military. He also predicted that
the revenues from military applications would decline to $10 million in a
few years and that turned out to be true. That happened shortly after the
end of the Cold War. Eventually this led to military applications being

only 5 percent of the company's current business. Hanley and his colleagues put together a marketing strategy that allowed them to achieve more than a 25 percent growth rate in the commercial business.

As a result of careful planning and execution, the company now has a vertically integrated business team structure, all the way from applications engineering to product development, manufacturing and marketing, focusing on high value-added products based on core technologies. All this happened, despite the fact that the number of employees has gone down from 630 in 1987 to 205 in 1996. Productivity has increased dramatically, both in sales dollars per employee and gross profits per employee. Finally, its customer base has diversified, going from four customers which accounted for 85 percent of sales to around two thousand customers, with the top twenty accounting for 70 percent of the business. It also should be mentioned that even if it had captured all the business in the military sector, it could have made revenues of around $45 million, whereas having undergone the conversion, it has created market opportunities that are greater than $800 million.

The most important factor contributing to the successful transition of the company was a well thought out, well understood, well articulated, and well executed strategy. This long-range strategic planning, which included worst case scenario preparation, was pivotal in the company's survival. Finally, careful cash management was also a very important contributing factor.

JAMES A. KOSHAK, HEXCEL CORPORATION

Hexcel Corporation is a manufacturer of advanced structural materials such as lightweight high performance composites and other materials which, until recently, were used exclusively in military and commercial aerospace applications. The company's current mission is to become the global leader in developing and selling these products to customers in the aerospace, space and defense, electronics/telecommunications, recreation and general industrial markets. In order to realize the benefits of the recent surge in commercial aerospace, Hexcel is taking innovative steps to leverage its leading-edge technology into new markets, products and manufacturing processes.

The composite industry has gone through major changes since its inception. Until the early 1990s composite materials had been seen more as a curiosity than as an industry pursued by some large chemical companies and a few small technology-oriented companies such as Hexcel. The composites market came into existence during and immediately after World War II. The primary application of Hexcel's products to this market was in military aircraft and vehicles due to high R&D costs. In the early 1990s two major events took place that highly influenced the composites industry. First the USSR collapsed and then the Berlin Wall was dismantled. This resulted in a dramatic cut in defense budgets. A significant portion of the military programs in the U.S. and Europe were either delayed or canceled outright. For example, the proposed production of the B-2 aircraft, which was to have used more composite materials than any other aircraft, fell from 75 and then to 21 after the fall of the USSR. In addition, the weakening economies worldwide led to a decline in air passenger travel and hammered the commercial airline industry. Boeing's peak delivery rate was approximately 720 planes in 1991, then fell below 400 by 1996. A number of airlines went bankrupt trying to attract more travelers by engaging in price wars. These changes led many of the diversified companies that had entered the composite business during the 1970s and 1980s to exit very quickly. Plants were padlocked, profits evaporated and many employees lost their jobs. All of these events indicate the composite industry was declining rapidly.

As the industry declined, Hexcel Corporation recognized too slowly that dramatic changes were underway. It had invested in too much capacity and had made large financial commitments. Hexcel developed a three stage strategy toward recovery. In the first stage, restructuring, the company re-focused its operations, reduced staffing, sold non-core assets, and repositioned its product lines. In the second stage, consolidation, it consolidated its resources and capacity more in line with demand and improved its profitability. In addition, it acquired the worldwide composites business of Ciba-Geigy and the composites products division of Hercules. The acquisitions extended its product lines upstream and downstream and expanded its market penetration. As a result, Hexcel is the only global vertically integrated company in the composites industry. The company also became a truly global player, going from a $300 million company in 1995 to an $800 million company in 1996 after its acquisitions.

The third stage, growth and expansion, is where Hexcel is today. It intends to use five strategies in this stage. The first strategy is to pursue further acquisitions and alliances. The company expects to continue to identify, explore and selectively enter into acquisitions and business alliances that will complement or extend its current technologies and product lines. Recent acquisitions of two assets of Fiberite Incorporated are examples of this strategy. Another strategy is to maintain a global perspective on operations. The old Hexcel focused primarily in the U.S. but today, as a result of the Ciba acquisition, it is a truly global company. A third strategy is to pursue vertical integration. The forward integration of the production of composite parts and carbon fiber operation from the Hercules acquisition provided Hexcel with excellent opportunities to work with its customers to quickly create innovative products that meet their needs from start to finish. The last two strategies are to create new processes, manufacture new materials through vigorous research and development, and maintain an engineering and research-oriented focus in order to further improve cost performance. For example, process development groups at Hexcel's plant in Arizona, which produces honeycomb core, focuses on new core processes.

Two measures of the company's success in reacting to changes in the market is how fast it can expand the total market demand for its products and grow effectively by customizing its products to the unique needs of its customers. It should also be emphasized that Hexcel is pursuing not only growth for itself, but also the expansion of the overall use of composite materials throughout the world.

The company's success in responding to the decline in its market is attributable to the fact that it never lost a couple of important competencies. One was its customer base and the other was its leadership position in technology. As a result, by undergoing a dramatic restructuring that consolidated the advanced structural materials industry, by pursuing aggressive strategies for expanding and diversifying its business, and by implementing a total overhaul of the company, Hexcel Corporation was able to radically change its business environment.

ALBERT SMITH, HARRIS CORPORATION[1]

The Harris Corporation is a Fortune 500 company with annual sales that exceed $3.5 billion and an employment level of approximately 26,000. The Harris Corporation has enjoyed double-digit earnings growth for the past four years by repositioning from a defense electronics company to a diversified company engaged in commercial high growth markets. Harris operates in four business sectors: communications products, semiconductors, office products and electronic systems. Mr. Smith discussed how the electronic systems sector has dealt with the significant downturn in defense spending, particularly the dramatic decreases in RDT&E and procurement budgets. He first indicated the results of the downturn on industrial consolidation. According to Mr. Smith, the U.S. aerospace/defense industry is nearing the endgame of a very successful consolidation phase initiated with the full support of Secretary of Defense Perry. In addition to dramatic savings to be realized, the consolidation will sustain the U.S. as the global leader in defense/aerospace systems. As reference, Mr. Smith stated that Europe is several years behind in consolidating their infrastructure for competitiveness.

In regards to Harris, the company has managed to sustain its U.S. government business through a strategy of focusing on its core markets and customers and penetrating adjacent markets. A deliberate strategy has also been employed to "pull back" from commercial diversification activities within the government cost structure. Mr. Smith pointed out several examples where diversification by government contractors has failed. Alternatively, he pointed out several techniques whereby technology and products can be moved into the commercial marketplace. Specifically, he discussed a concept of "channel partnering" to produce financial benefits for the Harris Corporation. These channel partnering techniques which include licensing, joint ventures, or movement of specific intellectual property, all involve marrying the government-developed technology to partners having commercial market knowledge, presence and leverage. Mr. Smith discussed three examples of channel partnering. They included the Harris/Lanier movement into healthcare information systems, the recently formed GE Harris railroad joint venture (which grew from $2 million to $60 million in annual sales within three years) and four venture spin-outs accomplished in a two year period.

As a result of these initiatives, the company's defense business has been able to stabilize revenue, realize a significant improvement in net income and support the overall Harris Corporation by acting as its R&D engine. Overall this strategy has resulted in significant enhancement in shareholder value for the company's stockholders.

RICHARD P. FLAM, ORBIT/FR

ORBIT/FR is a small company with less than 100 employees. It was formed through the merger of two companies, ORBIT and Flam and Russell. ORBIT was originally founded in Israel in 1950 as a supporter of military avionics. The merger which took place in 1996 led to a change in the main focus of the company from being a sole supporter of military avionics to manufacturing measurement systems for a new universe of commercial customers. Currently the business is evenly spread between military avionics and commercial customers.

Besides the manufacturing of measurement systems, which is the company's main line of business, it also has an RF systems line which, until recently, had traditionally been totally dedicated to military applications. Currently, however, the majority of the customers for such products are commercial users. A third business in which the company had been engaged was the production of specialized test measurement systems. The entire system was designed and built for a specific aerospace customer but since the company does not have any more customers in this field, it is no longer an ORBIT/FR business.

A key element of the company's philosophy is its effort toward focus on its core competencies. Through the acquisition of Advanced Electromagnetics, ORBIT/FR now has a new core competency in microwave and radar absorbing materials and anechoic chambers, which is complementary to the older business areas.

As indicated earlier, the company is undergoing a change in its customer base. Its new commercial customers, which comprise about 50 percent of its business, demand the same kinds of products as the military-based customers but they differ in that the commercial customers are basically in the world of wireless and satellite communications. So the company has been going through a cultural change from a project orientation to a product orientation over the last several years. In addition, the company currently exports approximately half of its production, with the largest portion going to the Pacific Rim, Japan, Korea, Taiwan, Mainland China and Singapore and a small portion going to Europe.

The company faces a number of major challenges as it reacts to change. One of these challenges is devising ways to attract customers from this new community of commercial customers. Each of these customers has distinct ways of doing business and different expectations such as the speed of response to demand. The company is facing the challenge of making its customers aware of its existence and the types of services it can offer them. This increases the importance of its sales representatives.

Another consequence of the change is that the company, although small, has to do business with a global perspective because a significant portion of its business is export-based. The challenge is to find ways to create global awareness of ORBIT/FR through aggressive marketing and sales efforts. The company has to be able to keep a very high customer service level in order to keep existing customers and to attract new ones.

A third challenge is competitiveness. The company has to perform its business faster and more efficiently to remain competitive in the market. It is currently generating revenues of $20 million a year with less than 100 employees. So as it grows, it has to focus more on human resources.

The company constructed some strategies to tackle these problems. One of these strategies is increasing value through vertical integration. This seems to be the contrary to today's common philosophy of outsourcing. However, it would be a greater challenge for the company if it pursued outsourcing because it is usually extremely difficult for a small company to generate the attention and the loyalty from the third-party providers.

Another strategy is to form strategic partnerships. An example is one that was formed with Hewlett-Packard. The partnership is designed so that when ORBIT/FR sells a product, so does Hewlett-Packard. Therefore, the partnership is beneficial for both parties. By utilizing this strategy, ORBIT/FR has been able to access a salesforce of 1800 engineers and make a significant percentage of its sales through Hewlett-Packard, particularly overseas.

One of the factors that has helped the company succeed has been the realization that in making the transition from military to dual-use products, it had to move from project-oriented to product-oriented work. This meant differences in planning for technical efforts, allocation of resources, and new product development funding. It also meant adjusting to different expectations, different ways of doing business, and a need to respond faster to customer needs. Additionally, focusing on the human capital of the company, and making sure that employees continue to develop and remain productive has been perhaps the most important management task.

JOHN MANUEL, LOCKHEED MARTIN CORPORATION[2]

Lockheed Martin is a $27 billion global enterprise with a core competency in systems integration of complex large systems for large customers, with a focus on national and global challenging issues. It is organized into five sectors which have relative autonomy. Each sector president is, by design, a president and chief operating officer. At the same time, there is teamwork at play. In other words, the corporate headquarters is not a holding company. Thus, competitive strategies, investment strategies and general guidance are devised at the headquarters level, while the sector presidents execute and deliver program strategy.

At Lockheed Martin, technology management and organizational practices are in support of the corporate strategy. The corporate strategy for a publicly held company such as Lockheed Martin is to make commitments to its shareholders and employees for continued growth. Since corporate strategy calls for continued growth, and the defense environment has been static, this translates into looking for new growth areas, while focusing on the company's core competencies.

Within this framework, the company has sought to consolidate and modernize against a benchmark of global competitiveness. In addition, corporate synergy with an emphasis on company-wide coordination is seen as a very important aspect of corporate strategy; it eliminates duplication and achieves economies of scale. The company coordinates its activities so that they complement each other. It is aware of the fact that in order to prosper as a company and as a nation, it has to view the marketplace externally and offer its customers the best that is available. This means that if the company is not capable of doing a part of its business, it has to purchase that part and outsource it to remain competitive. Within the company, this is known as external focus in vertical integration.

One important issue is globalization. As an example, the company is offering an American spread of proton launchers against the French, the Chinese and the Japanese. In addition to that, Lockheed Martin is also involved in communication and information systems that are all space-based systems. It also has imaging systems which will be used to launch the first commercial space imaging satellite in 1998.

In terms of dual-use technologies, the company has countless examples bridging the boundaries between defense and commercial technology. From working on a single stage to orbit vehicle, to commercial space imaging satellites and related communications and information applications, x-ray analysis and archiving for medical records, crop management using multi-spectral IR technology, even transforming simulation and training systems into third generation video arcade games, Lockheed Martin is involved in an array of exciting and promising projects.

GERALD BRASUELL, BEI SENSORS AND SYSTEMS

From the early 1950s, BEI's Systron Donner Inertial Division has been driven by the needs of the defense community and government. In 1992, the common belief within the company was that it was very successful. That year, the company had put into production the Maverick Missile program. However, the end of the Cold War hampered the existing programs. They were terminated one after another and that led the company to have a significant amount of obsolete product. The investment that had been made between 1987 and 1992 was underutilized. Within just two years, the division's income fell by 50 percent. In other words, the problem was that it was not able to make money anymore.

The above mentioned circumstances required the company to start focusing on commercial applications. That also called for a change in the way of thinking of how the company did business. It was so used to a mil-spec mentality, that is, reliance upon a given set of specifications without questioning their purpose or value. Nobody questioned whether those specifications were optimum for the division's production procedures or for the application for which the product would be used. All those mil-specs simply flowed down to the people involved in design and manufacturing.

The company has made a successful transition from being military-driven to producing commercial products, some of which it feeds back into the military. BEI is diversifying in markets and using its products as a base core to be used in both military and civilian applications. In 1992 the Donner Inertial Division was about 7 percent commercial. In 1995 this percentage went up to about 22 percent. Currently, 32 percent of its business is in the commercial arena. The goal for the next five years is to become 75 percent commercial.

The company first recognized the automotive industry as an emerging market opportunity. It developed a new technology for a solid-state quartz-rate sensor, which had applications in the automotive industry. Also as the company had chosen to get into the automotive industry, it had to get ISO 9000 certification along with a QS 9000 rating. It embarked on that task in January 1996 as part of its obligation under contracts with General Motors. In June 1996, it received the ISO 9001 certification and became the 235th company in the world to get a QS 9000 certification. The reason underlying this success was that the company already had well designed procedures from previously being a Mil-Q 9858A company. However, not all of the procedures were cost effective, so each procedure was reviewed to determine its necessity and its relationships with all other procedures.

Another area of change that the company pursued was the Single Process Initiative. Managers met with their regional defense contracting agency and went through an audit certification process which led to a documentation for the Single Process Initiative. This meant that the company would convert its existing and future defense contracts and use ISO 9001 in lieu of Q-9858A.

In the process of trying to grow effectively within its resource limits, it has been divesting itself of manufacturing tasks that are not within its core competencies, and has been investing heavily in areas that fall within its core competencies. Thus, it invested in automation of the quartz process and in the automation of the core nugget. It also has pursued strategic alliances with both international and domestic major automotive suppliers.

BEI also went into a mode of planned product obsolescence and divestiture. That is, in order to become more efficient, it planned for and got rid of products which were custom-built in small quantity and had only military application. As a result, it reduced its products from 168 in 1996 to 58 in 1997. Within this mind-set, it reduced the number of products developed specifically for unique military applications. Finally, it made a transition in R&D partnerships, from labs to universities, setting up relationships with Carnegie Mellon and Stanford, for example.

As a result, the decision to change the way of doing business resulted in a movement to a more profitable position.

RICHARD J. FARRELLY, BASE TEN SYSTEMS[3]

Base Ten Systems was founded in 1966 and is currently a publicly traded company. Historically it specializes in what it calls safety critical software, which manages the guns, rockets, missiles and gravity weapons of most NATO aircraft. It is also involved in the production of secure communications equipment endorsed by the National Security Agency. It manufactures defense products to specifications for any prime contractor. It also has another line of business in medical software, which has been growing in recent years. The company's sales hit a peak in 1991 at about $50 million. That was the year that it had completed its last major defense production contract. Since then, sales decreased to a minimum of $12 million before returning to its current level of $25 million. The reason for this fluctuation is that the company began modifying its strategy in 1991. During this period, Base Ten Systems decided to shut down a number of its foreign facilities and some of its domestic facilities.

In the last few years, the company began to emphasize commercial product development including medical screening, which is subject to FDA approval. The key to the company's success in this market is that it uses the same critical techniques that were developed in its traditional defense business. Manufacturing, quality assurance, documentation, field support and project management skills required for defense systems are very similar to those required to deal with major pharmaceutical companies, the FDA, and their foreign counterparts. Therefore, Base Ten Systems initially focused its commercial product development on the areas where it could apply the same skill set. Now its business environment has come to a point that the company is able to transfer technology more actively from the commercial area to the defense sector than the other way around.

When Base Ten founded a medical technology division, its first management decision was to use as much as possible of what it had in place, which included human resources, facilities, standards, procedures, organization structures, and policies. It then looked for the common threads of technologies and disciplines. It recognized that what it had in place was not adequate to serve the new market. Especially complex and dynamic were some regulatory affairs, which included what had to be done to be compliant with jurisdictions in the United States, Canada, European Union, and the rest of the world. The company did not have expertise in this area. The company made use of the services of an FDA-oriented law firm in Washington. Working closely with this law firm led to recognition that the requirements for the FDA and the medical industry were not all that different from those for the defense industry. The differences were related to the training and orientation towards the FDA. Remedies were developed for those areas where there were differences without incurring massive expenditures of personnel, facilities or associated things. The company hired a few key consultants from these new industries to give its people a dual capability both in the defense and the commercial sector.

A major difference, though, was in the finance area. The company had to overcome its cost-plus mentality and finance its own business. To adapt its financial systems, Base Ten Systems worked with Deloitte Touche as its public accountant and with Cowen & Company as its investment banker.

The company also now has collaborative relationships with a number of institutions such as Hershey Medical Center, The Pennsylvania State University, where most of the clinical tests are conducted. Base Ten Systems also works internationally with institutions such as the Wolfson Institute of Preventive Medicine and Saint Bartholomews and King's College hospitals in London. Those people have provisions to initiate and evaluate medical software. The company's own people are not allowed to render medical opinions or judgments, however.

Base Ten System's efforts to establish a presence in the medical sector also led to the development of a medical operating system called MED-DOS, which is a shell for use between an application system, a conventional operating system, and an applications program. This system, which allows one to convert a PC into a safety critical system, has been very successful. The company so far has received six United States patents in this area.

As a result of pursuing the medical sector, the company now has international facilities that can provide customer support and service in the pharmaceutical area. The company's customers include eleven of the top twenty pharmaceutical manufacturers in the world.

In the defense sector, things seem to be going pretty smoothly. In April 1996, Boeing selected Base Ten Systems to provide a maintenance data recorder for the Apache helicopter. The company also was selected for the engineering and the initial production of an interface blanken unit for the F-18 naval aircraft.

Briefly, thanks to its pursuit of the medical sector and adapting and utilizing its skills and competencies that it already had from the defense sector, Base Ten Systems has managed to remain in business as many competitors have dropped out one by one.

JAMES D. SCANLON, LOCKHEED MARTIN CONTROL SYSTEMS GROUP

During the last decade, Lockheed Martin's Control Systems Group has adjusted successfully to the rapidly changing defense industry environment. The company's business has recovered from four years of declining orders and employment to six consecutive years of double-digit growth and a forecast for continued growth at the same pace. Jim Scanlon and his colleagues have been able to re-focus their business from 90 percent military in 1987 to 50 percent today and a projection of 30 percent for the year 2002.

On the military side, the company continues to produce high tech fly-by-wire control systems for aircraft, such as the B-2, V22, F/A-18, C-17, and Sweden's JAS 39 Gripen Fighter. In addition, it has been developing a fly-by-wire control for India's light combat aircraft, as well as producing full-authority digital electronic controls for General Electric engines.

Examples of new markets that the company has entered include electronic controls for railroad locomotives, and hybrid electric vehicles including buses, trucks and taxis. Some of its customers in hybrid drives include Navistar International, which will be road-testing six trucks using Hybri-Drive™ drive-train technology, and Orion Bus Industries, which has received major orders from New York's Metropolitan Transit Authority as well as New Jersey's, and is using hybrids for its buses. In addition the company is working with the New York State Energy Research and Development Agency to develop a hybrid electric taxicab.

In order to adjust to the changing environment, the company devised a strategy to apply its hard-won core competencies in aerospace to closely related adjacent markets. Scanlon and associates found out that some of its core competencies in defense technology offered great potential for commercial application. However utilizing those competencies was not an easy task. In fact, with a reputation as an aerospace company, the company had to convince the world that it could design, develop and produce competitively priced products for the commercial marketplace, no matter how capable they were. Leveraging the relationship with General Electric and Lockheed Martin helped the company in that respect. Through working with local government and congressional representation, it was able to overcome some regulatory hurdles.

402   The Defense Industry in the Post-Cold War Era

One important issue the company faced during the transition period was that doing business commercially required a whole new way of thinking, especially for a defense contractor that was used to dealing with a single customer in a highly structured business environment. In order to attain this breakthrough way of thinking, Scanlon and colleagues put together a young entrepreneurial team that was not aware of the fact that defense companies were supposed to have a hard time competing in the commercial marketplace. This approach enabled the team to think "outside the box" and challenged its members to look for ways to apply some core competencies to non-aerospace markets. This innovative way of thinking brought with it some good discoveries such as recognition that the company's military factory could not be used to produce commercial electronics. Consequently, they built a factory within a factory devoted solely to their commercial non-critical flight products.

Through all these changes, the new product and new line, the company never lost sight of reducing its costs. It has taken its lowest cost commercial processes and brought them full circle back into the military product area. In this way, it was able to utilize not only its core competencies for the commercial area, but also new discoveries within the commercial area that could be fed back to its government business. For example, the company instituted the use of plastic encapsulated microcircuits (PEMC), a commercial grade part, to lower costs for its military customers. This process has been FAA certified and its facilities are commercially certified to such standards as ISO 9001. In addition, to ensure that its customers get the highest quality products, the company has adopted a six-sigma program that also helped lower costs.

A major benefit for the company during this transition was that all of its 1700 employees acquired a breakthrough-thinking spirit. Although the competition in the commercial marketplace is fierce, everyone in the company is fully persuaded that they will respond to the competition successfully.

JOHN V. SPONYOE, PRESIDENT, FEDERAL SYSTEMS GROUP, LOCKHEED MARTIN[4]

The Electronic Platforms Integration Group of Lockheed Martin successfully shifted its business focus from an almost exclusive involvement in the aerospace industry to a more balanced focus, which now also includes international business and postal and transportation systems. It went through this transition while an operating unit of three very different companies. For thirty-five years, it was part of IBM, then Loral Corporation for two years, and finally a group within Lockheed Martin for the last year-and-a-half.

John Sponyoe and colleagues at IBM-Owego were able to make this transition successful by focusing on their core competencies. They had an excellent infrastructure, good engineering and software skills, a very good investment stream and quality people. They believed that it was easier to grow a business from half a billion to a billion dollars than to start from zero and go up to half a billion. The issue here was whether they would be able to apply their competencies and resources to new areas which they thought had growth potential. One factor that helped them in that respect was that in the mid-1970s, they had some re-focusing experience as a business within IBM. This experience taught them that if they really wanted to grow, they had to consider doing things differently.

One of the company's strategies was to focus on systems integration. Following this strategy, the company also actively pursued international opportunities. Although IBM's name called for international systems integration, IBM-Owego was not allowed to bid outside the U.S. It only had a small slice of the declining defense business, primarily in aerospace and naval systems in the early 1990s. This triggered a desire to find out if there were any similar programs that were being bid outside of the U.S. and then make a strong effort to rely on the company's core competencies. This proactive approach helped them to win a program called Merlin in Great Britain, an ASW program similar to the LAMPS program in the U.S. The mission assigned to this program was slightly different from that in the U.S., but basically utilized the same technology. Merlin also required collaboration with British industry if it were to be successful.

Another area the company entered was software development for federal systems integration. In areas such as logistics, supply, maintenance, transportation, finance, safety, security, etc., management was not confident that it had the right rate structure to compete for these kinds of activities. The strategy it followed was to team up with those companies that had been more competitive in these activities and built the relationships they needed in order to be able to bid on these programs. Its initial success was teaming with AT&T on a program called Sustaining Base Information Services to re-architect all the camps, posts and stations throughout the U.S. Army. This strategy of combining the core strengths of many companies continues to pay dividends to Owego.

Another area the company chose to pursue was postal systems. It looked at three areas within this sector. One was postal sorting equipment, in which it had never been involved before. However, Endicott, a facility within IBM, had many years' experience with check sorting equipment. By taking the skills in Endicott and teaming up with AEG, a German company, the company bid a program as a leader-follower, AEG being the leader, and won. Today, the postal business is one of the largest at Owego and continues to grow and diversify.

There were some issues that enabled Lockheed Martin to make things happen. It had the business processes to enable it to focus its business. It believed that it was very fortunate to start this activity early. As a result,

404 The Defense Industry in the Post-Cold War Era

it was able to take significant costs out of the business by analyzing and maximizing its fundamental processes. In addition, the company looked at the issue of quality as a means of survival. It knew that customer satisfaction was key to success when re-focusing a business.

In summary, the transition was successful because the company tried to utilize its core competencies most effectively, train its people, team up with like-minded companies, improve appropriate business processes, and focus on customer needs.

NOTES

1. Mr Smith is currently President, Sanders, a Lockheed Martin Company.
2. Mr Manuel retired from Lockheed Martin in 1998.
3. Mr Farrelly retired from Base Ten Systems in 1998.
4. Mr. Sponyoe is currently Chief Executive Officer, Lockheed Martin Global Telecommunications.

ACKNOWLEGDEMENT

These case presentations have been reproduced with kind permission from their authors.

INDEX